Mutations of Alternative Algebras

Mathematics and Its Applications

Volume 278

Mutations of Alternative Algebras

by

Alberto Elduque
Department of Mathematics,
University of Zaragoza,
Zaragoza, Spain

and

Hyo Chul Myung
Department of Mathematics,
University of Northern Iowa,
Cedar Falls, Iowa, U.S.A.

KLUWER ACADEMIC PUBLISHERS
DORDRECHT / BOSTON / LONDON

A C.I.P. Catalogue record for this book is available from the Library of Congress.

ISBN 0-7923-2735-7

Published by Kluwer Academic Publishers,
P.O. Box 17, 3300 AA Dordrecht, The Netherlands.

Kluwer Academic Publishers incorporates
the publishing programmes of
D. Reidel, Martinus Nijhoff, Dr W. Junk and MTP Press.

Sold and distributed in the U.S.A. and Canada
by Kluwer Academic Publishers,
101 Philip Drive, Norwell, MA 02061, U.S.A.

In all other countries, sold and distributed
by Kluwer Academic Publishers Group,
P.O. Box 322, 3300 AH Dordrecht, The Netherlands.

Printed on acid-free paper

Printed in the Netherlands

A Pili y Eva

To Karen, Peggy, Jane and Michael

TABLE OF CONTENTS

PREFACE

In recent years, nonassociative algebras, other than Lie algebras, have frequently appeared in other disciplines in sciences and mathematics, and, in particular, in physics. Around 1978, two classes of nonassociative algebras emerged as algebraic models in theoretical physics and mechanics. One class consisted of flexible Lie–admissible algebras, whose introduction was originally due to A.A. Albert in 1948 [**A2**]. The objective was to extend the framework of quantization problems based on associative algebras to a more general one using flexible Lie–admissible algebras [**O1**]. A book detailing the mathematical exposition in this subject was published in 1986 under the title *Malcev–Admissible Algebras* [**My4**].

Another class comprised nonassociative algebras obtained from "mutations" of associative algebras, which originated from a quantized version of generalized Hamilton's equations [**Sa1**]. Thus, the introduction of these algebras arose from an attempt to generalize classical mechanics as well as quantum mechanics. The present book concerns the mathematical side of the theory of this class of algebras. Included in this book are mutations of alternative algebras which are a natural generalization of mutations of associative algebras and which are mainly distinguished by mutations of Cayley–Dickson algebras. This fact provided the title of this book *Mutations of Alternative Algebras*.

Let A be an alternative algebra with multiplication xy over a field F. For fixed elements $p, q \in A$, the (left) (p, q)–mutation $A(p, q)$ of A is defined as the algebra with multiplication $x * y = (xp)y - (yq)x$ defined on the vector space A. If A is associative, then the product $*$ becomes $x * y = xpy - yqx$, which is the original source of $*$. Although mutation algebras are both Jordan–admissible and Malcev–admissible, their structure theory takes on quite different approach from that in the aforementioned book on Malcev–admissible algebras, and that of other well known nonassociative algebras, such as Lie, alternative or Jordan algebras. This is due to the fact that the class of mutation algebras lacks the intrinsic characterization by means of identities because it does not constitute a variety; therefore, it seems impossible to identify a minimal set of identities that defines this class.

In general, the structure of $A(p, q)$ is largely governed by that of the underlying alternative algebra A as well as by the mutation parameters p and q. One of the main tasks in the study of mutation algebras was the search for an appropriate radical which leads to a meaningful structure theory for $A(p,q)$, regardless of the choice of p and q. Recently such a radical was found; namely, there exists the solvable radical $J(p,q)$ of $A(p,q)$ for any artinian alternative algebra A and for arbitrary fixed elements p and q in A.

The primary concern of this book is to present a self–contained and detailed account of the mathematical theory of mutation algebras. The major portion of the material is devoted to the structure theory for mutations of artinian alternative algebras. Therefore, basic structure theory about alternative algebras, which is summarized in Chapter I without proofs, plays a central role throughout this book, in particular, for Chapters III and IV. Virtually all results in the remainder of this book are presented with proofs in considerable detail. Thus, we have conscientiously tried to make the material accessible to nonspecialists in the subject, who wish to become acquainted with mutation algebras and their applications. Owing to the physical origins of these two classes of algebras, another concern was to offer this book as a counterpart to the book *Malcev–Admissible Algebras*, although they have quite different features of structure theory.

The primary objective of Chapter I is to provide background material pertaining to the ensuing chapters. Some basic results on modules, radical and central simplicity for nonassociative algebras are presented with proofs. The structure of composition algebras and alternative algebras is exposited in Section I.3, but we have made no attempt to provide the proofs which can be found in standard texts on nonassociative algebras [**J3**, **Sc3**,**ZSSS1**]. Two identities that arise naturally from the mutation product $*$ are Malcev–admissibility (Lie–admissibility for the associative case) and Jordan–admissibility. In essence, these are the only identities which play important roles for our discussions, although mutation algebras satisfy many other identities (Chapter V). The main concerns in Sections I.4 and I.5 are to present some basic results about Malcev–admissible algebras and Jordan–admissible algebras, including brief historical remarks. Also, included in these two sections are some classical results on Lie and Jordan ideals in a semiprime associative algebra, which are instrumental for later discussions.

Chapter II is the highlight of this book in that we set forth the framework for the structure theory of mutation algebras. For any artinian associative algebra A, the radicals $R(A)$ and $R(A(p,q))$ of A and $A(p,q)$ (in the sense of Section I.1) have the relation $R(A) \subseteq R(A(p,q))$. In general, these radicals are too far apart from each other, and do not seem to shed enough light on the structure of $A(p,q)$. However, it is possible to locate two ideals $R(p,q) \subseteq J(p,q)$, of $A(p,q)$, between $R(A)$ and $R(A(p,q))$, which interplay nicely with $R(A)$ and $R(A(p,q))$, and with the structure of $A(p,q)$ for any mutation parameters p and q. Moreover, it turns out that $R(p,q)$ is nilpotent in $A(p,q)$ and $J(p,q)$ is the solvable radical of $A(p,q)$. There exists another ideal, $R^0(p,q)$ of $A(p,q)$ sitting inside $R(p,q)$, which describes more explicitly the role of p and q.

The general mutation algebra $A(p,q)$ does not satisfy the third–power identity $(x*x)*x = x*(x*x)$ that is implied by well known identities, such as flexibility or power–associativity. In Section II.2 we discuss the relationships between these identities in $A(p,q)$ and conditions for them to be equivalent. Several earlier works on identity problems in $A(p,q)$ are obtained here as special cases from our general results in this section. The simplicity or primeness of $A(p,q)$ implies the same of A. The converse is not true without an additional condition; the investigation of this condition is the main topic in Section II.3. The ideal $R^0(p,q)$ plays a central role for this. Also, included in this section are the relationships among the nucleus, centers and centroids of $A(p,q)$ and A.

Section II.4 is devoted to the structure of $A(p,q)$ for any simple artinian associative algebra A. The principal result is that if A is simple artinian, then there exists a subalgebra S of $A(p,q)$ such that $A(p,q) = S \oplus R^0(p,q)$ and S is isomorphic to a "twisted" mutation algebra. If $p \neq q$, then S is simple. In case $p = q$, it is shown that either $A(p,p)$ is solvable, or $A(p,p)/J(p,p)$ is a semisimple Lie algebra whose commutator algebra is simple. Finally, in Section II.5 we establish the structure theorem: If A is an artinian associative algebra, then $A(p,q)/J(p,q)$ is a direct sum of simple twisted mutation algebras and semisimple Lie algebras cited above. The relationships among the radicals $R(A) \subseteq R(p,q) \subseteq J(p,q) \subseteq R(A(p,q))$ are also investigated in this section.

Chapter III concerns the structure of mutations of alternative algebras. Virtually all results in Chapter II are extended to the alternative case. Since a simple alternative algebra is either associative or a Cayley–Dickson algebra over its center, this chapter is in large devoted to mutations of Cayley–Dickson algebras C. For an artinian alternative algebra A, $A(p,q)/J(p,q)$ is shown to be a direct sum of simple algebras and semisimple Lie algebras. Those simple summands that do not arise from mutations of an associative algebra are of the form $C(p,q)/J(p,q)$. The main objective in Section III.4 is to determine the structure of $C(p,q)$ and $C(p,q)/J(p,q)$.

Chapter IV deals with the automorphism group $\operatorname{Aut} A(p,q)$ and the derivation algebra $\operatorname{Der} A(p,q)$ of $A(p,q)$. The central importance in this chapter is the structure of $\operatorname{Aut} A(p,q)$ and $\operatorname{Der} A(p,q)$ for central simple artinian alternative algebras A. Thus, if A is not associative, then A is a Cayley–Dickson algebra C. For a generalized quaternion algebra Q, the structure of $\operatorname{Aut} Q(p,q)$ and $\operatorname{Der} Q(p,q)$ is more explicitly described (Section IV.5). Similar results are obtained for $\operatorname{Aut} C(p,q)$ and $\operatorname{Der} C(p,q)$ when $p+q$ is invertible (Section IV.6). For associative algebras A and B, where B is prime with a unit element, it is possible to determine all isomorphisms of $A(p,q)$

to $B(a,b)$ in terms of isomorphisms or anti–isomorphisms of A to B, when (a,b) satisfies a certain normality condition (Section IV.3). This plays a principal role for the main results on $\operatorname{Aut} A(p,q)$.

The primary objective in Chapter V is to investigate identities satisfied by mutations of all associative algebras and representations for these mutation algebras. The T–ideal of identities satisfied by the mutation algebras is not only vastly larger than the identities resulting from Lie– and Jordan–admissibilities, but also it is a formidable task to find a generating set of that T–ideal. This situation makes it impossible to study the structure of mutation algebras and its representations within the intrinsic approach based on the theory of varieties. The first part of this chapter is intended to offer a general tool for the treatment of identity problems in mutation algebras. This is done by establishing a gradation for the T–ideal of these identities. All identities of degree ≤ 4 are determined in this gradation, which result in two new identities of degree 4 that are not consequences of Lie– and Jordan–admissibilities. The second part of Chapter V focuses on a particular type of representation, called here a strict representation, that arises naturally from modules for an associative algebra and from Peirce decompositions of a mutation algebra. Such irreducible representations are determined for mutations of any simple artinian associative algebra.

During the writing of this book, it became apparent that the pervading influence in the approach presented here can be attributed to the theory of artinian associative or alternative algebras, rather than those identities stemming from mutation algebras. Therefore, we have taken the view throughout that the material can be more appealing for applications of the theory of noncommutative associative or alternative algebras than a standard theory of nonassociative algebras. We have chosen not to explore the physical side of the theory of mutation algebras, not only because of our desire to keep the text within an algebraic framework but also because of our insufficient competence in the discipline of physical applications related to mutation algebras.

Since the writing of this book was first undertaken in the fall of 1989, the initial draft of the material went through several revisions and improvements, due to on–going research activities in the subject, notably in some Spanish schools of mathematics, in particular, at the University of Zaragoza and the University of Oviedo. Concurrently with our book project, F. Montaner was working on his doctoral thesis in this subject. He generously made his unpublished work available to us, which has played an indispensible role for the final organization of this book.

J.E. Bruha and R.B. Campbell assisted with this project by carefully reading part of the manuscript and offering several critical comments. It was challenging, as well as often frustrating, to prepare a manuscript two continents apart. On the

Spanish side, A. Elduque not only served as coauthor, but was the principal LaTeX–setter. On the United States side we are indebted to N. Sash and R.B. Campbell for assisting in placing our thoughts onto the printed page. We express our gratitude to all of these individuals for their assistance. The authors alone, however, accept the responsibility for any errors or omissions that may occur.

We wish to conclude with the following personal acknowledgments. A. Elduque is greatly indebted for the support from the Dirección General de Investigación Científica y Técnica (DGICYT, Spain) under the program "Perfeccionamiento y Movilidad del Personal Investigador (1989)", which made possible the visit to the University of Northern Iowa during the fall semester of 1989. He wishes to thank UNI for the generous hospitality during his stay; it was during that semester that this project was conceived. In subsequent years, Elduque also received additional support from DGICYT under the projects PS 87–0054 and PS 90–0129, and from the "Diputación General de Aragón" under the project PCB–6/91.

H.C. Myung wishes to thank the University of Northern Iowa for the Professional Development Leave during the fall semester of 1989, which provided the essential time necessary to engage in the early development of this writing project, and for a UNI Summer Research Grant in 1991 to continue this project. Myung is also grateful to UNI, the University of Zaragoza, and the University of Oviedo for support to travel to Spain in 1990 and 1992. In particular, he is most appreciative to S. González, who made the arrangements and provided the support through these two universities. These visits were quite helpful in keeping this writing project on schedule. The authors wish to gratefully acknowledge all this support without which this project could not have been completed in its present form.

We also wish to acknowledge the wonder of "electronic mail", without which our writing task would have been considerably more difficult. This modern "jinni" literally transported all parts of our manuscript across the Atlantic, from one continent to the other, in seconds, often on a daily basis.

Zaragoza and Cedar Falls A. Elduque and H.C. Myung

CHAPTER I

NONASSOCIATIVE ALGEBRAS

1. Definitions and basic results

The primary aim of this chapter is to present some basic results about nonassociative algebras which will be important background for later investigations. No attempt is made to provide a detailed account of any particular topic with great generality. The discussion focuses on definitions and elementary results about some well known nonassociative algebras, and relationships between them. It is also intended to make the presentation accessible to nonspecialists in nonassociative algebras who wish to make contact with mutation algebras and their applications. Thus, deliberate efforts are made to provide proofs for some known results whenever possible, without unnecessary difficulty.

In Section I.2 we discuss some basic facts about centroids and centrality of simple algebras, and prove two fundamental results on this subject: (1) A simple algebra A over a field F with center $Z(A) \neq 0$ has a unit element, and $Z(A) = \Gamma 1$, where the centroid Γ of A is a field extending F. (2) A simple algebra over a field F with centroid Γ is central simple over F if and only if $\Gamma = F$. In Section I.3, some classical results about alternative algebras are surveyed with particular emphasis on the structure of composition algebras. The central topic in this section is the structure of Cayley–Dickson algebras which are the only simple alternative algebras (over the center) that are not associative. No efforts are made to provide proofs of these results that may be found in standard texts in nonassociative algebras, e.g., [**Sc3,J2,3,ZSSS1**].

Mutations of any alternative algebra A are contained in the classes of Malcev–admissible algebras and of Jordan–admissible algebras. If, in particular, A is associative, then its mutations are Lie–admissible. The material in Sections I.4 and I.5 is designed to offer a very brief introduction to the structure of these two classes of algebras. Proofs are not given of all of the topics discussed here; however, a number of results are presented with proofs, whenever possible without digression from the main subject, which include some results on ideals in the Lie algebra A^- and Jordan algebra A^+ for an associative algebra A. Since a meaningful structure theory of the two classes of algebras focuses on flexible Malcev–admissible algebras and flexible Jordan–admissible algebras (also called noncommutative Jordan algebras), and since a "small" subclass of mutation algebras is flexible (Chapters II and III), a few results cited in Sections I.4 and I.5 are independent of the theory of mutation algebras.

In this section we give some basic definitions, and prove elementary results about modules, semisimplicity and radicals for nonassociative algebras.

A *nonassociative algebra* A over a field F or simply an *algebra* A over F with multiplication denoted by xy is a vector space over F such that the map $A \times A \to$

$A\ :\ (x,y) \mapsto xy$ is bilinear over F, that is,

$$(\alpha x + \beta y)z = \alpha(xz) + \beta(yz), \quad x(\alpha y + \beta z) = \alpha(xy) + \beta(xz)$$

for $\alpha, \beta \in F$ and $x, y, z \in A$. A (nonassociative) *ring* R is an additive abelian group with a bi–additive multiplication $R \times R \to R\ :\ (x,y) \mapsto xy$. Thus, any algebra A over a field F is regarded as a ring. Although it is commonly understood that a ring satisfies an additional condition, the *associative law*,

$$(xy)z = x(yz),$$

throughout this book 'algebra' refers to a nonassociative algebra; this means that the associative law is not being assumed.

The definitions of terms, subalgebra, ideal, direct sum of ideals, unit element, homomorphism, kernel, isomorphism, quotient algebra, etc., do not involve associativity and hence are applicable to any algebra. The following fundamental homomorphism theorem can be easily proven for any algebra.

Theorem 1.1. *For algebras A, A', let $f : A \to A'$ be a homomorphism, and let* $\ker f$ *and* $\operatorname{Im} f$ *denote the kernel and the image of f. Let I and J be ideals of A.*

(i) *There exists a homomorphism $\bar{f} : A/I \to A'$ with $f = \bar{f} \circ \pi$ if and only if $I \subseteq \ker f$, where π denotes the natural homomorphism $: A \to A/I$.*

(ii) $\operatorname{Im} f$ *is a subalgebra of A' and* $A/\ker f \cong \operatorname{Im} f$.

(iii) *$I + J$ and $I \cap J$ are ideals of A and $(I+J)/I \cong J/(I \cap J)$.*

(iv) *If $I \subseteq J$ then J/I is an ideal of A/I and $(A/I)/(J/I) \cong A/J$.* □

Theorem 1.1(ii) means that the homomorphic images of an algebra A are determined by the ideals in itself. If A has no proper ideals ($\neq 0, A$) or no proper homomorphisms then $A^2 = A$ or $A^2 = 0$, and, in the latter, A is one–dimensional over F and is uninteresting. Thus, an algebra A over F is defined to be *simple* over F if it has no proper ideals and $A^2 \neq 0$. The simple algebras in a class of algebras over F play the role of building blocks, so that "nice" algebras in that class are direct sums of simple algebras. One wants to define such algebras to be semisimple.

Definition 1.1. An algebra A over F is called *semisimple* if it is the direct sum of finitely many simple ideals of A. □

The semisimplicity in some well known algebras of finite dimension is defined to have zero radical, where the radical is the unique maximal ideal relative to a certain radical property (e.g., solvability, nility), and is shown to be equivalent to Definition

1.1. For these algebras, the radical is also the unique minimal ideal relative to semisimple quotient algebras. Since not every algebra has a radical property which leads to a meaningful radical theory (see [**My4**, p.272]), the latter description is more convenient for the general algebra. We describe such a unique minimal ideal in terms of maximal ideals.

Definition 1.2. Let S be a set and let V be a vector space over a field F. Then, V is called an *S–module* over F if there is a map $S \times V \to V \ : \ (s,x) \mapsto sx$ such that $s(\alpha x + \beta y) = \alpha(sx) + \beta(sy)$ for $\alpha, \beta \in F$, $x, y \in V$ and $s \in S$. A subspace W of V is called an *S–submodule* of V if $SW \subseteq W$. □

If W is an S–submodule of V then the quotient space V/W is made into an S–module, called the quotient module, under $s(x+W) = sx+W$, $s \in S$, $x \in V$. An S–module homomorphism $f : V \to V'$ is a linear map over F such that $f(sx) = sf(x)$, $s \in S$, $x \in V$. It is easy to see that Theorem 1.1 holds for S–modules and S–module homomorphisms.

Definition 1.3. An S–module V is called *completely reducible* if for every S–submodule W there is an S–submodule W' such that $V = W \oplus W'$. An S–submodule W of V is called *irreducible* if it has no proper S–submodules of V. □

Let A be an algebra over a field F. Denote

$$\text{Left multiplication } \ L_x : A \to A \ : \ y \mapsto xy,$$

$$\text{Right multiplication } \ R_x : A \to A \ : \ y \mapsto yx.$$

Let $L_A = \{L_x : x \in A\}$, $R_A = \{R_x : x \in A\}$ and denote by $Hom_F A$ the associative algebra over F of all linear transformations on A. The *multiplication algebra* $M(A)$ of A is defined to be the associative subalgebra of $Hom_F A$ generated by $L_A \cup R_A$. We regard A as an $M(A)$–module. Thus, $B \subseteq A$ is an ideal of A if and only if B is an $M(A)$–submodule of A, i.e., stable under $M(A)$. Note that a nonzero trivial irreducible $M(A)$–submodule W of A (i.e., $M(A)W = 0$) is one–dimensional, and hence A is simple over F if and only if it is a nonzero nontrivial irreducible $M(A)$–module.

Lemma 1.2. *Let S be a set and V be a completely reducible S–module over F.*

(i) *For any S–submodule W of V, W and V/W are completely reducible.*

(ii) *If $V \neq 0$ then V contains a nonzero irreducible S–submodule.*

Proof. (i) Let W_0 be an S–submodule of W. Since V is completely reducible, there is an S–submodule W_0' of V such that $V = W_0 \oplus W_0'$. But $W = W \cap V =$

$W \cap (W_0 \oplus W_0') = W_0 \oplus (W \cap W_0')$, where $W \cap W_0'$ is an S–submodule of W. Thus, W is completely reducible. Let $\bar{V} = V/W$ and $\bar{P}$ be an S–submodule of $\bar{V}$. If P denotes the inverse image of $\bar{P}$ under the natural homomorphism : $V \to \bar{V}$ then we obtain $V = P \oplus P'$ for some S–submodule P' of V. But $\bar{V} = \bar{P} \oplus \bar{P}'$ since $P \supseteq W$, thus $\bar{V}$ is completely reducible.

(ii) Let x be a nonzero element of V. If every nonzero S–submodule of V contains x, then V must be irreducible. Assume that there is a nonzero S–submodule not containing x. By Zorn's lemma we choose a maximal S–submodule W with $x \notin W$. There exists an S–submodule $W' \neq 0$ of V such that $V = W \oplus W'$. We assert that W' is irreducible. Suppose that W' is reducible. Since W' is completely reducible by part (i), there exist proper S–submodules W_1, W_2 of W' such that $W' = W_1 \oplus W_2$, but $x \in (W \oplus W_1) \cap (W \oplus W_2)$ by the maximality of W. This implies $x \in W$, a contradiction. Hence, W' is irreducible. □

We prove the following standard characterization of complete reducibility, which is also useful for later investigations.

Theorem 1.3. *An S–module $V \neq 0$ for a set S is completely reducible if and only if V is the direct sum of irreducible S–submodules of V.*

Proof. Assume V is completely reducible. Let $\{V_i\}$ be the set of all irreducible S–submodules of V. The set is nonempty by Lemma 1.2. It follows from Lemma 1.2(ii) that $V = \sum_i V_i$. By Zorn's lemma, there exists a maximal subset $\{V_k\}$ of $\{V_i\}$ such that $\sum \oplus V_k$ is a direct sum. Suppose $V \neq \sum \oplus V_k$. Then, $V_i \not\subseteq \sum \oplus V_k$ for some $V_i \in \{V_i\}$ and hence $V_i \cap \sum \oplus V_k = 0$ by irreducibility of V_i. Thus, $V_i \oplus \sum \oplus V_k$ is direct, a contradiction, by the maximality of $\{V_k\}$. This shows $V = \sum \oplus V_k$. Now, let $V = \sum \oplus V_k$ for irreducible S–submodules V_k and let W be an S–submodule of V. By Zorn's lemma, we choose a maximal S–submodule W' with $W \cap W' = 0$. For each k, $(W \oplus W') \cap V_k = 0$ or V_k. For the former, $W \oplus W' \oplus V_k$ is direct and $W \cap (W' \oplus V_k) = 0$, so $V_k = 0$ by the maximality of W'. In the latter, we have $V_k \subseteq W \oplus W'$ and $V = W \oplus W'$. □

Corollary 1.4. *If A is a semisimple algebra over F then for any proper ideal I of A there is an ideal J of A such that $A = I \oplus J$. If A is finite–dimensional and $A^2 = A$ then the converse holds.*

Proof. If A is semisimple then A is a completely reducible $M(A)$–module by Theorem 1.3. Assume that A is a completely reducible $M(A)$–module. Then any ideal of an ideal of A is also an ideal of A. Thus, if A is finite-dimensional and $A^2 = A$ then by induction on $\dim A$ it follows that A is semisimple. □

An ideal M of A is called *maximal* if $M \neq A$ and whenever $M \subseteq J \subseteq A$ for ideals J of A then $M = J$ or $J = A$. The maximal ideals of A are the maximal

$M(A)$–submodules of A. The existence of a maximal ideal is guaranteed when A is finite–dimensional or has a unit element. In this case, every proper ideal of A is contained in a maximal ideal of A.

Definition 1.4. The *radical* $R(A)$ of any algebra A over F is defined to be the intersection of all maximal ideals M of A such that $A^2 \not\subseteq M$. If such maximal ideals do not exist then we let $R(A) = A$. □

This radical is sometimes called the *Albert radical*.

Proposition 1.5. *If B and C are ideals of an algebra A over F such that $A = B \oplus C$, then $R(A) = R(B) \oplus R(C)$.*

Proof. Let M be any maximal ideal of A not containing A^2. Suppose that $B \not\subseteq M$ and $C \not\subseteq M$. Since M is maximal, $A = B + M$ and so $C^2 = AC = BC+BM = MC \subseteq M$. Similarly, $B^2 \subseteq M$ and $A^2 = B^2+C^2 \subseteq M$, a contradiction. Thus, it must be that either $B \subseteq M$ or $C \subseteq M$. It follows from this that the maximal ideals M of A not containing A^2 are obtained either by $M = B \oplus N$ for a maximal ideal N of C with $N \not\supseteq C^2$, or by $M = N \oplus C$ for a maximal ideal N of B with $N \not\supseteq B^2$. By Definition 1.4, this implies $R(A) = R(B) \oplus R(C)$. □

We show that if A is finite–dimensional then $R(A)$ is the unique minimal ideal of A such that $A/R(A)$ is semisimple or $A/R(A) = 0$. The following Chinese Remainder Theorem is useful.

Theorem 1.6. *Let $B_1, \ldots, B_r$ be ideals of A such that $B_i + (\cap_{j\neq i} B_j) = A$ $(1 \leq j \leq r)$. Then, as algebras,*

$$A/\cap_{i=1}^{r} B_i \cong A/B_1 \oplus \cdots \oplus A/B_r.$$

Proof. By induction on r it suffices to prove for $r = 2$. Let $f : A \longrightarrow A/B_1 \oplus A/B_2 \,:\, x \mapsto (x+B_1, x+B_2)$. Clearly, f is a homomorphism and $\ker f = B_1 \cap B_2$, so $A/\, B_1 \cap B_2 \cong \operatorname{Im} f$. To show f is onto, let $(x+B_1, y+B_2) \in A/B_1 \oplus A/B_2$. From $A = B_1 + B_2$, we have $x = b_1 + b_2$ and $y = c_1 + c_2$ for $b_1, c_1 \in B_1$ and $b_2, c_2 \in B_2$. Since $x+B_1 = b_2+B_1$ and $y+B_2 = c_1+B_2$, $f(b_2+c_1) = (b_2+c_1+B_1, b_2+c_1+B_2) = (b_2+B_1, c_1+B_2) = (x+B_1, y+B_2)$. □

Theorem 1.7. *If A is a finite–dimensional algebra over a field F then the radical $R(A)$ of A is the unique minimal ideal of A such that the quotient algebra $A/R(A)$ is semisimple or zero.*

Proof. Assume $R(A) \neq A$. Let $\{M_i\}$ be the set of maximal ideals of A which do not contain A^2. By finite–dimensionality, we can choose a minimal finite set

$\{1,\ldots,n\}$ of indices such that $R(A) = M_1 \cap \ldots \cap M_n$. Thus, $R(A) \neq R_k = M_1 \cap \ldots \cap M_{k-1} \cap M_{k+1} \cap \ldots \cap M_n$ for all $k = 1,\ldots,n$. By the maximality of M_k, either $M_k + R_k = M_k$ or $M_k + R_k = A$. In the former, $R_k \subseteq M_k$ and so $R(A) = M_k \cap R_k = R_k$, a contradiction. Thus, $A = M_k + R_k$ and by Chinese Remainder Theorem $A/R(A) \cong A/M_1 \oplus \cdots \oplus A/M_n$. Since the M_k's are maximal and $A^2 \not\subseteq M_k$, $A = M_k + A^2$, hence $(A/M_k)^2 = A/M_k$ is simple. This shows that $A/R(A)$ is semisimple.

Let B be an ideal of A such that $A/B = \bar{A}_1 \oplus \cdots \oplus \bar{A}_r$ for simple ideals $\bar{A}_i$ of A/B. Then, $\bar{A}_1 \oplus \cdots \oplus \bar{A}_{k-1} \oplus \bar{A}_{k+1} \oplus \cdots \oplus \bar{A}_n = U_k/B$ for an ideal $U_k \supseteq B$ of A. Since $A/U_k \cong (A/B)/(U_k/B) \cong \bar{A}_k$ is simple, each U_k is a maximal ideal of A with $A^2 \not\subseteq U_k$ and $0 = \cap_{k=1}^r (U_k/B) = (\cap_{k=1}^r U_k)/B$. Thus, $R(A) \subseteq \cap_{k=1}^r U_k = B$. □

Corollary 1.8. *If A is semisimple then $R(A) = 0$.* □

Theorem 1.9. *Let A be a finite-dimensional algebra over F and let $T(A)$ denote the intersection of all maximal ideals of A. Then, $T(A) \subseteq A^2$ and $A/T(A) = R(A)/T(A) \oplus A^2/T(A)$ where $R(A)^2 \subseteq T(A)$ and $A^2/T(A)$ is semisimple.*

Proof. If $A = A^2$ then clearly $T(A) \subseteq A^2$. Assume $A \neq A^2$. Since any subspace of codimension 1 containing A^2 is a maximal ideal of A, it is easy to see that A^2 is the intersection of some maximal ideals of codimension 1, and hence $T(A) \subseteq A^2$. As in the proof of Theorem 1.7 we choose a minimal set $\{M_1,\ldots,M_n\}$ of maximal ideals $M_1,\ldots,M_n$ such that $T(A) = M_1 \cap \ldots \cap M_n$, and $A/T(A) \cong A/M_1 \oplus \cdots \oplus A/M_n$. We may further assume that $A^2 \subseteq M_i$ for $i = 1,\ldots,r$ and $A^2 \not\subseteq M_i$ for $i = r+1,\ldots,n$ where r is an integer with $0 \le r \le n$. Thus, $(A/M_1 \oplus \cdots \oplus A/M_r)^2 = 0$ and $A/M_{r+1} \oplus \cdots \oplus A/M_n$ is semisimple. There exist ideals N and P containing $T(A)$ such that $A/T(A) = N/T(A) \oplus P/T(A)$ with $N^2 \subseteq T(A)$ and $P/T(A)$ semisimple. Since $T(A) \subseteq A^2$ and $A^2/T(A) = (A/T(A))^2 = (N/T(A))^2 + (P/T(A))^2 = (P/T(A))^2 = P/T(A)$, $A^2 = P$ and $N = \{x \in A : xA + Ax \subseteq T(A)\}$. Since $A/N \cong (A/T(A))/(N/T(A)) \cong A^2/T(A)$ is semisimple, by Theorem 1.7 $R(A) \subseteq N$ and hence $R(A)^2 \subseteq T(A)$. If there is a maximal ideal M of A with $A^2 \not\subseteq M$ and $N \not\subseteq M$ then $A = N + M$ and so $A^2 \subseteq N^2 + M \subseteq T(A) + M = M$, a contradiction, to show $N = R(A)$. □

Corollary 1.10. *Let A be a finite-dimensional algebra over F.*

(i) $R(A) = \{x \in A : xA + Ax \subseteq T(A)\}$.

(ii) $R(A) = T(A)$ *if and only if* $A^2 = A$.

(iii) *If* $R(A) = 0$ *then* $A^2 = A$ *and* A *is semisimple.* □

Note that if A has a unit element then $A^2 = A$. Recall that an algebra A is called *nilpotent* if there is a positive integer n such that all products of any n elements of A in any association are zero.

Definition 1.5. For a subalgebra B of an algebra A over F, define the *derived series* $B = B^{(0)} \supseteq B^{(1)} \supseteq \cdots \supseteq B^{(m)} \supseteq \cdots$, recursively by $B^{(m)} = B^{(m-1)}B^{(m-1)}$ for $m = 1, 2, \ldots$. If there is a positive integer m such that $B^{(m)} = 0$ then B is called *solvable*. If there is a unique maximal solvable ideal, $\operatorname{Rad} A$, of A which contains all solvable ideals of A then $\operatorname{Rad} A$ is called the *solvable radical* of A. □

It is easy to see that if B is a solvable ideal of A and C is a solvable subalgebra of A then $B + C$ is solvable. Thus, if A is finite–dimensional then $\operatorname{Rad} A$ exists and equals a solvable ideal of A with maximal dimension. For a class of algebras A of finite dimension where $\operatorname{Rad} A = 0$ implies the semisimplicity (in the sense of Definition 1.1), we have $R(A) = \operatorname{Rad} A$. For, since $\operatorname{Rad}(A/\operatorname{Rad} A) = 0$ [**Sc3**, p.18], by Theorem 1.7 $R(A) \subseteq \operatorname{Rad} A$ whereas $(R(A) + \operatorname{Rad} A)/R(A)$ is a solvable ideal of $A/R(A)$, being semisimple, thus $R(A) = \operatorname{Rad} A$. Well known examples of such algebras are Lie, Malcev ($\operatorname{char} F = 0$), Jordan ($\operatorname{char} F \neq 2$) and alternative algebras [**J2,3,Sc3,ZSSS1**]. For the last two algebras, $R(A) = \operatorname{Rad} A$ is nilpotent also [**J3,ZSSS1**].

If A is finite–dimensional over F then it is well known that the radical $R(M(A))$ of $M(A)$ is solvable and so nilpotent, i.e., $R(M(A))^m = 0$ for some positive integer m. There is a close relationship between $R(M(A))$ and $T(A)$.

Theorem 1.11. *If A is a finite–dimensional algebra over F then $T(A) = R(M(A))A$.*

Proof. Denote $S = R(M(A))$. Let M be a maximal ideal of A. If $SA + M = A$ then $A = S(SA + M) + M \subseteq S^2A + M$ and $A \subseteq S^mA + M = M$ for some $m > 0$, a contradiction. Thus, $A \neq SA + M$, and since $SA + M$ is an ideal of A, we have $SA + M = M$ by the maximality of M. Hence $SA \subseteq T(A)$. For the inclusion $SA \supseteq T(A)$, we regard A/SA as a module for the semisimple associative algebra $M(A)/S$. By a well known result for associative algebras, A/SA is a completely reducible $M(A)/S$–module and hence is the direct sum $A/SA = B_1/SA \oplus \cdots \oplus B_n/SA$ of irreducible $M(A)/S$–submodules B_i/SA, being ideals of A/SA. Since each B_i is an ideal of A, as in the proof of Theorem 1.7 there is a maximal ideal $U_k \supseteq SA$ of A such that $U_k/SA = B_1/SA \oplus \cdots \oplus B_{k-1}/SA \oplus B_{k+1}/SA \oplus \cdots \oplus B_n/SA$. Thus, $0 = \cap_{k=1}^n (U_k/SA) = (\cap_{k=1}^n U_k)/SA$ and so $T(A) \subseteq \cap_{k=1}^n U_k \subseteq SA$. This shows that $T(A) = SA = R(M(A))A$. □

Corollary 1.12. *If $A^2 = A$ then $R(A) = R(M(A))A$.* □

2. Simple and central simple algebras

The structure of certain class U of algebras over a field F not only depends on the type of algebras in U but also on the base field F. For example, in order to obtain a basis of an algebra in U which yields a nicest possible multiplication, the base field F needs to be large enough to contain all roots of certain polynomial equations over F, e.g., an algebraically closed field. This problem also arises in the classification of simple algebras in U and its representations [**J2,3**]. In general, it is more convenient to study those simple algebras which remain simple after any extension of scalars. In this section we briefly discuss some basic relationships between simple algebras and their scalar extensions.

Let A be an algebra over a field F and K be an extension field of F. The tensor product algebra $K \otimes_F A$ over F is made into an algebra over K under the scalar multiplication

$$\beta\left(\sum_i \alpha_i \otimes x_i\right) = \sum_i \beta\alpha_i \otimes x_i$$

for $\alpha_i, \beta \in K$ and $x_i \in A$. The K–algebra so obtained is called the *scalar extension* of A to K and is denoted by A_K $(= K \otimes_F A)$.

Definition 2.1. An algebra A over F is called *central simple* over F if A_K is simple over K for any extension field K of F. □

For an algebra A over F, denote:

$$\text{(2.1)}\qquad \begin{aligned} \text{Commutator} &: \quad [x, y] = xy - yx, \\ \text{Associator} &: \quad (x, y, z) = (xy)z - x(yz). \end{aligned}$$

Definition 2.2. The *nucleus* $N(A)$ of A is defined to be the set

$$N(A) = \{x \in A : (x, A, A) = (A, x, A) = (A, A, x) = 0\}.$$

The subset $Z(A) = \{c \in N(A) : [c, A] = 0\}$ is called the *center* of A. The *centroid* $\Gamma = \Gamma(A)$ of A is defined as the set

$$\Gamma = \{\gamma \in Hom_F A : [\gamma, M(A)] = 0\}.$$

□

Viewing an algebra A over F as a ring, we let $Hom_{\mathbb{Z}} A$ denote the associative ring of additive maps : $A \to A$. Clearly, $Hom_F A \subseteq Hom_{\mathbb{Z}} A$. It follows easily that the multiplication algebra $M(A)$ is the subring of $Hom_{\mathbb{Z}} A$ generated by $L_A \cup R_A$

and that $\Gamma = \Gamma(A)$ is a subring of $Hom_F A$. By a direct expansion, we obtain the identities

$$\begin{aligned}[xy,z]-x[y,z]-[x,z]y &= (x,y,z)-(x,z,y)+(z,x,y),\\ (xy,z,w)-(x,yz,w)+(x,y,zw) &= x(y,z,w)+(x,y,z)w\end{aligned} \tag{2.2}$$

holding for any algebra A. It follows from (2.2) that $N(A)$ and $Z(A)$ are subalgebras of A and hence $Z(A)$ is a commutative associative subalgebra of A.

For any subfield K of F, the algebra A over F is also regarded as an algebra over K by restricting the scalar multiplication to K. The resulting algebra, denoted by A^K, is called the K–descent of A. If F is a finite extension field of K of degree m and A is finite–dimensional over F then $\dim_K A^K = m \dim_F A$. The following results are often convenient.

Lemma 2.1. *Let A be an algebra over F.*

(i) *A is simple over F if and only if $M(A)x = A$ for all $0 \neq x \in A$.*

(ii) *A is simple as an algebra if and only if A is simple as a ring.*

(iii) *If K is a subfield of F then A is simple over F if and only if the K–descent A^K is simple over K.*

Proof. (i) If A is simple over F then for any $0 \neq x \in A$, $M(A)x$ is a nonzero ideal of A as a ring or as an algebra and hence $M(A)x = A$. The converse is clear since any ideal of A is stable under $M(A)$. Parts (ii) and (iii) follow from (i), since the relation $M(A)x = A$ does not depend on the scalar multiplication in A. □

Lemma 2.1(iii) generalizes a result in [**My4**, p. 189] where it is proven for a finite Galois extension F of K.

Lemma 2.2. *If $A^2 = A$ then Γ is a commutative associative algebra with 1, and furthermore, $\Gamma = \{\gamma \in Hom_{\mathbb{Z}} A : [\gamma, L_A] = [\gamma, R_A] = 0\}$.*

Proof. For $\gamma, \delta \in \Gamma$ and $x, y \in A$, we have

$$\begin{aligned}\gamma\circ\delta(xy) &= \gamma\circ\delta(L_x y) = \gamma\circ L_x\circ\delta(y) = \gamma(x\delta(y))\\ &= \gamma\circ R_{\delta(y)}(x) = R_{\delta(y)}\circ\gamma(x) = \gamma(x)\delta(y)\\ &= L_{\gamma(x)}\circ\delta(y) = \delta\circ L_{\gamma(x)}(y) = \delta\circ R_y(\gamma(x))\\ &= \delta\circ R_y\circ\gamma(x) = \delta\circ\gamma(xy).\end{aligned}$$

Since $A^2 = A$, this gives $\gamma\circ\delta = \delta\circ\gamma$ on A. For the second part, it suffices to show that all $\gamma \in Hom_{\mathbb{Z}} A$ with $[\gamma, L_A] = [\gamma, R_A] = 0$ are F–linear. For $\alpha \in F$, $\gamma(\alpha(xy)) = \gamma\circ L_{\alpha x}(y) = \alpha L_x\circ\gamma(y) = \alpha(\gamma\circ L_x(y)) = \alpha(\gamma(xy))$ and hence γ is F–linear. □

For $f, g \in Hom_F A$, we write $f \circ g = fg$. If $\gamma \in \Gamma$ then for $x, y \in A$, $\gamma(xy) = \gamma R_y(x) = R_y(\gamma(x)) = L_{\gamma(x)} y$ and hence

$$\gamma L_x = L_{\gamma(x)}, \quad \gamma R_x = R_{\gamma(x)} \tag{2.3}$$

for all $\gamma \in \Gamma$ and $x \in A$. For a simple algebra, we obtain a stronger result:

Theorem 2.3. *Let A be a simple algebra over F and Γ be its centroid.*

(i) *Γ is a field extending F.*

(ii) *If $Z(A) \neq 0$ then A has a unit element 1 and $Z(A) = \Gamma 1$.*

Proof. (i) By Lemma 2.2 it suffices to prove that each $0 \neq \gamma \in \Gamma$ has the inverse $\gamma^{-1} \in \Gamma$. Since $\operatorname{Im} \gamma = \gamma A \neq 0$ and $\ker \gamma$ are easily seen to be ideals of A, $\operatorname{Im} \gamma = A$ and $\ker \gamma = 0$. Thus, γ has the inverse γ^{-1} in $Hom_F A$. To show $\gamma^{-1} \in \Gamma$, we use (2.3) to obtain

$$\gamma^{-1} R_x = \gamma^{-1} R_{\gamma\gamma^{-1}(x)} = \gamma^{-1} \gamma R_{\gamma^{-1}(x)} = R_{\gamma^{-1}(x)},$$

and similarly, $R_x \gamma^{-1} = \gamma^{-1} R_x$ and $\gamma^{-1} L_x = L_x \gamma^{-1}$. Thus, $\gamma^{-1} \in \Gamma$ and Γ is a field. For $\alpha \in F$, the scalar map : $A \to A$: $x \mapsto \alpha x$ is clearly in Γ, and identifying this map with α, we have $F \subseteq \Gamma$.

(ii) If $c \in Z(A)$ then $L_c = R_c$, $L_c L_x = L_x L_c = L_{cx}$ and $R_c R_x = R_x R_c = R_{cx}$ for all $x \in A$, implying $R_{Z(A)} \subseteq \Gamma$. Note that if $0 \neq c \in Z(A)$ then $R_c \neq 0$ and so R_c is invertible in Γ. There is an element $1 \in A$ with $R_c 1 = L_c 1 = c$. For $x \in A$, $(x1)c = x(1c) = xc$, hence $R_c(x1 - x) = 0$ and $x1 = x$. Similarly, we have $1x = x$ to show that 1 is the unit element of A. From the above, $Z(A) = Z(A)1 = R_{Z(A)} 1 \subseteq \Gamma 1$, and clearly $\Gamma 1 \subseteq Z(A)$, thus $Z(A) = \Gamma 1$. □

Any simple algebra A over F is regarded as an algebra over the centroid Γ under $\gamma x = \gamma(x)$ for $x \in A$ and $\gamma \in \Gamma$. Before proving the standard characterization of central simplicity, we first note:

Lemma 2.4. *If K is a proper extension field of the field F then $K \otimes_F K$ is not simple over K.*

Proof. Consider the multiplication map $f : K \otimes_F K \to K$: $\alpha \otimes \beta \mapsto \alpha\beta$ for $\alpha, \beta \in K$. Clearly, f is a K-algebra homomorphism, so that $\ker f$ is an ideal of $K \otimes_F K$. Since $\ker f$ contains a nonzero element $\alpha \otimes 1 - 1 \otimes \alpha$ for $\alpha \in K$ with $\alpha \notin F$, it is a proper ideal of $K \otimes_F K$. □

Theorem 2.5. *A simple algebra A over a field F with centroid Γ is central simple over F if and only if $\Gamma = F$.*

Proof. Suppose $\Gamma = F$. Since A is simple over F, $M(A)$ acts irreducibly on A. By Jacobson's density theorem, for any linearly independent elements $x_1, \ldots, x_n \in A$ and for any elements $y_1, \ldots, y_n \in A$ there is a $\rho \in M(A)$ with $\rho(x_i) = y_i$, $i = 1, \ldots, n$. Let K be an extension field of F and consider $A_K = K \otimes_F A$. Any nonzero element $x \in A_K$ is expressed as $x = \alpha_1 \otimes x_1 + \cdots + \alpha_n \otimes x_n$ for some nonzero scalars $\alpha_1, \ldots, \alpha_n \in K$ and linearly independent elements $x_1, \ldots, x_n \in A$. Choosing a $\rho \in M(A)$ with $\rho(x_1) = x_1$, $\rho(x_2) = \cdots = \rho(x_n) = 0$ and using $M(A)x_1 = A$ (by Lemma 2.1), we have

$$\begin{aligned} M(K \otimes_F A)x &\supseteq M(K \otimes_F A)(1 \otimes \rho)x = M(K \otimes_F A)(\alpha_1 \otimes x_1) \\ &= (\alpha_1 K) \otimes_F M(A)x_1 = K \otimes_F A. \end{aligned}$$

Thus, by Lemma 2.1(i) $K \otimes_F A$ is simple over K.

Assume that A is central simple over F. Suppose $F \neq \Gamma$. Since A is also an algebra over Γ, $A \cong \Gamma \otimes_\Gamma A$ and

$$\Gamma \otimes_F A \cong \Gamma \otimes_F (\Gamma \otimes_\Gamma A) \cong (\Gamma \otimes_F \Gamma) \otimes_\Gamma A. \tag{2.4}$$

By Lemma 2.4, $\Gamma \otimes_F \Gamma$ has a proper ideal Δ over Γ, and $\Delta \otimes_\Gamma A$ is a proper ideal of $(\Gamma \otimes_F \Gamma) \otimes_\Gamma A$. Thus, $\Gamma \otimes_F A$ is not simple over Γ, a contradiction. □

Let A be a simple algebra over F and K be a proper subfield of F. Then, the K–descent A^K is simple over K and $A^K \cong F \otimes_F A^K$, since A^K is an algebra over F. As in (2.4),

$$(A^K)_F = F \otimes_K A^K \cong F \otimes_K (F \otimes_F A^K) \cong (F \otimes_K F) \otimes_F A^K.$$

As argued above, $(A^K)_F$ is not simple over F. This shows that there exist a large number of simple algebras which are not central simple. When F is a finite Galois extension of K, a stronger result has been obtained in [**My4**, p.187]; $(A^K)_F$ is semisimple over F.

Corollary 2.6. *Any simple algebra A over F is central simple over its centroid Γ.*

Proof. By Lemma 2.1 (iii) A is simple over Γ and by Lemma 2.2 the centroid of A over Γ is Γ. It follows from this and Theorem 2.5 that A is central simple over Γ. □

Let A be a finite–dimensional simple algebra over an algebraically closed field F, e.g., the field of complex numbers. Since the centroid Γ of A is a finite extension field of F, $\Gamma = F$, and hence A is central simple over F. Let A be a semisimple algebra with decomposition $A = S_1 \oplus \cdots \oplus S_n$ into simple ideals S_i of A. It is easy to see that as algebras,

$$\begin{aligned} M(A) &= M(S_1) \oplus \cdots \oplus M(S_n), \\ Z(A) &= Z(S_1) \oplus \cdots \oplus Z(S_n), \\ \Gamma(A) &= \Gamma(S_1) \oplus \cdots \oplus \Gamma(S_n). \end{aligned}$$

3. Alternative algebras

An algebra A over a field F with multiplication xy is called *alternative* if it satisfies the left and right alternative laws

(3.1) $$(x, x, y) = 0 = (y, x, x)$$

for all $x, y \in A$. In what follows, the base field F is assumed to be of characteristic $\neq 2$. Linearizing (3.1), we find that A is alternative if and only if (x, y, z) is an alternating function in $x, y, z \in A$; that is,

(3.2) $$(x, y, z) = -(y, x, z) = (y, z, x) = -(z, y, x)$$

for all $x, y, z \in A$. In particular, (3.2) with $x = z$ gives the *flexible identity*

(3.3) $$(x, y, x) = 0 \quad \text{or} \quad (xy)x = x(yx).$$

An algebra is called *flexible* if it satisfies (3.3).

The following Moufang identities, which are consequences of (3.1)–(3.3), play fundamental roles for the structure of alternative algebras [**Sc3,J3,ZSSS1**]:

(3.4) $$(xax)y = x(a(xy)),$$

(3.5) $$y(xax) = ((yx)a)x,$$

(3.6) $$(xy)(ax) = x(ya)x$$

for all $a, x, y \in A$. Using the Moufang identities, it is not difficult to verify the following theorem of Artin [**Sc3**]:

Theorem 3.1. *Any two elements in an alternative algebra generate an associative subalgebra.* □

Thus, alternative algebras are nonassociative algebras which are "sufficiently" close to associative algebras. The first discovery of an alternative algebra that is not associative was independently due to Cayley (1845) and Graves (1844), commonly called the *Cayley algebra* (over the field $\mathbb{R}$ of real numbers). The Cayley algebra is closely related to the classical Hurwitz problem (1898) which was evolved into the study of composition algebras. The Cayley algebra can also be constructed through the so–called *Cayley–Dickson process*. In this construction, alternative algebras are obtained only for dimension $1, 2, 4$ or 8 which is also the only possible dimension of a composition algebra. Such algebras of dimension 8 are called *Cayley–Dickson algebras* (or *octonion algebras*). It is a well known structure theorem that Cayley–Dickson algebras are the only simple alternative algebras that are not associative [**Sc3,ZSSS1**]. Thus, in a sense, the theory of alternative algebras begins and ends with the Cayley algebra. Cayley–Dickson algebras and, in particular, the Cayley algebra have many striking relationships with other nonassociative algebras as well as with geometry, topology and physics.

It is the main concern in this section to review some basic definitions and structure theorems of alternative algebras.

Definition 3.1. Let A be an algebra over a field F. A quadratic form $n : A \to F$ is said to be *nondegenerate* if the associated bilinear form $(x, y) = \frac{1}{2}(n(x+y) - n(x) - n(y))$ is nondegenerate. We say that n *permits composition* in A if the composition law

$$n(xy) = n(x)n(y) \tag{3.7}$$

is satisfied for all $x, y \in A$. A *composition algebra* is an algebra A over F with a unit element 1 which has a nondegenerate quadratic form permitting composition. □

The Hurwitz problem is to determine all nondegenerate quadratic forms permitting composition on an n–dimensional Euclidean space. The problem for arbitrary field is the generalized Hurwitz problem which is closely linked with the Cayley–Dickson process.

Definition 3.2. An algebra A over F with a unit element 1 is called *quadratic* over F if every $x \in A$ satisfies the quadratic equation

$$x^2 - t(x)x + n(x)1 = 0 \tag{3.8}$$

where $t(x)$ and $n(x)$ are scalars in F with $t(\alpha) = 2\alpha$ and $n(\alpha) = \alpha^2$ for $\alpha \in F$. In this case, $t(x)$ and $n(x)$ are called the *trace* and *norm*, respectively, of x in A. □

An *involution* on an algebra A over F is a linear map $A \to A : x \mapsto \bar{x}$ such that $\bar{\bar{x}} = x$ and $\overline{xy} = \bar{y}\bar{x}$ for all $x, y \in A$. Involutions play important roles in the structure theory of alternative algebras. The following results establish relationships between composition algebras and alternative quadratic algebras [**ZSSS1**].

Theorem 3.2. (i) *If A is a quadratic algebra over F, then the trace t is a linear form on A and the norm n is a quadratic form on A.*

(ii) *The norm of any alternative quadratic algebra over F permits composition.*

(iii) *Any simple alternative quadratic algebra over F is a composition algebra.*

(iv) *Any composition algebra A over F is an alternative quadratic algebra with involution $x \mapsto \bar{x}$ such that $\bar{x} = t(x)1 - x$, $t(x)1 = x + \bar{x} \in F1$ and $n(x) = x\bar{x} \in F$ for all $x \in A$.*

(v) *Any composition algebra over F is either simple or isomorphic to the direct sum $F \oplus F$ with involution $\overline{(\alpha, \beta)} = (\beta, \alpha)$.* □

Therefore, the main task is to classify the algebras described in Theorem 3.2. In fact, these algebras all arise from the Cayley–Dickson process. We begin with an algebra A over F with a unit element 1 and with an involution $a \mapsto \bar{a}$ such that $a + \bar{a}, a\bar{a} \in F$ for all $a \in A$ (identify $\alpha 1 = \alpha$, $\alpha \in F$), hence A is a quadratic algebra over F with trace $a + \bar{a}$ and norm $a\bar{a}$. The *Cayley–Dickson process* is a doubling process to construct a new algebra of dimension $2 \dim A$ from A ($\dim A < \infty$), which contains A as a subalgebra and has an involution.

Let $\alpha \neq 0$ be a fixed scalar in F and let vA be a linear isomorphic copy of A (put $v = v1$). We define a multiplication in the vector space direct sum $(A, \alpha) \equiv A \oplus vA$ by

$$(a_1 + va_2)(a_3 + va_4) = (a_1 a_3 + \alpha a_4 \bar{a}_2) + v(\bar{a}_1 a_4 + a_3 a_2) \tag{3.9}$$

for $a_i \in A$, $i = 1, 2, 3, 4$, where the distinctive element v satisfies $v^2 = \alpha$. For $x = a_1 + va_2 \in (A, \alpha)$, let

$$\bar{x} = \bar{a}_1 - va_2, \; t(x) = t(a_1), \; n(x) = n(a_1) - \alpha n(a_2), \tag{3.10}$$

where $t(a) = a + \bar{a}$ and $n(a) = a\bar{a}$ for $a \in A$. It is easy to see that (A, α) is a quadratic algebra over F with involution $x \mapsto \bar{x}$, trace and norm given by (3.10), and that if n is nondegenerate on A, then so is n on (A, α) [**Sc3,ZSSS1**]. Thus, starting from a composition algebra A of dimension m, we can continue the Cayley–Dickson process to obtain an algebra of dimension $2^t m$. However, only a few composition or alternative algebras are gained by this process. The following result can be easily proven by a direct calculation using (3.9) and (3.10) [**Sc3,ZSSS1**].

Lemma 3.3. (i) *If A is a composition algebra, then (A, α) is a composition algebra if and only if A is associative.*

(ii) *If A is an algebra over F with a unit element and with involution $a \mapsto \bar{a}$ such that $a + \bar{a}$, $a\bar{a} \in F$ for all $a \in A$, then (A, α) is alternative if and only if A is associative.* □

When the Cayley–Dickson process begins with a field F, the composition law collapses after the first three processes, thereby producing only three additional classes of composition algebras of dimensions 2, 4 and 8. The first two are associative, and the last one, called a *Cayley–Dickson algebra*, is not associative but is a central simple alternative algebra over F [**Sc3,ZSSS1**].

Denote by v_1, v_2, v_3 the distinctive elements in the first three steps of the Cayley–Dickson process, so that $v_1^2 = \alpha$, $v_2^2 = \beta$ and $v_3^2 = \gamma$ for nonzero scalars $\alpha, \beta, \gamma \in F$. Letting $e_1 = v_1$, we obtain a basis $\{1, e_1\}$ of (F, α) with $e_1^2 = \alpha (= \alpha 1)$. This is extended to a basis $\{1, e_1, e_2, e_3\}$ of $Q(\alpha, \beta) \equiv (F(\alpha), \beta)$ by putting $e_2 = v_2$ and $e_3 = v_1 v_2 = e_1 e_2$. It follows immediately from (3.9) that $e_3^2 = -\alpha\beta$, $e_1 e_2 = e_3 = -e_2 e_1$, $e_2 e_3 = -\beta e_1 = -e_3 e_2$ and $e_3 e_1 = -\alpha e_2 = -e_1 e_3$. This is a standard basis for the *generalized quaternion algebra* $Q(\alpha, \beta)$. Similarly, letting $e_4 = v_3, e_5 = e_1 e_4 = v_1 v_3, e_6 = e_2 e_4 = v_2 v_3, e_7 = (e_1 e_2) e_4 = (v_1 v_2) v_3$, we obtain a standard basis $\{1, e_1, e_2, e_3, e_4, e_5, e_6, e_7\}$ for a Cayley–Dickson algebra $C(\alpha, \beta, \gamma) = (Q(\alpha, \beta), \gamma)$. The following multiplication table for $C(\alpha, \beta, \gamma)$ is easily obtained from (3.9).

TABLE I.1

	1	e_1	e_2	e_3	e_4	e_5	e_6	e_7
1	1	e_1	e_2	e_3	e_4	e_5	e_6	e_7
e_1	e_1	$\alpha 1$	e_3	αe_2	e_5	αe_4	$-e_7$	$-\alpha e_6$
e_2	e_2	$-e_3$	$\beta 1$	$-\beta e_1$	e_6	e_7	βe_4	βe_5
e_3	e_3	$-\alpha e_2$	βe_1	$-\alpha\beta 1$	e_7	αe_6	$-\beta e_5$	$-\alpha\beta e_4$
e_4	e_4	$-e_5$	$-e_6$	$-e_7$	$\gamma 1$	$-\gamma e_1$	$-\gamma e_2$	$-\gamma e_3$
e_5	e_5	$-\alpha e_4$	$-e_7$	$-\alpha e_6$	γe_1	$-\alpha\gamma 1$	γe_3	$\alpha\gamma e_2$
e_6	e_6	e_7	$-\beta e_4$	βe_5	γe_2	$-\gamma e_3$	$-\beta\gamma 1$	$-\beta\gamma e_1$
e_7	e_7	αe_6	$-\beta e_5$	$\alpha\beta e_4$	γe_3	$-\alpha\gamma e_2$	$\beta\gamma e_1$	$\alpha\beta\gamma 1$

In Table I.1, the blocks containing $\{1, e_1\}$ and $\{1, e_1, e_2, e_3\}$ are the multiplication tables for (F, α) and $Q(\alpha, \beta)$. If α is not a square in F, then (F, α) is isomorphic to a separable quadratic extension field $F(s)$, and otherwise, $(F, \alpha) \cong F \oplus F$. Let $x = \sum_{i=0}^{7} x_i e_i$, $(e_0 = 1,\ x_i \in F)$, be any element of $C(\alpha, \beta, \gamma)$. Using $n(x) = x\bar{x}$,

(3.10) and Table I.1, we easily find

(3.11)
$$\begin{aligned} n(x) &= x_0^2 \quad \text{for } F; \\ n(x) &= x_0^2 - \alpha x_1^2 \quad \text{for } (F,\alpha); \\ n(x) &= x_0^2 - \alpha x_1^2 - \beta x_2^2 + \alpha\beta x_3^2 \quad \text{for } Q(\alpha,\beta); \\ n(x) &= x_0^2 - \alpha x_1^2 - \beta x_2^2 + \alpha\beta x_3^2 \\ &\quad -\gamma x_4^2 + \alpha\gamma x_5^2 + \beta\gamma x_6^2 - \alpha\beta\gamma x_7^2 \quad \text{for } C(\alpha,\beta,\gamma). \end{aligned}$$

The algebras constructed above are essentially all about composition algebras, as shown by the generalized Hurwitz theorem [**Sc3,ZSSS1**]:

Theorem 3.4. *Let A be an algebra over F (not necessarily finite–dimensional). Then, A is a composition algebra if and only if A is isomorphic to one of the algebras: $F\oplus F$, $F(s)$ (a separable quadratic extension field), $Q(\alpha,\beta)$ and $C(\alpha,\beta,\gamma)$.* □

Definition 3.3. Let A be an algebra over F. A nonzero element $a \in A$ is called a *zero divisor* in A if there exists an element $b \neq 0$ in A such that $ab = 0$ or $ba = 0$. The algebra A is called a *division algebra* if for any elements $a, b \in A$ with $a \neq 0$ the equations $ax = b$ and $ya = b$ have a unique solution, or equivalently, the linear maps L_a, R_a are invertible in $Hom_F A$. □

If A is finite–dimensional, then it is clear that A is a division algebra if and only if it has no zero divisors. For a composition algebra, this can be described in terms of norm:

Lemma 3.5. *A composition algebra A with norm n has no zero divisors if and only if $n(x) \neq 0$ for $x \neq 0$ in A if and only if it is a division algebra.* □

This is easily seen from $n(x) = x\bar{x}$. A composition algebra which does not satisfy one of the conditions in Lemma 3.5 is called *split*. Split composition algebras are easier to describe [**Sc3,ZSSS1**]:

Theorem 3.6. *There exists a unique split composition algebra over F for each of dimensions 2, 4 and 8. It is isomorphic to one of the algebras: $F \oplus F$, the algebra $M_2(F)$ of 2×2 matrices over F, or the split Cayley-Dickson algebra $C(F)$ with multiplication given by*

(3.12)
$$\begin{aligned} &e_i^2 = e_i \ (i = 1, 2), \ e_1 u_i = u_i = u_i e_2, \ e_2 v_i = v_i = v_i e_1, \\ &u_i u_j = \epsilon_{ijk} v_k, \ v_i v_j = \epsilon_{ijk} u_k, \ u_i v_j = -\delta_{ij} e_1, \ v_i u_j = -\delta_{ij} e_2 \end{aligned}$$

for $i, j, k = 1, 2, 3$, *and all other products are zero, where* $\{e_1, e_2, u_i, v_i : i = 1, 2, 3\}$ *is a basis of* $C(F)$ *and* ϵ_{ijk} *is a totally skew symmetric tensor in* $1, 2, 3$ *with* $\epsilon_{123} = 1$. □

We note that $C(F) \cong C(1, 1, 1)$. The algebra $C(F)$ is also called the *Zorn vector matrix algebra* over F, because it can be described in terms of 2×2 matrices of scalars and vectors [**ZSSS1**,**My4**]. The descriptions of non–split or division composition algebras A of dimension ≥ 4 are more subtle and are given in terms of norm n. A quadratic form n on A is said to be *similar* to a quadratic form n' on A' if there is a bijective linear map $f : A \to A'$ such that $n'(f(x)) = n(x)$ for all $x \in A$. Note that the similarity relation is an equivalence relation on the norms in composition algebras of the same dimension. The following theorem is due to Jacobson (for a proof, see [**Sc3**, p.70]):

Theorem 3.7. *Let A and A' be composition algebras of the same dimension with norms n and n', respectively. Then, A is isomorphic to A' if and only if n is similar to n'.* □

Let $F(s)$ be a separable quadratic extension field of F. If $x = a_1 + va_2 \in Q(\alpha, \beta) = F(s) \oplus vF(s)$ is a nonzero element with $n(x) = 0$ then $a_2 \neq 0$ and by (3.10) $\beta = n(a_1)n(a_2)^{-1} = n(a_1 a_2^{-1})$, thus the equation $n(x) = \beta$ has a solution x in $F(s)$. A similar observation is made with $Q(\alpha, \beta)$ and $C(\alpha, \beta, \gamma)$. Thus, we have

Theorem 3.8. *A quaternion algebra $Q(\alpha, \beta) = (F, \alpha) \oplus v(F, \alpha)$ is a division algebra if and only if $(F, \alpha) = F(s)$ is a separable quadratic extension field of F and the equation $n(x) = \beta$ has no solution in $F(s)$. A Cayley–Dickson algebra $C(\alpha, \beta, \gamma) = Q(\alpha, \beta) \oplus vQ(\alpha, \beta)$ is a division algebra if and only if $Q(\alpha, \beta)$ is a division algebra and the equation $n(x) = \gamma$ has no solution in $Q(\alpha, \beta)$.* □

If, in particular, F is algebraically closed, then the equations above have solutions by (3.11) and hence any composition algebra over F of dimension ≥ 2 is split. Thus, by Theorem 3.6 we have

Corollary 3.9. *Any composition algebra over an algebraically closed field F is isomorphic to one of the algebras: F, $F \oplus F$, $M_2(F)$ or $C(F)$.* □

For the field $\mathbb{R}$ of real numbers we have a stronger result [**ZSSS1**, p.40]:

Theorem 3.10. *Any division composition algebra over $\mathbb{R}$ is isomorphic to one of the algebras: $\mathbb{R}$, $\mathbb{C}$ (the field of complex numbers), $Q(-1, -1)$ or $C(-1, -1, -1)$.* □

The algebra $Q(-1, -1)$ is known as the division ring of *real quaternions* and $C(-1, -1, -1)$ is the classical algebra of *Cayley numbers*. More generally, $\mathbb{R}$, $\mathbb{C}$,

$Q(-1,-1)$ and $C(-1,-1,-1)$ are the only finite–dimensional real alternative algebras without zero divisors (the generalized Frobenius theorem [**ZSSS1**, p.40]). Linearizing the quadratic equation (3.8) and using Table I.1, we easily obtain [**ZSSS1**, p.41]:

Theorem 3.11. *Let A be a composition algebra over F.*

(i) *If $\dim_F A \geq 4$ then the center $Z(A) = F1$, and $N(C(\alpha,\beta,\gamma)) = Z(C(\alpha,\beta,\gamma)) = F1$, where $N(A)$ denotes the nucleus of A (Definition 2.2).*

(ii) *The trace t of A is symmetric and associative:*

$$t(xy) = t(yx), \quad t((xy)z) = t(x(yz))$$

for all $x, y, z \in A$. □

Thus, by Theorems 2.3 and 2.5, any composition algebra of dimension ≥ 4 is central simple, and $C(\alpha,\beta,\gamma)$ is the only finite–dimensional central simple alternative algebra that is not associative [**Sc3,ZSSS1**]:

Theorem 3.12. *A finite–dimensional central simple alternative algebra over a field F is either a Cayley–Dickson algebra or the algebra $M_n(D)$ of $n \times n$ matrices over a central division associative algebra D over F.* □

For a finite–dimensional alternative algebra A over F, let $\operatorname{Rad} A$ denote the solvable radical of A. As noted in Section I.1, $\operatorname{Rad} A = R(A)$, and A is semisimple if and only if $R(A) = 0$. Moreover, if $R(A) \neq A$ then $A/R(A)$ is semisimple (Theorem 1.7). An algebra A over F is called *separable* if for any extension field K of F, $A_K = K \otimes_F A$ is semisimple. Wedderburn principal theorem shows how an alternative algebra decomposes as a semi–direct sum of a semisimple subalgebra and its radical [**Sc3**]:

Theorem 3.13. *Let A be a finite-dimensional alternative algebra over a field F. If $A/R(A)$ is separable, then A has a semi–direct sum decomposition $A = S + R(A)$, where S is a semisimple subalgebra of A and $S \cong A/R(A)$.* □

Wedderburn principal theorem holds for certain broader classes of algebras, including Jordan algebras of characteristic $\neq 2$ [**J3**]. The decomposition above is well known for Lie and Malcev algebras of characteristic 0, and is called a *Levi decomposition* [**J3,Ku2**]. In this case, all semisimple algebras are separable.

The structure of simple alternative algebras A over a field F of arbitrary dimension is more subtle, and the principal result in this case is Kleinfeld's theorem that if A is not associative then it is a Cayley–Dickson algebra over its center $Z(A)$. The proof of this case can be achieved by several steps. First, consider the ideal $\langle [N,A] \rangle$

of A generated by $[N, A]$ where $N = N(A)$ denotes the nucleus of A. Clearly, $N = Z$, the center of A, if and only if $\langle [N, A] \rangle = 0$. The ideal $\langle [N, A] \rangle$ can be described by $\langle [N, A] \rangle = [N, A] + [N, A]A = [N, A] + A[N, A]$ ([**ZSSS1**, p.140]). If A is simple then $\langle [N, A] \rangle = 0$ or A, and if in addition A is not associative then it is not difficult to show $\langle [N, A] \rangle \neq A$, so that [**ZSSS1**, p.141]:

Theorem 3.14. *If A is a simple alternative algebra that is not associative then $N(A) = Z(A)$.* □

If A is simple and not associative then it must not be commutative, because of the following result of Zhevlakov [**ZSSS1**, p.143]:

Theorem 3.15. *A simple commutative alternative algebra is a field.* □

A striking feature of alternative algebras is that they satisfy many useful identities involving commutators and associators. Among those is the identity, due to Shestakov [**ZSSS1**, p.148],

$$n_1(x, y)x^2 - n_2(x, y)x + n_3(x, y) = 0 \tag{3.13}$$

which is satisfied by any alternative algebra A, where n_i $(i = 1, 2, 3)$ given by

$$\begin{aligned} n_1(x, y) &= [x, y]^4, \\ n_2(x, y) &= [x, y]([x, y][x, yx] + [x, yx][x, y]), \\ n_3(x, y) &= [x, y]^2[x, yx]^2 \end{aligned} \tag{3.14}$$

are contained in $N(A)$.

In what follows, assume that A is simple over F and not associative. Thus, if $Z(A) \neq 0$ then by Theorem 2.3 $Z(A)$ is a field and A has a unit element 1, so that A satisfies (3.13) with coefficients in a field. In fact, it is shown that there exist elements $a, b \in A$ such that $0 \neq [a, b]^4 = n_1(a, b) \in N = Z$ [**ZSSS1**, p.151]. Thus, the main concern is to show that (3.13) induces a quadratic equation $x^2 + \alpha(x)x + \beta(x)1 = 0$ with $\alpha(x), \beta(x) \in Z$. But, this follows from the following stronger result which can be easily proven by only using $0 \neq [a, b]^4$ and (3.13) [**ZSSS1**, p.150].

Lemma 3.16. *Let A be an alternative algebra with a unit element 1 over an infinite field F. If $n_i(x, y) \in F$ for all $x, y \in A$ and $i = 1, 2, 3$ and $n_1(a, b) = [a, b]^4 \neq 0$ for some $a, b \in A$ then A is a quadratic algebra over F.* □

We now regard A as an algebra over Z and consider $A_K = K \otimes_Z A$ for some infinite extension field K of Z. Thus, A_K is central simple over K, so $N(A_K) = Z(A_K) = K$. Moreover, since A is not commutative, A_K satisfies the conditions in

Lemma 3.16 and hence is a quadratic algebra over K. But then, A is quadratic over Z [**Sc3**, p.50]. By Theorem 3.2 (iii) A is a composition algebra over Z, which must be a generalized quaternion or Cayley–Dickson algebra over Z. Therefore, if A is simple and not associative then by Theorem 3.15 A is not commutative, so it must be a Cayley–Dickson algebra. In this way we arrive at Kleinfeld's theorem:

Theorem 3.17. *If A is a simple alternative algebra that is not associative then the center $Z(A)$ is a field and A is a Cayley–Dickson algebra over $Z(A)$.* □

4. Malcev–admissible algebras

An algebra L over a field F with multiplication denoted by $[x,y]$ is called a *Lie algebra* if it is anticommutative, i.e., $[x,x]=0$ for all $x \in L$, and satisfies the Jacobi identity

$$J(x,y,z) \equiv [[x,y],z]+[[y,z],x]+[[z,x],y]=0 \tag{4.1}$$

for all $x,y,z \in L$. For an algebra A over F with multiplication xy, let A^- denote the algebra with multiplication $[x,y]$,

$$[x,y]=xy-yx$$

defined on the same vector space as A. The left side $J(x,y,z)$ of (4.1) is called the *Jacobian* in A. The basic source of Lie algebras comes from A^- for associative algebras A, i.e., $J(A,A,A)=0$ (this is easily seen by expanding $J(x,y,z)$ and using the associative law). Thus, any subalgebra of A^- for an associative algebra A is a Lie algebra. For example, let $A=M_n(F)$ or A be an associative algebra over F with an involution $x \mapsto \bar{x}$. Then, the set of trace zero elements in A or of skew symmetric elements in A (i.e., $\{x \in A : \bar{x}=-x\}$) is a subalgebra of A^- and hence is a Lie algebra. In fact, by Poincaré–Birkhoff–Witt Theorem, any Lie algebra is isomorphic to a subalgebra of A^- for some associative algebra A. However, the associative algebras are not the only source to obtain Lie algebras from A^-.

Definition 4.1. An algebra A over a field F is called *Lie–admissible* if A^- is a Lie algebra, i.e., $J(A,A,A)=0$. □

Lie-admissible algebras were first introduced by Albert [**A3**] in 1948 and originated from one of the defining identities for a standard algebra. Lie and associative algebras are clearly Lie–admissible. In any algebra A, let $S(x,y,z)$ denote

$$S(x,y,z)=(x,y,z)+(y,z,x)+(z,x,y).$$

Lemma 4.1. *Let A be an algebra over F.*

(i) *A is Lie–admissible if and only if*

$$S(x, y, z) = S(x, z, y) \tag{4.2}$$

for all $x, y, z \in A$.

(ii) *If* $S(A, A, A) = 0$, *then A is Lie–admissible.*

(iii) *If A is Lie–admissible of characteristic* $\neq 2$ *and satisfies the third–power identity* $(x, x, x) = 0$, *then* $S(A, A, A) = 0$.

Proof. (i) The identity

$$[xy, z] + [yz, x] + [zx, y] = S(x, y, z) \tag{4.3}$$

holds for any algebra. Hence

$$J(x, y, z) = S(x, y, z) - S(x, z, y)$$

for all $x, y, z \in A$. It follows from this that A is Lie–admissible if and only if (4.2) holds for A.

(ii) is immediate from (i).

(iii) Linearizing $(x, x, x) = 0$ twice, we obtain

$$[xy + yx, z] + [yz + zy, x] + [zx + xz, y] = 0$$

for all $x, y, z \in A$, which is equivalent to the identity $S(x, y, z) + S(x, z, y) = 0$ by (4.3). Hence, by (4.2) $S(A, A, A) = 0$ since the characteristic of F is not two. □

A linear transformation δ on A is called a *derivation* of A if

$$\delta(xy) = \delta(x)y + x\delta(y)$$

for all $x, y \in A$. It is straightforward to check that the set Der A of all derivations of A is a Lie subalgebra of $(Hom_F A)^-$, called the *derivation algebra* of A. The study of derivation algebras is a central topic in the theory of nonassociative algebras.

For any algebra A over F, define the adjoint map ad_x by

$$ad_x : A \to A \ : \ y \mapsto ad_x y = [x, y] \tag{4.4}$$

for $x, y \in A$. For flexible algebras, we have:

Lemma 4.2. *Let A be an algebra over F of characteristic* $\neq 2$.

(i) *A is a flexible Lie–admissible if and only if*

$$(x, y, z) + (z, x, y) - (x, z, y) = 0 \tag{4.5}$$

for all $x, y, z \in A$.

(ii) *A is flexible Lie–admissible if and only if*

$$[xy, z] = x[y, z] + [x, z]y \tag{4.6}$$

for all $x, y, z \in A$ if and only if $ad_A \subseteq \mathrm{Der}\, A$.

Proof. (i) Linearizing the flexible identity $(x, y, x) = 0$ gives

$$(x, y, z) + (z, y, x) = 0 \tag{4.7}$$

for all $x, y, z \in A$. It follows easily from (4.7) that if A is flexible, then (4.5) is equivalent to $S(A, A, A) = 0$. Flexibility is obtained from (4.5) with $x = z$, and hence the result follows from Lemma 4.1.

(ii) Using the first identity of (2.2), we find that (4.5) and (4.6) are equivalent, that is, A is flexible Lie–admissible if and only if $ad_A \subseteq \mathrm{Der}\, A$. □

An algebra A over F is called *standard* if it satisfies (4.5) and

$$(x, y, wz) + (w, y, xz) + (z, y, wx) = 0$$

for all $x, y, z, w \in A$ [**A3**]. Hence, by Lemma 4.2, a standard algebra is flexible Lie–admissible. The identity above arises from a linearization of the defining identities of a Jordan algebra [**A3**], and hence a standard algebra is a noncommutative Jordan algebra for char $F \neq 2, 3$ [**A3,Sc4**] (see Definition 5.2). The introduction of standard algebras was the beginning of a series of generalizations for Jordan– and alternative–like theories during 1960-70's (e.g., [**Sc4,T1,2**]).

A *power–associative algebra* is an algebra in which every element generates an associative subalgebra (e.g., associative algebras and alternative algebras by Artin's theorem). A systematic study of Lie–admissible algebras, in particular, of flexible Lie–admissible algebras was aroused from Albert's problem [**A3**] posed in 1948: Determine all power–associative flexible Lie–admissible algebras A over F such that A^- is semisimple. Albert's problem was first solved in 1962 [**LT1**] by showing that if A is finite–dimensional and F is algebraically closed of characteristic 0, then A is a Lie algebra isomorphic to A^-. This result was extended to the case of char $F = p > 2$ when A^- is a classical Lie algebra or a generalized Witt algebra [**My1,4**].

Around 1980, a unifying approach to the study of flexible Lie–admissible algebras surged using representations of Lie algebras. To briefly illustrate the underlying idea of this approach, let L be a Lie algebra over F and let A be an L–module with composition ax for $a \in L$ and $x \in A$. Thus, it *affords* a representation $\phi : L \to Hom_F A$ by $\phi(a)x = ax$. Let $Hom_L(A \otimes_F A, A)$ denote the F–space of all L–module homomorphisms of $A \otimes_F A$ into A (see Section I.1).

Proposition 4.3. *Let L be a Lie algebra over F and A be an L–module with representation ϕ afforded by the L–module. There exists a bijective correspondence between the set $Hom_L(A \otimes_F A, A)$ and the set of bilinear multiplications xy defined on A such that $\phi(L) \subseteq \operatorname{Der} A$. The correspondence is given by*

$$f(x \otimes y) = xy \tag{4.8}$$

for $f \in Hom_L(A \otimes_F A, A)$ and $x, y \in A$.

Proof. If $f \in Hom_L(A \otimes_F A, A)$, then we define a multiplication xy in A by (4.8). It follows that

$$\begin{aligned} a(xy) &= a(f(x \otimes y)) = f(a(x \otimes y)) \\ &= f((ax) \otimes y) + f(x \otimes (ay)) \\ &= (ax)y + x(ay) = (\phi(a)x)y + x(\phi(a)y) \end{aligned}$$

for $x, y \in A$ and $a \in L$, and hence $\phi(L) \subseteq \operatorname{Der} A$. Conversely, if A is an algebra over F with multiplication xy and with $\phi(L) \subseteq \operatorname{Der} A$ then the map $f : A \otimes_F A \to A$ defined by (4.8) is easily seen to be an L–module homomorphism of $A \otimes_F A$ into A. □

Proposition 4.3 means that, for a fixed subalgebra L of $(Hom_F A)^-$, the determination of all algebra multiplications defined on A with $L \subseteq \operatorname{Der} A$ reduces to the problem of determining $Hom_L(A \otimes_F A, A)$. For example, in view of Lemma 4.2 (ii) one easily finds that an algebra A is flexible Lie–admissible if and only if the map $x \otimes y \mapsto xy$ is an A^-–module homomorphism of $A \otimes_F A$ into A under the adjoint action (4.4). Therefore, to determine all flexible Lie–admissible products on A such that A^- is a prescribed Lie algebra, it suffices to determine $Hom_{A^-}(A \otimes_F A, A)$ for the adjoint module A. The latter problem is intimately related to the decomposition of $A \otimes_F A$ into irreducible submodules, as shown by the following result [**EMy1**]:

Proposition 4.4. *Let L be a finite–dimensional Lie algebra over a field F and let U, V be finite–dimensional L–modules over F such that $U = U_1 \oplus \cdots \oplus U_n$ is the direct sum of irreducible L–modules U_i and V is an absolutely irreducible L–module over F (i.e., any scalar extension of V is irreducible). Then, $\dim_F Hom_L(U, V)$ equals the number of U_i's isomorphic to V.*

Proof. Let m be such that $U_i \cong V$ for $1 \leq i \leq m$ and $U_i \not\cong V$ for $m+1 \leq i \leq n$. Hence, $Hom_L(U_i, V) = 0$ for $m + 1 \leq i \leq n$ and

$$Hom_L(U, V) \cong Hom_L(V, V) \oplus \cdots \oplus Hom_L(V, V) \ (m \text{ copies}),$$

$$K \otimes_F Hom_L(U, V) \cong Hom_{L_K}(V_K, V_K) \oplus \cdots \oplus Hom_{L_K}(V_K, V_K) \ (m \text{ copies})$$

where K is the algebraic closure of F. Since V_K is irreducible over K, by Schur's lemma, $Hom_{L_K}(V_K, V_K) \cong K$. This implies $\dim Hom_L(U, V) = m$. □

The decomposition of $A \otimes_F A$ into irreducible L–submodules is a difficult problem for the general case, although various special cases have been studied in the representation theory of Lie algebras (e.g., [**BO2,K1,MMP1,My4,OM1**]). For example, let A^- be a finite–dimensional simple Lie algebra over an algebraically closed field F of characteristic 0. For the adjoint module A for A^-, it is well known that

$$(4.9) \qquad \dim_F Hom_{A^-}(A \otimes_F A, A) = \begin{cases} 1, & \text{if } A^- \text{ is not of type } A_n \ (n \geq 2), \\ 2, & \text{if } A^- \text{ is of type } A_n \ (n \geq 2) \end{cases}$$

[**BO2,My4**]. Hence if A^- is not of type A_n $(n \geq 2)$, then the map $x \otimes y \mapsto [x, y]$ is a basis of $Hom_{A^-}(A \otimes_F A, A)$. For type A_n $(n \geq 2)$, letting $A^- = sl(n+1, F)$, we define $x \# y$ by

$$(4.10) \qquad x\#y = xy + yx - \frac{2}{n+1}(\operatorname{trace} xy)I$$

where xy denotes the matrix product and I is the identity matrix. It is easy to see that $x \otimes y \mapsto x\#y$ is a nonzero symmetric element in $Hom_{A^-}(A \otimes_F A, A)$. Using this, Albert's problem can be easily resolved without the assumption of power–associativity.

Theorem 4.5. *Let A be a finite–dimensional flexible Lie–admissible algebra over an algebraically closed field F of characteristic 0 such that A^- is simple. Then, A is isomorphic to an algebra with multiplication xy given by*

$$(4.11) \qquad xy = \frac{1}{2}[x, y] + \beta x\#y$$

with $\beta = 0$ for A^- not of type A_n $(n \geq 2)$. If A^- is semisimple, then A is isomorphic to the direct sum of simple algebras given by (4.11).

Proof. If A^- is simple, then by the foregoing remarks, (4.9) and (4.10) the product xy in A is given by $xy = \alpha[x, y] + \beta x\#y$ for all $x, y \in A$ with $\alpha, \beta \in F$. But,

$$[x, y] = xy - yx = \alpha[x, y] + \beta x\#y - \alpha[y, x] - \beta y\#x = 2\alpha[x, y]$$

for all $x, y \in A$, since "#" is commutative, and this gives (4.11). Assume that A^- is semisimple and $A^- = A_1 \oplus \cdots \oplus A_n$ for simple ideals A_i of A^-. It is shown in [**My4**; Theorem 1.7] that each A_i is also a simple ideal of A. Hence, the result follows from the first part. If A^- is simple and is not of type A_n $(n \geq 2)$, then by (4.9) we have $\beta = 0$ in (4.11). □

If $A^- = sl(n+1, F)$ and $\beta \neq 0$ in (4.11), then by direct calculation one easily finds that the fourth–power identity $x^2x^2 = x^3x$ is not satisfied by A and hence A is not power–associative. For the general case, let R be the solvable radical of A^-, so that A^- has a Levi decomposition $A^- = S \oplus R$ for a semisimple subalgebra S of A^- [**J2**]. Thus, R is regarded as an ad_S–module. For the two special cases where $[S, R] = 0$ and R is abelian, i.e., $[R, R] = 0$, the structure of A has been determined [**BBO1,BO2,My4,OM1**]. The multiplication in A is described in terms of certain commutative products, bilinear and linear maps and linear functionals defined on irreducible ad_S–submodules of R. A book account of detail for these may be found in [**My4**].

Generalizing the discussion above to a broader class of algebras, we now turn our attention to Malcev–admissible algebras.

Definition 4.2. An algebra M with multiplication $[x, y]$ over a field F is called a *Malcev algebra* if it is anticommutative, i.e., $[x, x] = 0$ for any $x \in M$ and satisfies the Malcev identity

$$[[x, y], [x, z]] = [x, y, z, x] + [y, z, x, x] + [z, x, x, y] \tag{4.12}$$

for all $x, y, z \in M$, where $[x, y, z, t] = [[[x, y], z], t]$. An algebra A over F is called *Malcev–admissible* if A^- is a Malcev algebra. □

Identity (4.12) was first noted by Malcev in 1955 in analytic loop theory [**M1**]. If an analytic loop is Moufang, i.e., (3.6) is satisfied, then the tangent space at the unit element satisfies (4.12). In particular, if an analytic loop is a Lie group, then the tangent space at the unit element becomes a Lie algebra. Thus, the Malcev identity is regarded as a natural generalization of the Jacobi identity. This can also be seen from a certain equivalent form of (4.12), as shown by the following result [**S1**].

Lemma 4.6. *Let M be an anticommutative algebra with multiplication $[x, y]$ over a field F.*

(i) *M is a Malcev algebra if and only if it satisfies the identity*

$$J(x, y, [x, z]) = [J(x, y, z), x]. \tag{4.13}$$

In particular, any Lie algebra is a Malcev algebra.

(ii) *If the characteristic of F is not 2, then the Malcev identity (4.12) is equivalent to the multilinear identity:*

$$\begin{aligned} m(x, y, z, t) \equiv\ & [[x, z], [y, t]] - [x, y, z, t] - [y, z, t, x] \\ & - [z, t, x, y] - [t, x, y, z] = 0. \end{aligned} \tag{4.14}$$

Proof. (i) Using anticommutativity in M, we have

$$\begin{aligned} J(x,y,[x,z]) &= [[x,y],[x,z]] + [[y,[x,z]],x] + [[[x,z],x],y] \\ &= [[x,y],[x,z]] + [z,x,y,x] - [z,x,x,y], \\ [J(x,y,z),x] &= [x,y,z,x] + [y,z,x,x] + [z,x,y,x] \end{aligned}$$

for all $x,y,z \in M$. Hence, (4.12) is equivalent to (4.13).

(ii) If (4.14) holds for M, then $m(x,x,y,z) = 0$ gives (4.12). Assume that (4.12) holds for M, i.e., $m(x,x,y,z) = 0$ for all $x,y,z \in M$. We first note that $m(x,y,z,t) = m(t,x,y,z)$ for all x,y,z,t in any anticommutative algebra. But, linearizing $m(x,x,y,z) = 0$ gives the identity $m(x,y,z,t) = -m(y,x,z,t)$. Making repeated use of these two identities, we obtain

$$\begin{aligned} m(x,y,z,t) &= m(t,x,y,z) = -m(x,t,y,z) = -m(z,x,t,y) \\ &= m(x,z,t,y) = m(y,x,z,t) = -m(x,y,z,t). \end{aligned}$$

Hence, $2m(x,y,z,t) = 0$ or $m(x,y,z,t) = 0$, giving (4.14). □

As an immediate consequence of Lemma 4.6, we have:

Corollary 4.7. *An algebra A over a field F of characteristic $\neq 2$ is Malcev–admissible if and only if A^- satisfies (4.13) or (4.14).* □

Lemma 4.8. *A flexible algebra A over a field F is Malcev–admissible if and only if A satisfies the identity*

$$(4.15) \qquad 2S(x,y,[x,z]) = 2[S(x,y,z),x].$$

In particular, any flexible algebra of characteristic 2 is Malcev–admissible.

Proof. Flexibility of A implies (4.7), hence $S(x,y,z) = -S(x,z,y)$ for all $x,y,z \in A$. From the proof of Lemma 4.1 (i), we get $2S(x,y,z) = J(x,y,z)$. Hence, the result follows from (4.13). □

The primary source of Malcev–admissible algebras considered in this book originates from alternative algebras.

Theorem 4.9. *Let A be an alternative algebra over a field F.*

(i) *A is flexible Malcev–admissible.*

(ii) *A is Lie–admissible if and only if either A is associative or* $\operatorname{char} F = 2$ *or* 3.

Proof. (i) Using (3.2) and the Moufang identity (3.5) we obtain

$$\begin{aligned}(x,y,xz) &= (y,xz,x) = (y(xz))x - y((xz)x) \\ &= (y(xz))x - ((yx)z)x = -(y,x,z)x = (x,y,z)x, \\ (x,y,zx) &= -(x,zx,y) = -(x(zx))y + x((zx)y) \\ &= -x(z(xy)) + x((zx)y) = x(z,x,y) = x(x,y,z)\end{aligned}$$

for all $x, y, z \in A$. It follows from this that any alternative algebra satisfies the identity

(4.16) $$(x, y, [x, z]) = [(x, y, z), x].$$

Since (x, y, z) is an alternating function, we have $S(x, y, z) = 3(x, y, z)$, which combines with (4.16) and (4.15) to imply that A is flexible Malcev–admissible.

(ii) Since $J(x, y, z) = S(x, y, z) - S(x, z, y) = 3(x, y, z) - 3(x, z, y) = 6(x, y, z)$ for all $x, y, z \in A$, the result follows. □

Unlike Lie algebras, it is an open problem whether any Malcev algebra is embedded into A^- for some alternative algebra A. Cayley–Dickson algebras $C(\alpha, \beta, \gamma)$ of characteristic $\neq 2, 3$ are basic examples of flexible Malcev–admissible algebras that are not Lie–admissible. Denote $C \equiv C(\alpha, \beta, \gamma)$ and let

(4.17) $$M(\alpha, \beta, \gamma) = \{x \in C : t(x) = 0\}.$$

Since t is symmetric (Theorem 3.11), $M(\alpha, \beta, \gamma) = [C, C] \cong C^-/F1$ is a seven–dimensional Malcev algebra over F under the product $[\,,\,]$. Using the standard basis $\{e_1, \ldots, e_7\}$ in Table I.1, we obtain the multiplication table of $M(\alpha, \beta, \gamma)$ under the product $\frac{1}{2}[\,,\,]$ as given by Table I.2.

TABLE I.2

	e_1	e_2	e_3	e_4	e_5	e_6	e_7
e_1	0	e_3	αe_2	e_5	αe_4	$-e_7$	$-\alpha e_6$
e_2	$-e_3$	0	$-\beta e_1$	e_6	e_7	βe_4	βe_5
e_3	$-\alpha e_2$	βe_1	0	e_7	αe_6	$-\beta e_5$	$-\alpha\beta e_4$
e_4	$-e_5$	$-e_6$	$-e_7$	0	$-\gamma e_1$	$-\gamma e_2$	$-\gamma e_3$
e_5	$-\alpha e_4$	$-e_7$	$-\alpha e_6$	γe_1	0	γe_3	$\alpha\gamma e_2$
e_6	e_7	$-\beta e_4$	βe_5	γe_2	$-\gamma e_3$	0	$-\beta\gamma e_1$
e_7	αe_6	$-\beta e_5$	$\alpha\beta e_4$	γe_3	$-\alpha\gamma e_2$	$\beta\gamma e_1$	0

For $u, v \in M(\alpha, \beta, \gamma)$, let $u \times v$ denote the projection of uv onto $M(\alpha, \beta, \gamma)$. From a linearization of (3.8) it follows easily that

$$uv = u \times v - (u, v)1 = \frac{1}{2}[u, v] - (u, v)1 \tag{4.18}$$

for the bilinear form (,) associated with the norm n (Definition 3.1). Let U be a nonzero ideal of $M(\alpha, \beta, \gamma)$. By (4.18), $CU = U + F1$ and hence there is a $0 \neq a \in U$ with $n(a) \neq 0$; for, otherwise, $n(xa) = 0$ for any $x \in C$ and $a \in U$, so $n(1) = 0$. For $u \in M(\alpha, \beta, \gamma)$, we have $(ua)a = ua^2 = -n(a)u \in CU = U + F1$ and $u \in (U + F1) \cap M(\alpha, \beta, \gamma) = U$, hence $U = M(\alpha, \beta, \gamma)$. Thus, $M(\alpha, \beta, \gamma)$ is simple over F. Since Table I.2 can be extended to Table I.1 for C, being central over F, $M(\alpha, \beta, \gamma)$ is also central simple over F. Conversely, it is proven in **[Ku1]** that any finite–dimensional central simple, non–Lie, Malcev algebra over F of characteristic $\neq 2, 3$ is isomorphic to $M(\alpha, \beta, \gamma)$ for some $C(\alpha, \beta, \gamma)$. The same result has been obtained for arbitrary dimension **[F2]**. Moreover, any simple Malcev algebra over F of characteristic 3 turns out to be a Lie algebra **[F1]**. We summarize these in:

Theorem 4.10. *An algebra A over a field F of characteristic $\neq 2, 3$ is a central simple, non–Lie, Malcev algebra over F if and only if it is isomorphic to $M(\alpha, \beta, \gamma)$ with $\alpha\beta\gamma \neq 0$ for some Cayley-Dickson algebra $C(\alpha, \beta, \gamma)$. Moreover, any simple Malcev algebra of characteristic 3 is a Lie algebra.* □

A Malcev algebra $M(\alpha, \beta, \gamma)$ is said to be *split* if its associated $C(\alpha, \beta, \gamma)$ is split. If $M(\alpha, \beta, \gamma)$ is split over F, then we denote $M(F) = M(\alpha, \beta, \gamma)$. In view of Theorems 3.6 and 4.9, there exists a unique central simple, non–Lie, split Malcev algebra over the field F, which is isomorphic to $M(F) = M(\alpha, \beta, \gamma)$ with $C(F) =$

$C(\alpha, \beta, \gamma)$ given by (3.12). Letting $h = e_1 - e_2$ and replacing $-u_i$ by u_i in (3.12), we have a standard basis $\{h, u_i, v_i : i = 1, 2, 3\}$ for $M(F)$ with multiplication given by

$$(4.19)\qquad \begin{gathered}[h, u_i] = 2u_i, \ [h, v_i] = -2v_i, \ [u_i, v_j] = -\delta_{ij}h, \\ [u_i, u_j] = 2\epsilon_{ijk}v_k, \ [v_i, v_j] = 2\epsilon_{ijk}u_k\end{gathered}$$

for $i, j, k = 1, 2, 3$.

Definition 4.3. Let A be an algebra over a field F. The Lie subalgebra $\mathcal{L}(A)$ of $(Hom_F A)^-$ generated by $L_A \cup R_A$ is called the *Lie multiplication algebra* of A. Denote

$$\mathrm{IDer}\, A = \mathcal{L}(A) \cap \mathrm{Der}\, A.$$

Each element of $\mathrm{IDer}\, A$ is called an *inner derivation* of A. □

Let A be a finite–dimensional algebra over a field F. It is well known that if A is a semisimple Lie or Malcev algebra over F and $\mathrm{char}\, F = 0$, then $\mathrm{Der}\, A$ is inner, i.e., $\mathrm{IDer}\, A = \mathrm{Der}\, A$ [**J2,S1,Y1**]. If A is a separable (thus, semisimple) alternative or Jordan algebra, then $\mathrm{Der}\, A = \mathrm{IDer}\, A$ with $\mathrm{char}\, F \neq 2, 3$ for the alternative case [**Sc3**, p.87] and with $\mathrm{char}\, F = 0$ or other restrictions on $\mathrm{char}\, F$ for the Jordan case [**J3**, p.293]. If A is a central simple associative algebra over F, then by the theorem of Noether–Skolem, $\mathrm{Der}\, A = \mathrm{IDer}\, A$ [**P1**, p.230]. For the general case, by a theorem of Jacobson, we have that if A is a semisimple algebra (in the sense of Definition 1.1) with a one–sided unit element over F of characteristic 0, then $\mathrm{Der}\, A = \mathrm{IDer}\, A$ [**Sc3**, p.22]. Some extensions of this result are given in [**EM2**]. On the other hand, there exist finite–dimensional simple algebras over F with derivations that are not inner [**W1**].

For an algebra A over F, we put $S_1 = L_A \cup R_A$ and $S_{i+1} = [S_i, S_1]$, $i = 1, 2, \ldots$. It is not difficult to show that

$$(4.20)\qquad \mathcal{L}(A) = \sum_{i=1}^{\infty} S_i$$

[**Sc3**, p.19] and that $\mathrm{IDer}\, A$ is an ideal of $\mathrm{Der}\, A$ [**Sc3**, p.21]. If A is a Lie algebra, then since the Jacobi identity is equivalent to $[ad_x, ad_y] = ad_{[x,y]}$, by (4.20) we have $\mathcal{L}(A) = \mathrm{IDer}\, A = ad_A$. For the description of $\mathrm{IDer}\, M$ for a Malcev algebra M, let N denote the *J–nucleus* of M defined by

$$(4.21)\qquad N = \{a \in M : J(M, M, a) = 0\}.$$

It is clear that $a \in N$ if and only if $ad_a \in \mathrm{IDer}\, M$ and that $N = M$ if and only if M is a Lie algebra.

Theorem 4.11. *Let M be a Malcev algebra with multiplication $[x,y]$ over a field F of characteristic $\neq 2$.*

(i) *The J-nucleus N of M is an ideal of M.*

(ii) *For any $x,y \in M$, the linear map $d(x,y) : M \to M$ defined by*

$$d(x,y) = ad_{[x,y]} + [ad_x, ad_y] \tag{4.22}$$

is an inner derivation of M, and

$$\mathrm{IDer}\, M = ad_N + d(M,M). \tag{4.23}$$

If M is simple and is not Lie, then $\mathrm{IDer}\, M = d(M,M)$.

Proof. (i) Any Malcev algebra satisfies the identity

$$2J([a,x],y,z) = [y, J(z,a,x)] + [J(y,a,x),z] - J([y,z],a,x)$$

for all elements $a,x,y,z \in M$ [**S1**, p.431]. It follows from this that $J([N,M],M,M) = 0$, and hence N is an ideal of M.

(ii) We note that the identity $m(x,y,z,t) = 0$ of (4.14) is equivalent to the identity

$$ad_{[[x,y],z]} = ad_{[x,z]}ad_y + ad_x ad_{[y,z]} - ad_y ad_x ad_z + ad_z ad_y ad_x.$$

Interchanging y and z, and x and z in this, respectively, gives

$$\begin{aligned} ad_{[[x,z],y]} &= ad_{[x,y]}ad_z + ad_x ad_{[z,y]} - ad_z ad_x ad_y + ad_y ad_z ad_x, \\ ad_{[[z,y],x]} &= ad_{[z,x]}ad_y + ad_z ad_{[y,x]} - ad_y ad_z ad_x + ad_x ad_y ad_z. \end{aligned}$$

Adding these three relations, we obtain

$$ad_{d(x,y)z} = [ad_{[x,y]}, ad_z] + [[ad_x, ad_y], ad_z] = [d(x,y), ad_z],$$

since $d(x,y)z = [[x,y],z] + [[x,z],y] + [[z,y],x]$. Thus, $d(M,M) \subseteq \mathrm{IDer}\, M$.

For (4.23), in light of (4.22), it suffices to verify $\mathrm{IDer}\, M \subseteq ad_N + d(M,M)$. It follows from (4.22) that

$$\begin{aligned} [[ad_x, ad_y], ad_z] &= [d(x,y), ad_z] - [ad_{[x,y]}, ad_z] \\ &= ad_{d(x,y)z} + ad_{[[x,y],z]} - d([x,y],z) \end{aligned}$$

for all $x,y,z \in M$. Thus, by induction and (4.20) we obtain

$$\mathcal{L}(M) = ad_M + d(M,M).$$

Let $\delta = ad_b + \sum_i d(x_i, y_i)$ be a derivation of M. Then, ad_b is a derivation of M and hence, by the definition of N, $b \in N$, as desired. If M is simple, then $N = M$ or $N = 0$ by part (i). If M is not Lie, then $N = 0$, so $\mathrm{IDer}\, M = d(M,M)$. □

If $C(\alpha,\beta,\gamma)$ is a Cayley–Dickson algebra over F ($\operatorname{char} F \neq 2,3$), then by the theorem of Cartan–Jacobson $\operatorname{Der} C(\alpha,\beta,\gamma)$ is a 14–dimensional central simple Lie algebra of type G_2 [**Sc3**, p.82]. From (4.17) and (4.18) it follows easily that $\operatorname{Der} M(\alpha,\beta,\gamma) \cong \operatorname{Der} C(\alpha,\beta,\gamma) \cong G_2$ (see Section IV.6). Let $M = M(\alpha,\beta,\gamma)$ and $\operatorname{Der} M = G_2$. It is shown in [**EMy2**] that

$$\dim_F Hom_{G_2}(M \otimes_F M, M) = 1, \tag{4.24}$$

and hence the map $x \otimes y \mapsto [x,y]$ is a basis of $Hom_{G_2}(M \otimes_F M, M)$. This is well known for an algebraically closed field of characteristic 0, since the decomposition of $M \otimes_F M$ contains exactly one copy of the irreducible G_2–module M (see Proposition 4.4 and, e.g., [**My4**]).

For an algebra A over F of characteristic $\neq 2$, let A^+ denote the algebra with multiplication $x \circ y$,

$$x \circ y = \frac{1}{2}(xy + yx) \tag{4.25}$$

defined on the same vector space as A. We show that flexibility of A is equivalent to the identity

$$[x, y \circ z] = y \circ [x,z] + [x,y] \circ z, \tag{4.26}$$

that is, $ad_A \subseteq \operatorname{Der} A^+$. In fact, the expansion of (4.26) gives

$$(x,y,z) + (z,y,x) + (x,z,y) + (y,z,x) = (y,x,z) + (z,x,y)$$

for all $x,y,z \in A$, which is implied by the flexible identity (4.7). Conversely, (4.26) with $x = z$ implies $(xy)x = x(yx)$ for all $x,y \in A$.

Using the discussion above, Theorem 4.5 is easily extended to flexible Malcev–admissible algebras.

Theorem 4.12. *Let A be a finite–dimensional flexible Malcev–admissible algebra over an algebraically closed field F of characteristic 0 such that A^- is simple. Then, A is isomorphic to an algebra with multiplication xy defined by (4.11) where $x\#y$ is given by (4.10) and $\beta = 0$ for A^- not of type A_n ($n \geq 2$). If A^- is semisimple, then A is isomorphic to the direct sum of simple algebras given by (4.11).*

Proof. It remains to treat the case where A^- is non–Lie, Malcev. Hence, $A^- \cong M(F)$ is split and $\operatorname{Der} A^-$ is identified with $G_2 = d(A^-, A^-)$ (Theorem 4.11). But, since A is flexible, by (4.26) $ad_A \subseteq \operatorname{Der} A^+$ which implies $G_2 = d(A^-,A^-) \subseteq \operatorname{Der} A$ (see (4.22)). Therefore, the map $x \otimes y \mapsto xy$ (the product in A) is an element of $Hom_{G_2}(A \otimes_F A, A)$, being one–dimensional over F by (4.24). This gives the desired multiplication for A. If A^- is semisimple, then since A^- is a direct sum of simple ideals of A^- [**Ku1,2**], the result follows from the same proof as for Theorem 4.5. □

In Theorem 4.12, the only algebra A with A^- simple that does not appear in Theorem 4.5 is the seven–dimensional central simple, non–Lie, split Malcev algebra $M(F)$ given by (4.19). As for flexible Lie–admissible algebras, if A^- is not semisimple, then we consider a Levi decomposition $A^- = S \oplus R$ [**Ku2**]. The structure of A has been determined for the two special cases where $[S, R] = 0$, or R is abelian [**My4**].

Albert's problem for the Malcev–admissible case is easily resolved from Theorem 4.12.

Corollary 4.13. *If A is a finite-dimensional power-associative, flexible Malcev-admissible algebra over F of characteristic 0 such that A^- is semisimple, then A is a Malcev algebra isomorphic to A^-.*

Proof. Since any scalar extension of A^- is also semisimple, we may assume that F is algebraically closed. Thus, it follows from Theorem 4.12 that A is a direct sum of simple algebras A_i given by (4.11). But, A_i is a Malcev algebra isomorphic to A_i^-, unless A_i^- is a Lie algebra of type A_n $(n \geq 2)$. If A_i^- is of type A_n $(n \geq 2)$ and $\beta \neq 0$, then A_i cannot be power–associative. Thus, A is a direct sum of Malcev algebras A_i, and so $A \cong A^-$. □

Alternative algebras provide another class of Malcev–admissible algebras which are not in general flexible or power–associative. The investigation of the structure of these algebras is the primary subject in this book. For any algebra, we first define its mutations as:

Definition 4.4. Let A be an algebra over a field F with multiplication xy. For a fixed element $a \in A$, the *(left) a–homotope* of A, denoted by $A(a)$, is defined to be an algebra over F with multiplication $x \cdot y$ given by

$$x \cdot y = (xa)y \tag{4.27}$$

for $x, y \in A$. For fixed $p, q \in A$, the algebra $A(p,q)$ with multiplication $x * y$ given by

$$x * y = (xp)y - (yq)x \tag{4.28}$$

is called the *(left) (p,q)–mutation* of A. The right (p,q)–mutation of A is similarly defined using the right a–homotope product $x(ay)$. □

Let A^{op} denote the *opposite algebra* of A with multiplication $x \odot y = yx$ for $x, y \in A$. Since

$$x(py) - y(qx) = (x \odot (-q)) \odot y - (y \odot (-p)) \odot x,$$

the right (p,q)–mutation of A is the left $(-q,-p)$–mutation of A^{op}. Therefore, it suffices to study left mutations of an algebra.

It follows easily from (4.28) that the commutator $[x,y]^* = x*y - y*x$ in $A(p,q)$ is given by

$$[x,y]^* = (x(p+q))y - (y(p+q))x \tag{4.29}$$

and hence

$$A(p,q)^- = A(p+q)^-. \tag{4.30}$$

If p and q are chosen from the nucleus $N(A)$ of A, then the left and right (p,q)–mutations of A coincide and the product (4.28) becomes

$$x * y = xpy - yqx. \tag{4.31}$$

If, in particular, A is associative, then for any fixed $p,q \in A$, the multiplication in $A(p,q)$ is given by (4.31).

For alternative algebras, we have [**My3**]:

Proposition 4.14. *Let A be an alternative algebra over F. For any $a \in A$, the left or right a–homotope $A(a)$ of A is alternative, and for any $p,q \in A$, the left or right (p,q)–mutation $A(p,q)$ of A is Malcev–admissible.*

Proof. For $x,y \in A$, we have

$$(x \cdot x) \cdot y = ((xax)a)y = (xa)^2 y = (xa)((xa)y) = x \cdot (x \cdot y)$$

using (3.1) and Artin's theorem, and

$$y \cdot (x \cdot x) = (ya)(xax) = (((ya)x)a)y = (y \cdot x) \cdot x$$

by the Moufang identity (3.5). Hence, the left a–homotope $A(a)$ is alternative, and is Malcev–admissible by Theorem 4.9. It follows from this and (4.30) that the left (p,q)–mutation $A(p,q)$ of A is Malcev–admissible. The result follows similarly for the right homotope and mutation of A. □

If A has a unit element 1 and a is invertible in A with inverse a^{-1} (i.e., $aa^{-1} = a^{-1}a = 1$), then $A(a)$ is called the left or right *a–isotope* of A. If a is invertible in $N(A)$ (i.e., $a, a^{-1} \in N(A)$), then the map $A \to A(a)\ :\ x \mapsto xa^{-1}$ is an isomorphism since $(xy)a^{-1} = ((xa^{-1}a)y)a^{-1} = (((xa^{-1})a)y)a^{-1} = ((xa^{-1})a)(ya^{-1})$ for all $x,y \in A$. This is not true for the general case, even for alternative algebras [**Mc2**]. However, any isotope of a composition algebra A is shown to be isomorphic to A (Proposition III.2.2).

If A is an associative algebra, then since any homotope of A is also associative, it follows from (4.30) that $A(p,q)$ is Lie–admissible. The introduction of mutations of an associative algebra originated from a generalization of classical Hamiltonian mechanics given in, e.g., [**Sa1**]. A time–development equation of the form $i(dx/dt) = xrh - hsx$ was introduced from a quantum mechanical version of this classical mechanics, where xr is the product in an associative algebra A of operators in a physical system, h is a Hamiltonian, and r, s are in general non–hermitian nonsingular operators in A which represent nonself–adjoint forces [**Sa1**]. The right side of the above equation arises, as a special case, from the product (4.31) defined on any associative algebra with arbitrary fixed elements p, q [**My5**].

In a different view from the above approach, flexible Lie–admissible algebras have been used in [**O1**] to generalize the framework of the consistent canonical quantization procedure in the conventional theory. A quantization is called consistent if the Hamiltonian equation of motion $dQ/dt = i[H,Q]$ can reproduce the original Lagrange equation. Such a quantization can be done in a flexible Lie–admissible algebra A of operators in a physical system, based only on the canonical commutation relations and flexible Lie-admissibility (4.6) [**O1**]. The Heisenberg equation $dx/dt = i[H,x] = i(Hx - xH)$ for some $iH \in A$ is shown to be essentially the most general time–development equation in A for the system, where Hx denotes the product in A [**O1,My5**].

Let A be an associative algebra over F. For any fixed $p, q \in A$, we obtain from (4.31) the associator $(x,y,z)^*$ in $A(p,q)$ as

$$(x,y,z)^* = xpzqy - yqxpz + ypzqx - zqxpy. \tag{4.32}$$

Two special cases of this yield

$$(x,y,x)^* = x(pxq - qxp)y + y(pxq - qxp)x, \tag{4.33}$$

$$(x,x,x)^* = 2x(pxq - qxp)x. \tag{4.34}$$

If A has a unit element 1, then by (4.34) $(1,1,1)^* = 2(pq - qp)$. Thus, for $p, q \in A$ with $pq \neq qp$, $A(p,q)$ is not third power–associative. In particular, $A(p,q)$ is not in general flexible. In Section II.2, we investigate conditions for the third power and flexible identities to be equivalent in $A(p,q)$.

We close this section with some results on ideals of the algebras A^- and $[A,A]$ for simple alternative algebras A and semiprime or prime associative algebras A. These results will be used for the study of simple or prime mutation algebras (Chapters II–V). In general, the simplicity or primeness of A imposes strong restrictions on the ideals of A^-. The results for the associative case are in large due to [**H1**]. All algebras considered are over a field F of characteristic $\neq 2$. An algebra A over F is said to be *semiprime* if $I^2 = 0$ for any ideal I of A implies $I = 0$.

Lemma 4.15. *Let A be a semiprime associative algebra over F. If $ad_x^2 = 0$ for $x \in A$, then $ad_x = 0$, i.e., if $[[A,x],x] = 0$ for $x \in A$ then $x \in Z(A)$, the center of A.*

Proof. Assume $[[A,x],x] = 0$ for $x \in A$, and let $a, b \in A$. Making repeated use of (4.6), we have

$$\begin{aligned}[a,x][b,x] &= [[a,x]b,x] - [[a,x],x]b = [[a,x]b,x] \\ &= [[ab,x],x] - [a[b,x],x] = -[a[b,x],x] \\ &= -[a,x][b,x] + a[[b,x],x] = -[a,x][b,x].\end{aligned}$$

Thus, $[a,x][b,x] = 0$ and $[A,x][A,x] = [A,x]^2 = 0$. It suffices to show that $[A,x]$ is an ideal of A. For $a,b,c \in A$, by (4.6) $[a,x]c[b,x] = [a,x][cb,x] - [a,x][c,x]b = 0$, hence $[A,x]A[A,x] = 0$, which implies that the ideal $A[A,x]A$ squares to zero. By semiprimeness of A, $A[A,x]A = 0$, and this in turn shows that $A[A,x]$ and $[A,x]A$ are ideals of A, squaring to zero. Thus, $A[A,x] = [A,x]A = 0$, and $[A,x]$ is an ideal of A, so $[A,x] = 0$. □

Lemma 4.16. *Let A be a semiprime associative algebra over F, and let I be a subalgebra of A which is an ideal of A^-. Then, either $I \subseteq Z(A)$ or I contains a nonzero ideal of A.*

Proof. For $a \in A$ and $x, y \in I$, we have $a[x,y] = [ax,y] - [a,x]y \in [A,I] + I^2 \subseteq I$ and so $A[I,I] \subseteq I$. Hence, for $b \in A$,

$$a[x,y]b = [a[x,y],b] + ba[x,y] \in [I,A] + A[I,I] \subseteq I.$$

This shows that the ideal $A[I,I]A$ is contained in I. Assume $A[I,I]A = 0$. As in the proof of Lemma 4.15, semiprimeness of A implies $[I,I] = 0$. But, $[[A,I],I] \subseteq [I,I] = 0$, and $[[A,x],x] = 0$ for all $x \in I$, hence by Lemma 4.15 $I \subseteq Z(A)$. □

Theorem 4.17. *Let A be a simple alternative algebra over F.*

(i) *If I is an ideal of A^-, then either $I \subseteq Z(A)$ or $[A,A] \subseteq I$.*

(ii) *If A is not commutative, then the subalgebra of A generated by $[A,A]$ equals A.*

Proof. We first verify parts (i), (ii) for a simple associative algebra A. Let I be an ideal of A^- and let $T(I) = \{a \in A : [a,A] \subseteq I\}$. It is easy to see that $T(I)$ is an ideal of A^- containing I and is a subalgebra of A, since if $a, b \in T(I)$ and $c \in A$, then $[ab,c] = [a,bc] + [b,ca] \in I$. Since A is simple, by Lemma 4.16, either $I \subseteq T(I) \subseteq Z(A)$ or $T(I) = A$. In the latter case we have $[A,A] \subseteq I$. Let I now be the subalgebra of A generated by $[A,A]$. Clearly, I is an ideal of A^- and hence by Lemma 4.16 we get either $I = A$ or $I \subseteq Z(A)$. If $I \subseteq Z(A)$, then $[[A, A], A] = 0$ which by Lemma 4.15 implies $A \subseteq Z(A)$, so A is commutative. This proves parts (i) and (ii) for the associative case.

Assume then that A is a simple alternative algebra over F that is not associative. By Theorem 3.17 A is a Cayley–Dickson algebra over its center $Z(A)$, being an extension field of F. Thus, we have the decomposition $A^- = Z(A)1 \oplus [A, A]$ where $[A, A]$ is a simple Malcev algebra $M(\alpha, \beta, \gamma)$ over $Z(A)$ given by Table I.2. But, by Lemma 2.1 (iii), $[A, A]$ is also simple over F. If I is an ideal of A^- then $[A, I] = [[A, A], I] \subseteq I \cap [A, A]$, and hence $[A, I] = 0$ or $[A, I] = [A, A] \subseteq I$. This proves part (i). Since A is a quadratic algebra over $Z(A)$ and $[A, A]$ is the set of trace zero elements in A, there is an $x \in [A, A]$ with $0 \neq x^2 = \beta \in Z(A)$. For any $\alpha \in Z(A)$, we obtain $\alpha = (\alpha x)(\beta^{-1}x) \in [A, A]^2$ and $A = Z(A)1 + [A, A] \subseteq [A, A] + [A, A]^2$, showing part (ii). □

Lemma 4.18. *Let A be a simple associative algebra over F and let I be a nonzero ideal of the Lie algebra $[A, A]$.*

(i) *If $xI = 0$ for $x \in A$ then $x = 0$.*

(ii) *If $[I, I] \subseteq Z(A)$ then $I \subseteq Z(A)$.*

(iii) *If A is not commutative then $[[A, A], [A, A]] = [A, A]$.*

Proof. (i) Let $S = \{x \in A : xI = 0\}$. It suffices to verify $S = 0$. Suppose $S \neq 0$. Since SA is a nonzero ideal of A,

$$A = SA \subseteq [A, S] + AS \subseteq [A, A] + S,$$

hence $A = [A, A]+S$. If $x \in S$, $a \in [A, A]$ and $u \in I$, then $[x, a]u = xau = x[a, u] = 0$ and hence $[S, [A, A]] \subseteq S$. It follows from this that $S[A, A] \subseteq [A, A]S+[S, [A, A]] \subseteq S$ and so $A = SA = S([A, A] + S) \subseteq S$ since $A = [A, A] + S$. Therefore, we get $AI = 0$, and IA is an ideal of A, which in turn implies $I = 0$ by the simplicity of A, a contradiction.

(ii) We may assume that A is not commutative. Let u be any nonzero element of I. Assume first that $[u, I] = 0$, and define the linear map

$$\varphi : A \to I \ : \ x \mapsto [u, [u, x]]. \tag{4.35}$$

For $v \in I$, one has

$$\varphi(xv) = [u, [u, xv]] = [u, x[u, v]] + [u, [u, x]v] = [u, [u, x]]v$$

since $[u, v] = 0$. Hence, $\varphi(xv) = \varphi(x)v$ for all $v \in I$, and $\varphi([A, A]) = 0$ by the assumption and (4.35). If $x \in [A, A]$, $z \in A$ and $v \in I$, then by the above

$$0 = \varphi([x, zv]) = \varphi([x, z]v + z[x, v]) = \varphi([x, z])v + \varphi(z[x, v]) = \varphi(z)[x, v],$$

hence $\varphi(A)[[A,A],I]=0$. Since $[[A,A],I]$ is an ideal of $[A,A]$, it follows from this and part (i) that either $\varphi(A)=0$ or $[[A,A],I]=0$. In the former, by Lemma 4.15 and (4.35), $u\in Z(A)$. In the latter, we get $I\subseteq Z(A)$, since, by Theorem 4.17 (ii), A is generated by $[A,A]$, and therefore, $u\in Z(A)$.

Assume next that $[u,I]\neq 0$, and choose a $v\in I$ with $0\neq[u,v]=\alpha\in Z(A)$. Consider the linear map

$$\psi : A \to Z(A) \;:\; x\mapsto [v,[u,[u,x]]]=[v,\varphi(x)].$$

We assert that $\psi=0$. If $\psi\neq 0$, then $\psi(x)=\beta\neq 0$ is in $Z(A)$ for some $x\in A$. Since by (4.35) $\varphi(uy)=u\varphi(y)$ for all $y\in A$,

$$\psi(ux)=[v,\varphi(ux)]=[v,u\varphi(x)]=[v,u]\varphi(x)+u\psi(x)=-\alpha\varphi(x)+\beta u$$

is in $Z(A)$. Hence,

$$0=[-\alpha\varphi(x)+\beta u,v]=\alpha[v,\varphi(x)]+\beta\alpha=\alpha\psi(x)+\beta\alpha=2\alpha\beta,$$

a contradiction, since α and β are nonzero elements in the field $Z(A)$. Thus, $\psi=0$, as asserted. For $y\in A$, we obtain

$$0=\psi(uy)=[v,u\varphi(y)]=[v,u]\varphi(y)+u\psi(y)=-\alpha\varphi(y),$$

hence $\varphi(A)=0$. As before, this gives $u\in Z(A)$, showing part (ii).

(iii) Let $I=[[A,A],[A,A]]$. Since I is an ideal of A^- and of $[A,A]$, by Theorem 4.17 (i) either $I\subseteq Z(A)$ or $[A,A]\subseteq I$. In the latter case, $I=[A,A]$. If $I\subseteq Z(A)$, then by part (ii) $[A,A]\subseteq Z(A)$ and $[[A,x],x]=0$ for all $x\in A$. Hence, by Lemma 4.15 $A\subseteq Z(A)$, a contradiction. □

Theorem 4.19. *If A is a simple noncommutative associative algebra over F, then $[A,A]/\left(Z(A)\cap[A,A]\right)$ is a simple Lie algebra over F.*

Proof. It suffices to prove that any proper ideal of $[A,A]$ is contained in $Z(A)$. Let I be a proper ideal of $[A,A]$ and let

$$T(I)=\{x\in A:[x,A]\subseteq I\}.$$

Since $[ab,x]=[a,bx]+[b,xa]\in I$ for $a,b\in T(I)$ and $x\in A$, $T(I)$ is a subalgebra of A. Moreover,

$$\begin{gathered}[I,I]\subseteq T(I),\ [I,T(I)]\subseteq T(I),\\ [[T(I),T(I)],A]\subseteq[T(I),[T(I),A]]\subseteq T(I).\end{gathered}\tag{4.36}$$

Let $T(I)=T(I)^0\supseteq T(I)^{(1)}\supseteq\cdots\supseteq T(I)^{(m)}\supseteq\cdots$ be the derived series of $T(I)$ (Definition 1.5).

Let $a, b \in T(I)^{(1)}$ and $x \in A$. Then, by (4.36) $[a,b]x = [a,bx] - b[a,x] \in T(I)$ and hence, for $y \in A$,

$$y[a,b]x = [a,b]xy - [[a,b]x, y] \in T(I) + I.$$

Therefore, $AT(I)^{(2)}A \subseteq T(A) + I$. Since A is simple and $AT(I)^{(2)}A$ is an ideal of A, it must be that either $T(I)^{(2)} = 0$ or $A = T(I) + I$. In the latter, we have $[A,A] = [T(I)+I, T(I)+I] \subseteq [T(I),A] + [I,I] \subseteq I$, a contradiction. In the former, we have $I^{(3)} = 0$ by (4.36). By repeated applications of Lemma 4.18 (ii), we conclude $I \subseteq Z(A)$, as desired. □

Finally, we show that if the algebra A in Theorem 4.19 is prime or semiprime, then the Lie algebra $A^-/Z(A)$ is also prime or semiprime. This would follow immediately from the following lemma.

Lemma 4.20. *Let A be an associative algebra over F and let I, J be ideals of A^-.*

(i) *If A is semiprime and $[I,I] \subseteq Z(A)$, then $I \subseteq Z(A)$.*

(ii) *If A is prime, $I \not\subseteq Z(A)$ and a is an element of A such that $[a,I] = 0$, then $a \in Z(A)$.*

(iii) *If A is prime and $[I,J] \subseteq Z(A)$, then either $I \subseteq Z(A)$ or $J \subseteq Z(A)$.*

Proof. (i) We contend that $[I,I] = 0$. Suppose $[I,I] \neq 0$, and let a, b be elements of I with $0 \neq [a,b] = \alpha \in Z(A)$. Since $[a,[a,x]] \in [I,I] \subseteq Z(A)$ for any $x \in A$, we have

$$\begin{aligned}
[[a, b[a,ba]], b] &= [[a,b][a,ba], b] + [b[a,[a,ba]], b] \\
&= \alpha[[a,ba], b] + b[[a,[a,ba]], b] \\
&= \alpha[[a,ba], b] = \alpha[[a,b]a, b] = \alpha^3.
\end{aligned}$$

On the other hand, it follows that

$$\begin{aligned}
[[a, b[a,ba]], b] &= [[a,[a,b^2a]], b] - [[a,[a,b]ba], b] \\
&= -\alpha[[a,ba], b] = -\alpha[[a,b]a, b] = -\alpha^3.
\end{aligned}$$

Therefore, $\alpha^3 = 0$ and $(\alpha^2 A)^2 \subseteq \alpha^4 A^2 = 0$, so $\alpha^2 A = 0$ which implies $\alpha A = 0$ since A is semiprime. This shows that $F\alpha$ is an ideal of A with $(F\alpha)^2 = 0$, hence $\alpha = 0$, a contradiction. Therefore, $[I,I] = 0$ and $ad_x^2 = 0$ for all $x \in I$. It follows from Lemma 4.15 that $I \subseteq Z(A)$.

(ii) Let $S = \{x \in A : [x, I] = 0\}$. We may assume $S \neq 0$; otherwise, $a = 0$ and the assertion is trivial. Since S is an ideal of A^- and is a subalgebra of A, it follows from Lemma 4.16 that either $S \subseteq Z(A)$ or there is a nonzero ideal T of A contained in S. It suffices to rule out the latter case. In fact, we have

$$T[A, I] \subseteq [TA, I] + [T, I]A \subseteq [S, I] + [S, I]A = 0,$$

hence $T(A[A, I]A) = 0$. By primeness of A, this gives $[A, I] = 0$, a contradiction, since $I \not\subseteq Z(A)$.

(iii) If $[I, J] \subseteq Z(A)$ and $I \not\subseteq Z(A)$ then $[I, [J, J]] \subseteq [[I, J], J] \subseteq [Z(A), J] = 0$. Thus, by part (ii) we have $[J, J] \subseteq Z(A)$ which by part (i) implies $J \subseteq Z(A)$. □

As an immediate consequence of Lemma 4.20, we obtain:

Theorem 4.21. *Let A be a noncommutative associative algebra over F. If A is prime or semiprime, then the Lie algebra $A^-/Z(A)$ is prime or semiprime, respectively.* □

5. Jordan–admissible algebras

Throughout this section, the base field F considered is assumed to be of characteristic $\neq 2$. For an algebra A over F, let A^+ denote the commutative algebra with multiplication $x \circ y$ given by (4.25). For the (left) (p, q)–mutation $A(p, q)$ of A, the product in $A(p, q)^+$ will be denoted by $\{x, y\}^* = \frac{1}{2}(x * y + y * x)$. Thus, from (4.28) we find

$$\{x, y\}^* = \tfrac{1}{2}\left((x(p - q))y + (y(p - q))x\right), \tag{5.1}$$

$$A(p, q)^+ = A(p - q)^+. \tag{5.2}$$

Definition 5.1. An algebra J with multiplication denoted by $x \circ y$ over F is called a *Jordan algebra* if it is commutative, i.e., $x \circ y = y \circ x$ for all $x, y \in J$ and satisfies the *Jordan identity*

$$((x \circ x) \circ y) \circ x = (x \circ x) \circ (y \circ x) \tag{5.3}$$

for all $x, y \in J$. An algebra A over F is said to be *Jordan–admissible* if A^+ is a Jordan algebra. □

If A is an associative algebra, then expanding (5.3) and using associativity it is easily seen that A^+ is a Jordan algebra and so is any subalgebra of A^+. More generally, if A is an alternative algebra, then since (5.3) involves only two variables, by Artin's theorem any subalgebra of A^+ is a Jordan algebra. Therefore, by (5.2) $A(p, q)$ is Jordan–admissible for any fixed $p, q \in A$.

The Jordan algebras over a given field F are divided into two disjoint subclasses. The first subclass, denoted by SJ, consists of all Jordan algebras over F that are isomorphic to subalgebras of A^+ for associative algebras A over F. Each algebra in SJ is called *special*. The remaining subclass comprises those Jordan algebras not in SJ, called *exceptional*. Accordingly, a Jordan–admissible algebra A is said to be *special* or *exceptional* if A^+ is special or exceptional. Any alternative algebra A is special Jordan–admissible [**J3**, p.15], and so is $A(p,q)$ by (5.2).

Jordan algebras, named by Albert (1946), were first considered by physicist Pascual Jordan in the early 1930's in the hope that an infinite–dimensional algebraic setting which is essentially different from the standard theory of hermitian matrices may be introduced to generalize the formalism of quantum mechanics. Thus, this was the beginning of the search for infinite–dimensional exceptional Jordan algebras.

To briefly illustrate the source of exceptional Jordan algebras, consider an algebra A over a field F with a unit element and with an involution $x \mapsto \bar{x}$. If $M_n(A)$ denotes the set of $n \times n$ matrices with entries in A, then $M_n(A)$ becomes an algebra over F under the usual definitions of matrix operations. If we let $\bar{X} = (\bar{x}_{ij})^t$ for $X = (x_{ij}) \in M_n(A)$, then it is straightforward to see that the map $X \mapsto \bar{X}$ gives the "standard" involution on $M_n(A)$ associated with $x \mapsto \bar{x}$ $(x \in A)$. Let

$$H_n(A) = \{X \in M_n(A) : \bar{X} = X\},$$

i.e., the set of symmetric elements of $M_n(A)$ under the standard involution. Since $\overline{X \circ Y} = \frac{1}{2}\left(\overline{XY} + \overline{YX}\right) = \frac{1}{2}\left(\bar{Y}\bar{X} + \bar{X}\bar{Y}\right) = \frac{1}{2}(YX + XY) = X \circ Y$ for $X, Y \in H_n(A)$, $H_n(A)$ is a subalgebra of $M_n(A)^+$. A crucial result in the Jordan theory is that $H_n(A)$ for $n \geq 3$ is a Jordan algebra over F if and only if either A is associative, or $n = 3$ and A is alternative with symmetric elements of A in the nucleus $N(A)$ [**J3**, p.127, **ZSSS1**, p.54]. Thus, if $n \geq 4$, then A is associative and hence $H_n(A)$ is a special Jordan algebra.

Exceptional Jordan algebras arise when $n = 3$ and A is a Cayley–Dickson algebra C over F with the standard involution $x \mapsto \bar{x} = t(x)1 - x$ (Theorem 3.2). Thus, $H_3(C)$ is a Jordan algebra of dimension $1 + 1 + 1 + 8 + 8 + 8 = 27$. An algebra A over F is called an *Albert algebra* if A is an F–form of $H_3(C)$ for a Cayley–Dickson algebra C over K, an extension field of F; $K \otimes_F A \cong H_3(C)$. It is one of the most celebrated results in the Jordan theory that an algebra over an arbitrary field F of characteristic $\neq 2$ (not necessarily finite–dimensional) is a central simple exceptional Jordan algebra over F if and only if it is an Albert algebra over F (of dimension 27). The algebra $H_3(C)$ for the Cayley algebra C (over $\mathbb{R}$) first appeared in the fundamental paper of Jordan, von Neumann and Wigner [**JvonNW1**] in 1934. In the same year, Albert [**A1**] proved the exceptionality of $H_3(C)$, and later constructed an exceptional division algebra that is a form of $H_3(C)$ (so, a division

Albert algebra). The result above with finiteness condition has been proved in a number of papers up to the 1950's (see, e.g., [**Sc3,J3,BK1,Mc3,4**] and references therein). It was Zel'manov [**Z1**] in 1979 who first proved the result without finiteness condition (for survey, see [**Mc3,4**]). Therefore, Zel'manov's proof brought an end to the original problem of Jordan: There do not exist exceptional simple Jordan algebras other than Albert algebras.

If J is a Jordan algebra over F with product $x \circ y$ and if $[\ ,\ ]$ denotes any anticommutative product defined on the vector space J, then the algebra $(J, *)$ with multiplication $x * y = \frac{1}{2}[x, y] + x \circ y$ defined on J is Jordan–admissible such that $(J, *)^+ = J$. Conversely, every Jordan algebra can occur as the A^+ of a host of Jordan–admissible algebras A by simply changing the anticommutative product. This illustrates that Jordan–admissibility alone is too broad an identity to develop a meaningful structure theory.

Jordan–admissible algebras were first introduced in the Albert's 1948 paper [**A3**] along with Lie–admissible algebras, and its structure was investigated with the additional identity of flexibility. In fact, flexible Jordan–admissible algebras are regarded as a natural generalization of Jordan algebras. To see this, let A be an algebra over F with multiplication xy. Since $x \circ y = \frac{1}{2}(L_x + R_x)y$, it follows from (5.3) that Jordan–admissibility of A is equivalent to the identity

$$[L_{x^2} + R_{x^2}, L_x + R_x] = 0 \tag{5.4}$$

for all $x \in A$. In the absence of commutativity, the Jordan identity (5.3) has the following variants in A:

$$x(yx^2) = (xy)x^2; \ [L_x, R_{x^2}] = 0, \tag{5.5}$$
$$(x^2y)x = x^2(yx); \ [R_x, L_{x^2}] = 0, \tag{5.6}$$
$$(yx)x^2 = (yx^2)x; \ [R_x, R_{x^2}] = 0, \tag{5.7}$$
$$x^2(xy) = x(x^2y); \ [L_x, L_{x^2}] = 0. \tag{5.8}$$

If A is commutative, then (5.5)–(5.8) are all equivalent to the Jordan identity (5.3). More generally, for flexible algebras, we have [**A3,Sc1,2**]:

Proposition 5.1. *Let A be an algebra over F.*

(i) *If A is flexible, then (5.5)–(5.8) are all equivalent.*

(ii) *A is flexible and satisfies any one of (5.5)–(5.8) if and only if A is flexible Jordan–admissible.*

Proof. (i) Flexibility of A is written as

$$[R_x, L_x] = 0 \tag{5.9}$$

which is linearized to $[R_x, L_y] = [L_x, R_y]$. This with $y = x^2$ gives

$$[R_x, L_{x^2}] = [L_x, R_{x^2}], \tag{5.10}$$

hence (5.5) and (5.6) are equivalent. We write the flexible identity (4.7) as $L_{xy} - L_x L_y = R_{yx} - R_x R_y$ and put $y = x$ in this to obtain

$$L_{x^2} - L_x^2 = R_{x^2} - R_x^2 \tag{5.11}$$

for all $x \in A$. If A satisfies (5.7) and (5.9), then by (5.11) $[R_x, L_{x^2}] = 0$, giving (5.6). Similarly, (5.6) and (5.9) imply (5.7), and the equivalence of (5.5) and (5.8) follows from the same argument.

(ii) Jordan–admissibility (5.4) follows from (5.5)–(5.8). Hence, if A is flexible and satisfies any one of (5.5)–(5.8), then by part (i) A is flexible Jordan–admissible. Using (5.10) and (5.11), from (5.4) we obtain

$$\begin{aligned} 0 &= [L_{x^2} + R_{x^2}, L_x] + [L_{x^2} + R_{x^2}, R_x] \\ &= [2R_{x^2} - R_x^2 + L_x^2, L_x] + [2L_{x^2} - L_x^2 + R_x^2, R_x] \\ &= 2[R_{x^2}, L_x] + 2[L_{x^2}, R_x] \\ &= 4[R_x, L_{x^2}], \end{aligned}$$

giving (5.6). Hence, the result follows from part (i). □

Following [**Sc1**], we define:

Definition 5.2. An algebra A over a field F of arbitrary characteristic is defined to be *noncommutative Jordan* if A is flexible and satisfies (5.6). □

In view of Proposition 5.1, if char $F \neq 2$, then an algebra over F is noncommutative Jordan if and only if it is flexible Jordan–admissible. For char $F = 2$, there exist flexible (quadratic) Jordan–admissible algebras over F that are not noncommutative Jordan [**Mc1**]. The structure of flexible Jordan–admissible algebras is much better understood than that of flexible Lie–admissible algebras. If A is a finite–dimensional noncommutative Jordan algebra over F of characteristic 0, then the nil radical (the maximal nil ideal) of A is zero if and only if A is semisimple (in the sense of Definition 1.1) [**Sc3**, p.142]. Simple noncommutative Jordan algebras over an arbitrary field with descending chain condition on inner ideals have essentially been classified in [**Mc1**], and they are in large represented by simple associative, Jordan, Cayley–Dickson, quasiassociative algebras, and certain simple flexible algebras of degree 1 or 2. Algebras of interest for our discussion are the first four classes of algebras with only special Jordan algebras. In contrast, one can give a systematic construction of algebras to show that such structure theorems cannot be obtained for flexible Lie–admissible algebras [**My4**].

Proposition 5.2. *Every noncommutative Jordan algebra A over F* (char $F \neq 2$) *is power–associative.*

Proof. For $x \in A$, define x^k inductively by $x^{k+1} = x^k x$ for $k \geq 1$. We note that A is power–associative if and only if $x^j x^k = x^n$ for all $j, k \geq 1$ with $j+k = n = 2, 3, \ldots$. Since A is flexible, $x^2 x = x x^2$ for all $x \in A$ and hence this holds for $n = 3$. To prove the result by induction on n, assume that $x^j x^k = x^{j+k}$ is true for $j + k < n$ with $n \geq 4$. It suffices to verify

$$x^{n-p} x^p = x^n \tag{5.12}$$

for $p = 1, \ldots, n-1$. We prove (5.12) by induction on p $(1 \leq p \leq n-1)$. For $p = 1$, (5.12) clearly holds, and assume that $x^{n-k} x^k = x^n$ is true for $1 \leq k \leq p < n-1$.

Linearizing (5.6) twice, we obtain

$$(xz + zx, y, w) + (zw + wz, y, x) + (wx + xw, y, z) = 0.$$

Substituting $y = x^{n-(p+2)}, z = x, w = x^p$ into this and using induction hypothesis, we obtain

$$\begin{aligned} 0 &= 2(x^2, y, x^p) + 4(x^{p+1}, y, x) \\ &= 2(x^2 x^{n-(p+2)}) x^p - 2x^2 (x^{n-(p+2)} x^p) \\ &\quad +4(x^{p+1} x^{n-(p+2)}) x - 4x^{p+1} (x^{n-(p+2)} x) \\ &= 2x^{n-p} x^p - 2x^2 x^{n-2} + 4x^{n-1} x - 4x^{p+1} x^{n-(p+1)} \\ &= 4x^n - 4x^{p+1} x^{n-(p+1)}, \end{aligned}$$

since $x^2 x^{n-2} = x^2(x^{n-3}x) = (x^2 x^{n-3})x = x^n$ by (5.6) and $x^{n-p} x^p = x^n$ by the assumption. This shows that (5.12) is true for $p + 1$. □

As noted in Section I.4, flexible Lie–admissible algebras are not in general fourth power–associative ($x^3 x = x^2 x^2$). In fact, by a known result [**A2**], a flexible algebra over F of characteristic $\neq 2, 3, 5$ is fourth power–associative if and only if it is power–associative. We show in Section III.3 that the (p, q)–mutation $A(p, q)$ of an alternative algebra A is third power–associative if and only if it is power–associative and that if A contains an element that is not a zero divisor in A, then each of these two conditions is equivalent to flexibility of $A(p, q)$ or to the condition $p(xq) = q(xp)$ for all $x \in A$. If, in particular, A has a unit element, then the last condition forces $pq = qp$, imposing a strong restriction on p and q. Therefore, a "small" subclass of mutations of alternative algebras is noncommutative Jordan or flexible Malcev–admissible. An indepth study of identities satisfied by all $A(p, q)$ reveals that the structure theory of these mutation algebras cannot be developed by an intrinsic characterization in terms of identities (Chapter V). Instead, the structure of $A(p, q)$ is governed by that of A as well as by the fixed elements $p, q \in A$.

We turn our attention to quasiassociative algebras which are viewed as forms of certain mutations of associative algebras. Let A be any algebra over F (char $F \neq 2$) with multiplication xy. For fixed scalars $\lambda, \mu \in F$, let $A(\lambda, \mu)$ denote the algebra with multiplication

$$x * y = \lambda xy - \mu yx$$

defined on A. We call $A(\lambda, \mu)$ the (λ, μ)–mutation or a scalar mutation of A. If A has a unit element, then $A(\lambda, \mu)$ is a special case of the (p, q)–mutation $A(p, q)$. In particular, we denote $A(\lambda, \lambda - 1) = A(\lambda)$, unless there arises ambiguity with the λ–homotope of A (Definition 4.4). Thus, the product $x * y$ in $A(\lambda)$ is given by

$$x * y = \lambda xy + (1 - \lambda) yx. \tag{5.13}$$

In fact, if $\lambda - \mu \neq 0$, then the map

$$f : A(\lambda, \mu) \to A(\beta) \ : \ x \mapsto (\lambda - \mu)x$$

is an isomorphism, where $\beta = \lambda(\lambda - \mu)^{-1}$. To see this, we note

$$(\lambda - \mu)^{-1}(x * y) = \lambda(\lambda - \mu)^{-1} xy + (1 - \lambda(\lambda - \mu)^{-1}) yx = \beta xy + (1 - \beta) yx,$$

$$\begin{aligned} f(x * y) &= (\lambda - \mu)(x * y) = (\lambda - \mu)^2 \left((\lambda - \mu)^{-1} x * y\right) \\ &= \beta f(x) f(y) + (1 - \beta) f(y) f(x) = f(x) * f(y), \end{aligned}$$

showing f is an isomorphism. Therefore, it suffices to work with $A(\lambda)$.

The following relations are obtained easily from (5.13):

$$\begin{aligned} &(5.14) & [x, y]^* &= x * y - y * x = (2\lambda - 1)[x, y] \\ &(5.15) & \{x, y\}^* &= \frac{1}{2}(x * y + y * x) = x \circ y, \\ &(5.16) & (x, y, z)^* &= \lambda^2 (x, y, z) - (1 - \lambda)^2 (z, y, x) \\ &&& \quad + \lambda(1 - \lambda)\left((yx)z + z(xy) - x(zy) - (yz)x\right) \end{aligned}$$

for all $x, y, z \in A$.

Lemma 5.3. *Let A be an algebra with multiplication xy over F and let $\lambda \in F$ with $\lambda \neq \frac{1}{2}$.*

(i) *$A(\lambda)^+ = A^+$ and $A(\lambda)^- \cong A^-$.*

(ii) *$A = (A(\lambda))(\mu)$ with $\mu = \lambda(2\lambda - 1)^{-1}$.*

(iii) *A is flexible if and only if $A(\lambda)$ is flexible.*

(iv) *A is power–associative if and only if $A(\lambda)$ is power–associative.*

(v) *The ideals of* A *and* $A(\lambda)$ *coincide, and in particular,* A *is simple if and only if* $A(\lambda)$ *is simple.*

(vi) $R(A) = R(A(\lambda))$.

(vii) A *is semisimple if and only if* $A(\lambda)$ *is semisimple.*

Proof. Part (i) is immediate from (5.14) and (5.15). From the same relations, we obtain

$$xy = (2\lambda - 1)^{-1}(\lambda x * y + (\lambda - 1)y * x) = \mu x * y + (1 - \mu)y * x,$$

since $1 - \mu = (2\lambda - 1)^{-1}(\lambda - 1)$, verifying part (ii). If A is flexible, then by (5.16) $(x, y, x)^* = 0$ for all $x, y \in A$, giving flexibility of $A(\lambda)$. The converse follows from part (ii), and this proves (iii). Parts (iv)-(vi) are immediate consequences of (5.13) and part (ii) (see Definition 1.4). If A is semisimple, then $A = A_1 \oplus \cdots \oplus A_n$ for simple ideals A_i of A. But, $A(\lambda) = A_1(\lambda) \oplus \cdots \oplus A_n(\lambda)$ for simple ideals $A_i(\lambda)$ of $A(\lambda)$, and thus $A(\lambda)$ is semisimple. The converse is clear from part (ii). □

Definition 5.3. An algebra A over F (char $F \neq 2$) is called *quasiassociative* if A is an F–form of $B(\lambda)$ for some associative algebra B over an extension field K of F and some $\lambda \in K$; that is, $A_K = K \otimes_F A = B(\lambda)$. □

Let A be quasiassociative with $A_K = B(\lambda)$ $(\lambda \in K)$. If $\lambda = 0$ or 1, then A_K is anti–isomorphic or isomorphic to B, hence A is associative. If $\lambda = \frac{1}{2}$, then $A_K = B^+$ and A is a special Jordan algebra. Those quasiassociative algebras that are not covered by associative or Jordan algebras can be characterized by identities of degree 3 [**A3**].

Theorem 5.4. *Let* A *be an algebra over* F *which is neither associative nor Jordan. Then,* A *is quasiassociative if and only if* A *is flexible and satisfies the identity*

$$(5.17) \qquad (2\lambda_0 + 1)(x, y, z) = \lambda_0\left((yx)z + z(xy) - x(zy) - (yz)x\right)$$

for some $\lambda_0 \neq 0, -\frac{1}{4}$ *in* F.

Proof. Assume that A is quasiassociative with $A_K = B(\lambda)$. As remarked above, $\lambda \neq 0, 1, \frac{1}{2}$, and since B is associative, by Lemma 5.3 (iii) A is flexible. By Lemma 5.3 (ii), we have $B = A_K(\mu)$ with $\mu = \lambda(2\lambda - 1)^{-1} \in K$. Since B is associative and A is flexible, it follows from (4.7) and (5.16) that A satisfies

$$(5.18) \qquad \left(2(\mu^2 - \mu) + 1\right)(x, y, z) = (\mu^2 - \mu)\left((yx)z + z(xy) - x(zy) - (yz)x\right)$$

for all $x, y, z \in A$. Let $\lambda_0 = \mu^2 - \mu$. If $\lambda_0 \notin F$, then from (5.18) we would have $(x, y, z) = 0$ for all $x, y, z \in A$, that is, A is associative, a contradiction. Hence, $\lambda_0 \in F$ and this gives (5.17) with $\lambda_0 \neq 0, -\frac{1}{4}$, since $\mu \neq 0, 1, \frac{1}{2}$.

Conversely, assume that A is flexible and satisfies (5.17) with $\lambda_0 \neq 0, -\frac{1}{4}$ in F. Let μ be a zero of the polynomial $f(X) = X^2 - X - \lambda_0 \in F[X]$ and let $K = F(\mu)$. Then, A_K satisfies (5.18), and hence by (5.16) $(x, y, z)^* = 0$ for all $x, y, z \in A_K$ where $x * y$ denotes the product in $A_K(\mu)$. Therefore, $A_K(\mu)$ is associative, and by Lemma 5.3 (ii) $A_K = (A_K(\mu))(\lambda)$ with $\lambda = (2\mu - 1)^{-1}\mu$, since $\mu \neq 0, 1, \frac{1}{2}$. Thus, A is quasiassociative. □

Corollary 5.5. *If A is a quasiassociative algebra over F which is neither associative nor Jordan, then there exists a scalar $\lambda_0 \neq 0, -\frac{1}{4}$ in F such that $A_K(\mu)$ is associative, where $K = F(\mu)$ with $\mu^2 - \mu - \lambda_0 = 0$.*

Proof. Since A is flexible and satisfies (5.17) with $\lambda_0 \neq 0, -\frac{1}{4}$ in F, the result follows from the last part of the proof of Theorem 5.4. □

Corollary 5.6. *Let A be an algebra over F which is neither associative nor Jordan. If A is quasiassociative, then A satisfies the identity*

$$(x, y, z) = \lambda_0[[z, x], y] \tag{5.19}$$

for some $\lambda_0 \neq 0, -\frac{1}{4}$ in F. Conversely, if A satisfies (5.19) with $\lambda_0 \neq 0, -\frac{1}{4}, \frac{1}{2}$ in F, then it is quasiassociative.

Proof. Assume that A is quasiassociative. By Lemma 5.3 A is flexible Lie–admissible and by Lemma 4.2 (i)

$$S(x, y, z) \equiv (x, y, z) + (y, z, x) + (z, x, y) = 0$$

for all $x, y, z \in A$. Using this and flexibility (4.7) we obtain

$$\begin{aligned}
&x(zy) + (yz)x - (yx)z - z(xy) \\
&\quad = [[x, z], y] - (xz)y + (zx)y + y(xz) - y(zx) \\
&\qquad\quad + x(zy) + (yz)x - (yx)z - z(xy) \\
&\quad = [[x, z], y] - (x, z, y) + (z, x, y) - (y, x, z) + (y, z, x) \\
&\quad = [[x, z], y] + 2(y, z, x) + 2(z, x, y) \\
&\quad = [[x, z], y] - 2(x, y, z).
\end{aligned}$$

Hence, (5.17) becomes

$$-(2\lambda_0 + 1)(x, y, z) = \lambda_0 ([[x, z], y] - 2(x, y, z))$$

for $\lambda_0 \neq 0, -\frac{1}{4}$ in F, and this gives (5.19).

Suppose that A satisfies (5.19) for some $\lambda_0 \neq 0, -\frac{1}{4}, \frac{1}{2}$ in F. Flexibility of A follows from (5.19) with $z = x$. Permuting x, y, z cyclically in (5.19), we obtain

$\lambda_0 J(x,y,z) = S(x,y,z)$ for all $x,y,z \in A$. But by (4.3) $J(x,y,z) = 2S(x,y,z)$ since A is flexible, and hence $J(x,y,z) = S(x,y,z) = 0$ for all $x,y,z \in A$. This also shows that A is flexible Lie–admissible. We write (5.19) as $(x,y,z) = \lambda_0[[z,x],y] - 2\lambda_0 S(x,y,z)$, and expanding this gives

$$\begin{aligned}(2\lambda_0 + 1)(x,y,z) &= \lambda_0[[z,x],y] - 2\lambda_0(y,z,x) - 2\lambda_0(z,x,y) \\ &= \lambda_0\left((yx)z + z(xy) - x(zy) - (yz)x\right).\end{aligned}$$

Thus, by Theorem 5.4 A is quasiassociative. □

Generalizing Definition 5.3, we define an algebra A over F to be *quasialternative* if A is an F–form of $B(\lambda)$ for an alternative algebra B over an extension field K of F and for $\lambda \in K$. Thus, a quasialternative algebra is a form of a scalar mutation of an alternative algebra. Since B is alternative and $A_K = B(\lambda)$, it follows from (5.16) and Lemma 5.3 that any quasialternative algebra A is noncommutative Jordan and Malcev–admissible. We note that if $\lambda = 0, 1$, or $\frac{1}{2}$, then $B(0) = B^{op}$, $B(1) = B$ or $B(\frac{1}{2}) = B^+$ and hence A is alternative or (special) Jordan.

Let A be any algebra with multiplication xy over F, and let $\lambda \in F$. From (5.16) we obtain

$$\begin{aligned}(x,x,y)^* &= \lambda^2(x,x,y) - (1-\lambda)^2(y,x,x) \\ &\quad + \lambda(1-\lambda)\left(2x^2 \circ y - x(yx) - (xy)x\right)\end{aligned} \tag{5.20}$$

for all $x,y \in A$, where "$*$" and "$\circ$" denote the products in $A(\lambda)$ and A^+. We notice that a flexible algebra A is alternative if and only if $(x,x,y) = 0$ for $x,y \in A$.

The following is an analogous result of Theorem 5.4 for the alternative case.

Theorem 5.7. *Let A be an algebra with multiplication xy over F which is neither alternative nor Jordan. Then, A is quasialternative if and only if A is flexible and satisfies the identity*

$$(2\lambda_0 + 1)\left((x,x,y)\right) = 2\lambda_0\left(x^2 \circ y - (xy)x\right) \tag{5.21}$$

for $\lambda_0 \neq 0, -\frac{1}{4}$, where "$\circ$" is the product in A^+.

Proof. We proceed as for the proof of Theorem 5.4. Assume that A is quasialternative with $A_K = B(\lambda)$, where B is alternative over K. As noted above, A is flexible and $\lambda \neq 0, 1, \frac{1}{2}$. Since $B = A_K(\mu)$ with $\mu = \lambda(2\lambda - 1)^{-1}$ is alternative, from (4.7) and (5.20) we obtain

$$\left(2(\mu^2 - \mu) + 1\right)(x,x,y) = 2(\mu^2 - \mu)\left(x^2 \circ y - (xy)x\right)$$

for all $x,y \in A$. The remainder of the proof is the same as for Theorem 5.4. □

As for the associative case, if A is a quasialternative algebra over F which is neither alternative nor Jordan, then there exists an extension field $K = F(\mu)$ of F of degree ≤ 2 and $\lambda \in K$ such that $A_K = B(\lambda)$ for some alternative algebra B over K, where μ is a zero of $X^2 - X - \lambda_0$ with $\lambda_0 \neq 0, -\frac{1}{4}$ in F. If A is an alternative algebra with a unit element, then any scalar mutation $A(\lambda, \mu)$ of A or any quasialternative algebra is a special case of $A(p,q)$ with $p, q \in Z(A)$. In this case, $A(p,q)$ is said to be generalized quasialternative (Definition III.3.1). If $\alpha p + \beta q$ is invertible in the nucleus $N(A)$ for some $\alpha, \beta \in Z(A)$, then it can be shown that $A(p,q)$ is flexible or noncommutative Jordan if and only if $A(p,q)$ is generalized quasialternative (Theorem III.3.6).

We close this section with a proof of the result that an alternative algebra A is simple if and only if A^+ is simple. If A is associative, then this result is a consequence of the following lemma.

Lemma 5.8. *If A is a semiprime associative algebra over F, then any nonzero ideal of A^+ contains a nonzero ideal of A.*

Proof. Let I be a nonzero ideal of A^+, and let $x \in I$ and $a \in A$. Then, one gets $x \circ (xa - ax) = \frac{1}{2}(x^2 a - ax^2) \in I$, and since $x^2 = x \circ x \in I$, $x^2 \circ a = \frac{1}{2}(x^2 a + ax^2) \in I$, hence $ax^2, x^2 a \in I$. For $b \in A$, we have $ax^2 b = 2(ax^2 \circ b) - (ba)x^2 \in I$, so that $Ax^2 A \subseteq I$ for all $x \in I$. Thus, it suffices to verify $Ax^2 A \neq 0$ for some $x \in I$. Suppose $Ax^2 A = 0$ for all $x \in I$. By semiprimeness of A, $x^2 = 0$ for all $x \in I$ and hence $x \circ y = \frac{1}{2}\left((x+y)^2 - x^2 - y^2\right) = 0$ for any $x, y \in I$. But, for $0 \neq x \in I$ and $a \in A$, one has

$$xax = \frac{1}{2}\left(x(xa + ax) + (xa + ax)x\right) = \frac{1}{2}(x \circ (x \circ a)) = 0.$$

This implies that $(AxA)^2 \subseteq AxAxA = 0$ and hence $AxA = 0$, forcing $x = 0$, a contradiction. Therefore, I contains a nonzero ideal of A. □

Theorem 5.9. *An alternative algebra A over F is simple if and only if A^+ is simple.*

Proof. It is clear that if A^+ is simple, then so is A. If A is associative, then the result is immediate from Lemma 5.8. Assume that A is simple and not associative. By Theorem 3.6 A is a Cayley–Dickson algebra over the center $Z(A)$, an extension field of F. Hence, A satisfies the quadratic equation (3.8) over $Z(A)$. Letting $V = \{x \in A : t(x) = 0\}$, A decomposes as $A = Z(A)1 \oplus V$ (see (4.17)). Using (4.18), we find the multiplication "$\circ$" in A^+ to be given by

$$(\alpha 1 + x) \circ (\beta 1 + y) = (\alpha\beta - (x,y))1 + \alpha y + \beta x$$

for $\alpha, \beta \in Z(A)$ and $x, y \in V$. Therefore, A^+ is the Jordan algebra over $Z(A)$ associated with V and the symmetric bilinear form $(\ ,\ )$ [**J3**, p,13]. Since $\dim V > 1$

and (,) is nondegenerate, A^+ is simple over $Z(A)$ [**J3**, p.14]. In light of Lemma 2.1 (iii), A^+ is simple over F also, since F is a subfield of $Z(A)$. $\square$

CHAPTER II

MUTATIONS OF ASSOCIATIVE ALGEBRAS

1. Introduction

Throughout this chapter, we let A denote an associative algebra over a field F of characteristic $\neq 2$, and p and q denote fixed elements in A. Recall from Definition I.4.4 that the (p,q)–mutation $A(p,q)$ of A is the algebra with multiplication $x * y$ defined by

$$x * y = xpy - yqx \tag{1.1}$$

but with the same vector space as A, and that the commutator (Lie product) $[x,y]^* = x*y - y*x$ and the anticommutator (Jordan product) $\{x,y\}^* = \frac{1}{2}(x*y + y*x)$ in $A(p,q)$ are given by

$$[x,y]^* = x(p+q)y - y(p+q)x, \tag{1.2}$$

$$\{x,y\}^* = \tfrac{1}{2}[x(p-q)y + y(p-q)x]. \tag{1.3}$$

Consequently, we have

$$\begin{aligned} A(p,q)^- &= A(p+q)^-, \\ A(p,q)^+ &= A(p-q)^+ \end{aligned} \tag{1.4}$$

(see (I.4.29), (I.4.30), (I.5.1) and (I.5.2)), and hence $A(p,q)$ is both Lie– and Jordan–admissible (Definitions I.4.1 and I.5.1).

The (p,q)–mutation $A(p,q)$ will often be referred to as a mutation of A or a mutation algebra. Although the class of algebras that satisfy the same identities as mutation algebras is much larger than the class of associative algebras (see Chapter V), the structure of mutation algebras is closely related to the associative theory and consequently is governed by the structure of A as well as the fixed elements p and q in A. The earlier work on mutation algebras focused on finite–dimensional associative algebras [**My2**,**Os2**,**EGM1**,**MS1**]. More recently, mutation algebras have been studied for artinian associative algebras [**EM1**,**Mo1,3**]. In this section, we give some basic definitions and results which will be used for later investigations.

Definition 1.1. An algebra B over a field F is called *left (or right) artinian* if it satisfies the descending chain condition on left (or right) ideals of B, or equivalently, if any nonempty set of left (or right) ideals in B has a minimal element. □

Recall that the radical $R(B)$ of any algebra B is defined to be the intersection of all maximal ideals M of B such that $B^2 \not\subseteq M$ (Definition I.1.4). For a left (right) artinian associative algebra A, it is readily seen (Lemma 1.3) that $R(A)$ coincides with several different radicals of A that are commonly employed in the associative theory. For example, if A is left artinian, then the Jacobson radical $J(A)$ equals the prime radical $P(A)$ of A (the intersection of all prime ideals in A) and is the unique maximal nilpotent ideal of A which contains all nilpotent left ideals and nilpotent right ideals of A [**Hun1**,**J5**]. An associative algebra A is called *semisimple* if $J(A) = 0$ [**Hun1**,**J5**]. Much of our discussion in this chapter is based on the following Wedderburn–Artin theorem for artinian associative algebras [**Hun1**,**J5**].

Theorem 1.1. *Let A be a left (or right) artinian associative algebra over a field F.*

(i) *A is simple if and only if it is isomorphic to the algebra $M_n(D)$ of $n \times n$ matrices with entries in a division algebra D over F.*

(ii) *A is semisimple if and only if it is isomorphic to the direct sum $M_{n_1}(D_1) \oplus \ldots \oplus M_{n_r}(D_r)$ of matrix algebras $M_{n_i}(D_i)$ over division algebras D_i over F ($i = 1, \ldots, r$).* □

Henceforth, a left or right artinian algebra will be simply called artinian. We note that a semisimple associative algebra is left artinian if and only if it is right artinian [**Hun1**]. The following result is instrumental for the structure of $A(p,q)$ with A semisimple artinian [**Hun1**].

Proposition 1.2. *If A is a semisimple artinian associative algebra then any left (or right) ideal I of A is given by $I = Ae$ (or $I = eA$) for some idempotent $e \in A$ (i.e., $e^2 = e$).* □

Lemma 1.3. *Let A be an associative algebra over F and p, q be fixed elements in A. Let N be any nilpotent ideal of A. Then,*

(i) *N is contained in every maximal ideal M of A or $A(p,q)$ such that $A^2 \not\subseteq M$ or $A(p,q) * A(p,q) \not\subseteq M$.*

(ii) *If A is artinian then $J(A) = R(A) \subseteq R(A(p,q))$.*

Proof. (i) Let M be a maximal ideal of $A(p,q)$ such that $A(p,q) * A(p,q) \not\subseteq M$. Note from (1.1) that any ideal of A is an ideal of $A(p,q)$. Assume M does not contain N. Since M is maximal, $A(p,q) = N + M$ and $A(p,q) * A(p,q) = (N+M)*(N+M) \subseteq N * N + M \subseteq N^2 + M$, so $N^2 \not\subseteq M$ since $A * A \not\subseteq M$. Hence, $A(p,q) = N^2 + M$ and by induction $A(p,q) = N^m + M$ for any positive integer m. Since N is nilpotent, $A(p,q) = M$, a contradiction, so $N \subseteq M$. The proof is similar for the case where M is a maximal ideal of A with $A^2 \not\subseteq M$.

(ii) Since $J(A)$ is nilpotent, by part (i) and Definition I.1.4, $J(A) \subseteq R(A)$ and $J(A) \subseteq R(A(p,q))$. From the last part of the proof of Theorem I.1.7, it follows that if I is an ideal of A such that A/I is semisimple in the sense of Definition I.1.1, then $R(A) \subseteq I$. Since $J(A/J(A)) = 0$, by Theorem 1.1(ii) we have $R(A) \subseteq J(A)$ and hence $J(A) = R(A) \subseteq R(A(p,q))$. □

Hereafter, we will use $R(A)$ for the radical of A. Thus, for artinian A, $R(A)$ is the unique maximal nilpotent ideal of A. As we shall see later, $R(A) \neq R(A(p,q))$ in general. For example, $A(p,q)$ can be a radical algebra for $p = q$, i.e., $R(A(p,q)) = A(p,q)$, even when A is simple artinian. On the other hand, it is shown that if A is artinian and $A(p,q)$ has a unit element then $R(A) = R(A(p,q))$.

The primary aim in this chapter is to present a detailed account of the general structure of mutation algebras. The chapter is organized as follows. In Section II.2, we investigate conditions for $A(p,q)$ to satisfy some well known identities. Two principal identities which are not in general satisfied by $A(p,q)$ are third power–associativity (Lemma I.4.1) and flexibility (I.3.3). We give a condition for $A(p,q)$ to satisfy either one of these identities which are also shown to be equivalent in $A(p,q)$ under a slightly stronger condition. The nucleus, center and commutative center of $A(p,q)$ are explicitly described in terms of those of A.

The main task in the structure theory of mutation algebras has been to determine the structure of $A(p,q)$ for a simple associative algebra A. The crux of this problem is that even if A is simple, $A(p,q)$ is not in general simple. Thus, it is natural to investigate conditions for $A(p,q)$ to be simple for a simple associative algebra A. Let L_a and R_a denote the left and right multiplications in A by a. For fixed $p, q \in A$, we define $\Lambda(p,q)$ to be the subset of $Hom_F A$ given by

$$\Lambda(p,q) = \{L_pR_p, L_pR_q, L_qR_p, L_qR_q\}. \tag{1.5}$$

It follows immediately from (1.1) that the subsets $R^0(p,q)$ and $J(p)$ of A given by

$$\begin{aligned} R^0(p,q) &= \{x \in A : \Lambda(p,q)x = 0\}, \\ J(p) &= \{x \in A : x * A(p,p) \subseteq R^0(p,p)\} \\ &= \{x \in A : pxpyp = pypxp \text{ for all } y \in A\} \end{aligned} \tag{1.6}$$

are respectively ideals of $A(p,q)$ and of $A(p,p)$ with $R^0(p,p) \subseteq J(p)$. Note that $A(p,p) = A(p)^-$ is a Lie algebra and $R^0(p,p) = R^0(p,0)$. For simplicity, we denote $R^0(p) = R^0(p,p)$. Section II.3 is devoted to establishing the results that $A(p,q)$ for $p \neq q$ is simple (or prime) if and only if A is simple (or prime) and $R^0(p,q) = 0$, and that $A(p,p)$ is prime if and only if A is prime and $J(p) = 0$. On the other hand, a necessary and sufficient condition for $A(p,p)$ to be simple is that A is simple, $J(p) = 0$ and $A(p,p) * A(p,p) = A(p,p)$ (Theorem 3.14). Therefore, the ideals $R^0(p,q)$ and $J(p)$ are regarded as measurements of the simplicity or primeness of $A(p,q)$. It is also shown in Section II.3 that if $A(p,q)$ with $p \neq q$ is prime then the centroids of A and $A(p,q)$ coincide (see Definition I.2.2).

Section II.4 is the highlight of this chapter in that the structure of $A(p,q)$ is completely determined for any simple artinian associative algebra A. Let $M_{r\times s}(D)$

denote the D–vector space of all $r \times s$ matrices over D where D is a division algebra over F. We also regard $M_{r\times s}(D)$ as the D–vector space of all linear maps : $V \to W$ over D where V and W are vector spaces over D with $\dim_D V = r$ and $\dim_D W = s$. Each element of $M_{r\times s}(D)$ acts on V from the left while scalars in D acts on V and W from the right.

Definition 1.2. For fixed positive integers r, s, let $B = M_{r\times s}(D)$ where D is a division algebra over a field F. For fixed elements $u, v \in M_{s\times r}(D)$, define the operation "$*$" on B by

$$x * y = xuy - yux \tag{1.7}$$

for $x, y \in B$, where xu denotes the matrix product. The operation "$*$" converts $B = M_{r\times s}(D)$ into an F–algebra, which will be called the *twisted* (u, v)*–mutation* of B. We denote this algebra by $B(u, v)$ using the same notation as $A(p, q)$. □

For a fixed $a \in M_{s\times r}(D)$, the *twisted* a*–homotope* $B(a)$ of $B = M_{r\times s}(D)$ may be defined as the algebra over F with multiplication xay $(x, y \in B)$ defined on B. Clearly, $B(a)$ is associative, and the Jordan algebra $B(a)^+$ has been called the *Riccati algebra* [**LS1**] associated with a in the study of matrix Riccati equations. For the twisted (u, v)–mutation $B(u, v)$, as in (1.4) we have $B(u, v)^- = B(u + v)^-$ and $B(u, v)^+ = B(u - v)^+$. Note that if $r = s$ then a twisted mutation algebra is just a mutation algebra.

The main work in Section II.4 is to show that if A is simple artinian, so that $A = M_n(D)$, then there exists a subalgebra S of $A(p, q)$ such that

$$A(p, q) = R^0(p, q) \oplus S \tag{1.8}$$

and S is isomorphic to a twisted mutation algebra $M_{r\times s}(D)(u, v)$ for some $r, s \leq n$ and $u, v \in M_{s\times r}(D)$. Furthermore, if $p \neq q$ then $u \neq v$ and S is simple. In this case, $R^0(p, q)$ is shown to equal the radical $R(A(p, q))$ (Section II.5). Therefore, if A is simple artinian and $p \neq q$, then the decomposition (1.8) is regarded as an analog for $A(p, q)$ of the Wedderburn principal theorem for finite–dimensional alternative algebras (or other well known algebras) or of Levi's decompositions for Lie and Malcev algebras (of characteristic 0) [**Sc3**, **J2,3**,**Ku2**]. On the other hand, if $p = q$, so that $A(p, p)$ is a Lie algebra, then S is not in general simple and there is the unique maximal solvable ideal of $A(p, p)$ which contains the center $Z(S)$ of S and $R^0(p, q)$.

Finally, in Section II.5, we study the structure of the radical $R(A(p, q))$ for an artinian associative algebra A. The structure of $R(A(p, q))$ is determined relative to the radical $R(A)$. For this, the two ideals $R(p, q)$ and $J(p, q)$ of $A(p, q)$ defined by the following play important roles.

Definition 1.3. For an associative algebra A over F and fixed $p, q \in A$, define the subsets $R(p,q)$ and $J(p,q)$ of A by

$$R(p,q) = \{x \in A : \Lambda(p,q)x \subseteq R(A)\}, \tag{1.9}$$

$$J(p,q) = \{x \in A : x * y, \quad y * x \in R(p,q) \text{ for every } y \in A\}. \tag{1.10}$$

□

Lemma 1.4. *Let A, p and q be the same as in Definition 1.3. Then, $R(p,q)$ and $J(p,q)$ are ideals of $A(p,q)$ satisfying the relation*

$$R(A) \subseteq R(p,q) \subseteq J(p,q). \tag{1.11}$$

Proof. Since $R(A)$ is an ideal of A, clearly $R(A) \subseteq R(p,q)$. By Definition 1.3, it suffices to show that $R(p,q)$ is an ideal of $A(p,q)$. For $x \in R(p,q)$ and $y \in A$, $\Lambda(p,q)(x*y) = \Lambda(p,q)(xpy - yqx) = \Lambda(p,q)(xpy) - \Lambda(p,q)(yqx) \subseteq R(A)$ since $R(A)$ is an ideal of A. Similarly, $\Lambda(p,q)(y*x) \subseteq R(A)$ and hence $R(p,q)$ is an ideal of $A(p,q)$. □

Proposition 1.5. *Let A be an artinian associative algebra over F and let p, q be fixed elements of A. Then,*

(i) *we have the relation*

$$R(A) \subseteq R(p,q) \subseteq J(p,q) \subseteq R(A(p,q)). \tag{1.12}$$

(ii) *$J(p,q)$ is a solvable ideal of $A(p,q)$ such that $J(p,q)^{(3)} \subseteq R(A)$ where $J(p,q)^{(n)}$ $(n = 0, 1, \ldots)$ denotes the derived series (Definition I.1.5).*

(iii) *$R(p,q)$ is a nilpotent ideal of $A(p,q)$ such that if $R(A)^m = 0$ then $R(p,q)^{2m+1} = 0$ in $A(p,q)$.*

Proof. (i) By Lemma 1.4, we only need to verify the last inclusion. Suppose that $J(p,q)$ is not contained in $R(A(p,q))$. Then, there exists a maximal ideal M of $A(p,q)$ such that $A * A \not\subseteq M$ and $J(p,q) \not\subseteq M$. This implies $A = J(p,q) + M$ and so $A * A = J(p,q) * A + M * A \subseteq R(p,q) + M$ by the definition of $J(p,q)$. Since $A * A \not\subseteq M$ and $R(p,q)$ is an ideal of $A(p,q)$, by the maximality of M we have $A = R(p,q) + M$ and in particular, $R(p,q) \not\subseteq M$. From the definition of $R(p,q)$,

$$A * (R(p,q) * R(p,q)) = (R(p,q) * R(p,q)) * A \subseteq R(A), \tag{1.13}$$

and hence $R(p,q) * R(p,q) + R(A)$ is an ideal of $A(p,q)$. From this, we have $A = R(p,q) * R(p,q) + R(A) + M = R(p,q) * R(p,q) + M$ since $R(A) \subseteq M$ by Lemma

1.3 and since $A * A = (R(p,q) + M) * (R(p,q) + M) \subseteq R(p,q) * R(p,q) + M \subseteq R(p,q) * R(p,q) + R(A) + M$. Thus, by (1.13) and Lemma 1.3,

$$A * A = (R(p,q) * R(p,q) + M) * A \subseteq R(A) + M \subseteq M,$$

a contradiction. This implies $J(p,q) \subseteq R(A(p,q))$.

(ii) We have $J(p,q)^{(1)} = J(p,q) * J(p,q) \subseteq R(p,q)$ and $J(p,q)^{(3)} \subseteq (J(p,q)^{(1)} * J(p,q)^{(1)}) * A \subseteq (R(p,q) * R(p,q)) * A \subseteq R(A)$ by (1.13), so $J(p,q)^{(3)} \subseteq R(A)$. Since $R(A)$ is nilpotent in A (Lemma 1.3), it is also a nilpotent ideal of $A(p,q)$. Hence, $J(p,q)$ is a solvable ideal of $A(p,q)$.

(iii) Assume $R(A)^m = 0$. Then, any product in $A(p,q)$ of $2m+1$ elements in $R(p,q)$ is a sum of elements of the form

$$x = x_1 u_1 x_2 u_2 \ldots u_{2m} x_{2m+1}$$

where $x_i \in R(p,q)$, $i = 1, \ldots, 2m+1$, and $u_i = p$ or q, $i = 1, \ldots, 2m$. But ,

$$x = x_1(u_1 x_2 u_2) x_3 (u_3 x_4 u_4) \ldots x_{2m-1}(u_{2m-1} x_{2m} u_{2m}) x_{2m+1} \in R(A)^m = 0,$$

since each $u_i x_{i+1} u_{i+1} \in R(A)$ $(i = 1, \ldots, 2m-1)$ by the definition of $R(p,q)$. Thus, $R(p,q)$ is nilpotent in $A(p,q)$ with $R(p,q)^{2m+1} = 0$. □

Therefore, the structure of $R(A(p,q))$ relative to $R(A)$ largely depends on the behavior of $R(p,q)$ and $J(p,q)$. If A is artinian, then $A/R(A)$ is the direct sum of simple artinian algebras $A_1, \ldots, A_m$. In Section II.5, the quotient algebra $R(A(p,q))/R(A)$ is determined in terms of the radicals of mutation algebras of $A_1, \ldots, A_m$. Note that if $R(A) = 0$ then $R(p,q) = R^0(p,q)$ and $J(p,p) = J(p)$ by (1.6).

If D is a finite–dimensional associative division algebra over F then the center $Z(D) = \Gamma$ of D is a field containing F (Section I.2), and $\dim_\Gamma D = m^2$ where m is the dimension of a maximal subfield of D over Γ [**Hun1**, p.459]. Thus, $\dim_\Gamma M_n(D) = (mn)^2$, and mn is uniquely determined by $M_n(D)$.

Definition 1.4. If D is a finite–dimensional associative division algebra over F and $A = M_n(D)$, then the positive integer mn as above is called the *degree* of A. □

The following result is well known [**Hu1**, p.5].

Proposition 1.6. *Let D be the same as in Definition 1.4 and let mn be the degree of $A = M_n(D)$.*

(i) *If the characteristic of F does not divide mn, then*

$$A^- = Z(A) \oplus [A, A].$$

where $Z(A) = Z(D)I_n$.

(ii) *If the characteristic of* F *divides* mn, *then* $Z(A) \subseteq [A, A] \subset A$, *where* "$\subset$" *denotes proper containment.* □

2. Identities in mutation algebras

Mutation algebras do not satisfy some well known identities considered in the theory of nonassociative algebras, unless certain restrictions are imposed on the associative algebra A or on the fixed elements p and q. The primary concern in this section is to investigate conditions under which mutation algebras satisfy third power–associativity or flexibility, and its equivalence to other identities satisfied by mutation algebras. The present results generalize several earlier works on identities for mutation algebras [**My2**,**Os2**,**EGM1**,**GM1**,**2**,**E3**]. A systematic study of identities for mutation algebras will be presented in Chapter V. It is shown that there exist identities of degree 4 for mutation algebras which are not consequences of Lie– and Jordan–admissibilities (Theorem V.2.6). Some of the results in this section are extended to the variety generated by mutation algebras (Section V.3).

Throughout, the base field F is assumed to be of characteristic $\neq 2$. Let A be an associative algebra with multiplication xy over F and let p, q be fixed in A. Denote by $(x, y, z)^* = (x * y) * z - x * (y * z)$ the associator in $A(p, q)$. Recall from (I.4.32) – (I.4.34) the identities

$$(x, y, z)^* = xpzqy - yqxpz + ypzqx - zqxpy, \tag{2.1}$$

$$(x, y, x)^* = x(pxq - qxp)y + y(pxq - qxp)x, \tag{2.2}$$

$$(x, x, x)^* = 2x(pxq - qxp)x. \tag{2.3}$$

Theorem 2.1. *Let* A *be an associative algebra over* F *and let* p, q *be fixed in* A. *Then,* $A(p, q)$ *is third power–associative if and only if the identity*

$$x(pxq - qxp)x = 0 \tag{2.4}$$

holds for all $x \in A$ *if and only if* $A(p, q)$ *is power–associative.*

Proof. Note that $A(p, q)$ is third power–associative if and only if $(x, x, x)^* = 0$ for all $x \in A$. It follows from (2.3) that (2.4) is equivalent to third power–associativity of $A(p, q)$. Assume that (2.4) holds in A. Subtracting $xpxpx$ from (2.4) gives

$$\begin{aligned} x(p - q)xpx &= xpx(p - q)x, \\ x(p - q)xqx &= xqx(p - q)x \end{aligned} \tag{2.5}$$

for all $x \in A$. For a positive integer $k > 1$, define $x^{*k} = x^{*(k-1)} * x$. We verify the identity

$$x^{*r} * x^{*s} = [x(p-q)]^{r+s-1}x \tag{2.6}$$

for any positive integers r, s, which in particular implies power–associativity of $A(p,q)$. Note from (1.1) that $x * x = x(p-q)x$ for all $x \in A$. Assume (2.6) holds for r and s with $r+s < n$ where n is an integer with $n \geq 3$. If $r+s = n$, then using (2.5) and (2.6) with $r+s < n$ we obtain

$$\begin{aligned} x^{*r} * x^{*s} &= x^{*r}px^{*s} - x^{*s}qx^{*r} \\ &= [x(p-q)]^{r-1}xp[x(p-q)]^{s-1}x - [x(p-q)]^{s-1}xq[x(p-q)]^{r-1}x \\ &= [x(p-q)]^{r+s-2}xpx - [x(p-q)]^{r+s-2}xqx \\ &= [x(p-q)]^{n-1}x. \end{aligned}$$

Therefore, (2.6) holds for all $r, s>0$ and all $x \in A$. This implies power–associativity of $A(p,q)$. □

As to flexibility in $A(p,q)$, we have an identity similar to (2.4).

Theorem 2.2. *Let A be an associative algebra over F and let p, q be any fixed elements in A. Then, $A(p,q)$ is flexible if and only if*

$$x(pyq - qyp)x = 0 \tag{2.7}$$

for all $x, y \in A$.

Proof. Assume $A(p,q)$ is flexible, i.e., $(x,y,x)^* = 0$ for all $x, y \in A$. Linearizing (2.2) we obtain

$$x(pzq - qzp)y + z(pxq - qxp)y + y(pxq - qxp)z + y(pzq - qzp)x = 0$$

for all $x, y, z \in A$, which with $x = y$ implies

$$2x(pzq - qzp)x + z(pxq - qxp)x + x(pxq - qxp)z = 0.$$

Since the last two terms of this add to $(x,z,x)^* = 0$ by (2.2), we have (2.7) holding for $A(p,q)$. Assume that $A(p,q)$ satisfies (2.7). Linearizing it one gets $x(pyq - qyp)z + z(pyq - qyp)x = 0$ for all $x, y, z \in A$, which with $x = y$ implies $0 = x(pxq - qxp)z + z(pxq - qxp)x = (x,z,x)^*$ by (2.2). Thus, $A(p,q)$ is flexible. □

Corollary 2.3. *Let A and p, q be the same as in Theorem 2.2. Consider the following conditions:*

(i) $pxq = qxp$ *for all* $x \in A$.

(ii) $A(p,q)$ *is flexible.*

(iii) $A(p,q)$ *is a noncommutative Jordan algebra.*

(iv) $A(p,q)$ *is third power–associative.*

(v) $A(p,q)$ *is power–associative.*

Then, we have the implications: (i) $\Rightarrow$ (ii) $\Leftrightarrow$ (iii) $\Rightarrow$ (iv) $\Leftrightarrow$ (v).

Proof. Condition (i) clearly implies (2.7) and hence flexibility of $A(p,q)$. Recall from Proposition I.5.1 that an algebra over F (of characteristic $\neq 2$) is flexible Jordan–admissible if and only if it is noncommutative Jordan, and hence (ii) $\Leftrightarrow$ (iii). By Proposition I.5.2, any noncommutative Jordan algebra over F is power–associative. The equivalence of conditions (iv) and (v) is a consequence of Theorem 2.1. □

The following examples show that the implications (ii) $\Rightarrow$ (i) and (iv) $\Rightarrow$ (iii) are not valid in general.

Example 2.4. Let $B = F\langle X, Y\rangle$ denote the free associative algebra over F with generators X, Y, so that B is the associative algebra over F of polynomials without constant term in the noncommuting indeterminates X and Y. Let $A = B/B^5$ and write $x = X + B^5$ and $y = Y + B^5$. If we choose $p = x$ and $q = y$, then $pxq \neq qxp$ and condition (i) in Corollary 2.3 does not hold. However, since $A^5 = 0$, $u(pvq - qvp)u = 0$ for all $u, v \in A$ and hence by Theorem 2.2 $A(p,q)$ is flexible. □

Example 2.5. Let B be the same as in Example 2.4 and let I be the ideal of B generated by all monomials in B of degree five except X^2Y^2X and XYX^2Y, and by the element $X^2Y^2X - XYX^2Y$. It is readily seen that $B^6 \subseteq I$. For example, $(XYX^2Y)X = X(YX^2YX) \in I$ since $YX^2YX \in I$, $Y(XYX^2Y) = (YXYX^2)Y \in I$, etc. Let $A = B/I$ and write $x = X + I$ and $y = Y + I$. We choose $p = x$ and $q = y$ to form $A(p,q)$. Now, $x(pyq - qyp)x = x^2y^2x - xy^2x^2 = x^2y^2x \neq 0$ since $X^2Y^2X \notin I$ but $XY^2X^2 \in I$. Hence, by Theorem 2.2 $A(p,q)$ is not flexible. In order to show that $A(p,q)$ is third power–associative, we need to verify the identity $u(puq - qup)u = 0$ for all $u \in A$ (Theorem 2.1). Since $A^6 = 0$, it suffices to prove that the relation

$$w(\alpha,\beta) = (\alpha x + \beta y)[p(\alpha x + \beta y)q - q(\alpha x + \beta y)p](\alpha x + \beta y) = 0$$

holds for all $\alpha, \beta \in F$. Expanding this and using the definition of I, we find

$$w(\alpha,\beta) = \alpha^2\beta(x^2y^2x - xyx^2y) = 0$$

for all $\alpha, \beta \in F$. Thus, $A(p,q)$ is third power–associative. □

However, the conditions (i) – (v) in Corollary 2.3 are equivalent under some natural conditions on the associative algebra A or on the elements p and q.

Theorem 2.6. *Let A be an associative algebra over F and let p, q be any fixed elements in A. If A contains an element which is not a zero divisor, then the conditions (i) – (v) in Corollary 2.3 are equivalent.*

Proof. It suffices to verify the implication (iv) $\Rightarrow$ (i). Thus, assume that $A(p,q)$ is third power–associative, so that $x(pxq - qxp)x = 0$ for all $x \in A$. Let z be an element in A which is not a zero divisor. Substituting $x + z$ and $x - z$ for x in the above equation and adding the resulting equations we obtain

$$zpzqx + zpxqz + xpzqz = zqzpx + zqxpz + xqzpz \tag{2.8}$$

for all $x \in A$. Since $z(pzq - qzp)z = 0$ by (2.4) and z is not a zero divisor, $pzq - qzp = 0$ and hence by (2.8) $z(pxq - qxp)z = 0$. Thus, $pxq = qxp$ for all $x \in A$, giving condition (i). □

If A has a unit element 1, then the hypothesis in Theorem 2.6 is automatically satisfied. If $\alpha p + \beta q$ is invertible in A for some $\alpha, \beta \in Z(A)$, then $A(p,q)$ is isomorphic to a more specialized algebra:

Definition 2.1. An algebra over F is called *generalized quasiassociative* if it is isomorphic to a mutation algebra $A(p,q)$ of an associative algebra A with $p, q \in Z(A)$. □

Theorem 2.7. *Let A be an associative algebra with unit element 1 over F and let p, q be any fixed elements in A. If there are elements α, β in the center $Z(A)$ such that $\alpha p + \beta q$ is invertible in A, then the conditions (i) – (v) in Corollary 2.3 are equivalent to each of the following:*

(vi) *$(\alpha p + \beta q)^{-1}p$ and $(\alpha p + \beta q)^{-1}q$ are contained in $Z(A)$.*

(vii) *$A(p,q)$ is isomorphic to $A(r,s)$ for some $r, s \in Z(A)$ with $\alpha r + \beta s = 1$.*

(viii) *$A(p,q)$ is a generalized quasiassociative algebra.*

Proof. Note first that the conditions (i) – (v) are equivalent. Hence, it suffices to verify the implications (i) $\Rightarrow$ (vi) $\Rightarrow$ (vii) $\Rightarrow$ (viii) $\Rightarrow$ (i). Assume that $\alpha p + \beta q$ is invertible in A for some $\alpha, \beta \in Z(A)$. If $pxq = qxp$ for all $x \in A$, then

$$(\alpha p + \beta q)xq = qx(\alpha q + \beta q).$$

Substituting $x(\alpha p + \beta q)^{-1}$ for x into this and multiplying $(\alpha p + \beta q)^{-1}$ from the left, we obtain $x[(\alpha p + \beta q)^{-1}q] = [(\alpha p + \beta q)^{-1}q]x$ and so $(\alpha p + \beta q)^{-1}q \in Z(A)$. Similarly, it is seen that $(\alpha p+\beta q)^{-1}p \in Z(A)$. Let $r=(\alpha p+\beta q)^{-1}p$ and $s=(\alpha p+\beta q)^{-1}q$, and assume that r and s are in $Z(A)$. Then, the linear map ϕ given by

$$\phi : A(r,s) \to A(p,q) : x \to x(\alpha p+\beta q)^{-1} \tag{2.9}$$

is an algebra isomorphism, since

$$\begin{aligned} \phi(x*y) &= (xry-ysx)(\alpha p+\beta q)^{-1} \\ &= [x(\alpha p+\beta q)^{-1}py - y(\alpha p+\beta q)^{-1}qx](\alpha p+\beta q)^{-1} \\ &= \phi(x)*\phi(y). \end{aligned}$$

Thus, (vi) implies (vii), and by Definition 2.1 if (vii) holds, then $A(p,q)$ is generalized quasiassociative and thus is flexible by (2.2). □

The existence of a unit element in a mutation algebra gives a stronger condition and implies the equivalence of all the conditions discussed above.

Theorem 2.8. *Let A be an associative algebra over F and let p,q be any fixed elements in A. Then, $A(p,q)$ has a unit element e if and only if A has a unit element 1, $p-q$ is invertible in A and the equivalent conditions (i) – (v) in Corollary 2.3 hold. In this case, $e=(p-q)^{-1}$ and $A(s+1,s)\cong A(p,q)$ under the map ϕ given by (2.9) with $\alpha=-\beta=1$ where $s=(p-q)^{-1}q=eq$ is in $Z(A)$, so that $A(p,q)$ is generalized quasiassociative.*

Proof. Assume that $A(p,q)$ has a unit element e. Since $x*e=e*x=x$, $xpe-eqx=x=epx-xqe$ for all $x\in A$. From this, we have

$$e(p-q)x+x(p-q)e=2x \tag{2.10}$$

for all $x\in A$. Multiplying (2.10) by $(p-q)e$ from the right, we obtain

$$e(p-q)x(p-q)e+x(p-q)e(p-q)e=2x(p-q)e$$

which implies

$$e(p-q)x(p-q)e=x(p-q)e, \tag{2.11}$$

since $e(p-q)e=e$. Similarly, multiplying (2.10) by $e(p-q)$ from the left, one gets

$$e(p-q)x(p-q)e=e(p-q)x. \tag{2.12}$$

It follows from (2.10) – (2.12) that

$$e(p-q)x=x(p-q)e=x$$

holds for all $x \in A$. Thus, $e(p-q) = e(p-q)(p-q)e = (p-q)e$ is the unit element of A and so $e(p-q) = (p-q)e = 1$, or $e = (p-q)^{-1}$. On the other hand,

$$xp(p-q)^{-1} - (p-q)^{-1}qx = x \tag{2.13}$$

for all $x \in A$. Multiplying (2.13) by $p-q$ from both sides, (2.13) becomes equivalent to

$$(p-q)xp - qx(p-q) = (p-q)x(p-q). \tag{2.14}$$

Expanding (2.14), we find that (2.14) is equivalent to condition (i) in Corollary 2.3. Similarly, condition (i) is shown to be equivalent to the identity

$$px(p-q) - (p-q)xq = (p-q)x(p-q). \tag{2.15}$$

Conversely, assume that A has a unit element 1, $p-q$ is invertible in A and the equivalent conditions (i) – (v) in Corollary 2.3 hold. Let $e = (p-q)^{-1}$. Then, (2.14) and (2.15) are satisfied, and multiplying these by e from both sides gives $xpe - eqx = x = epx - xqe$ for all $x \in A$. Hence, e is a unit element of $A(p,q)$. Since $s+1 = (p-q)^{-1}p$, the remaining assertion follows from Theorem 2.8 with $\alpha = -\beta = 1$. □

In the remainder of this section, we describe the nucleus, center and commutative center of $A(p,q)$ in terms of those of A. For any algebra B, the *commutative center* $K(B)$ of B is defined to be the set:

$$K(B) = \{x \in B : [x,y] = xy - yx = 0 \text{ for all } y \in B\},$$

and recall that the center $Z(B)$ of B is the set $Z(B) = N(B) \cap K(B)$ where $N(B)$ is the nucleus of B (Definition I.2.2). The *double commutative center* $K_2(B)$ of B is defined by

$$K_2(B) = \{x \in B : [[x,B],B] = 0\}. \tag{2.16}$$

An algebra B is called *prime* if the product of any two nonzero ideals of B is not zero. A prime algebra is, in particular, semiprime (Section I.4). For $x \in B$, let L_x and R_x denote the left and right multiplications in B by x (Section I.1).

Lemma 2.9. *If B is a prime algebra, then for any $0 \neq \alpha \in Z(B)$, $L_\alpha = R_\alpha$ is injective, that is, $Z(B)$ contains no zero divisors.*

Proof. Let $L_\alpha a = 0$ for some $a \in B$. If $a \neq 0$, then αB and $\ker L_\alpha$ would be nonzero ideals of B which multiply to zero, a contradiction. □

Let A be an associative algebra over F and let α, β be fixed elements in the center $Z(A)$. In $A(\alpha,\beta)$, the commutator and associator become

$$[x,y]^* = (\alpha+\beta)[x,y] = [(\alpha+\beta)x, y], \tag{2.17}$$

$$(x,y,z)^* = \alpha\beta(xzy + yzx - yxz - zxy) = \alpha\beta[[x,z],y]. \tag{2.18}$$

It follows immediately from (2.17) that

$$K(A(\alpha,\beta)) = \{x \in A : (\alpha+\beta)x \in Z(A)\}. \tag{2.19}$$

If $x \in N(A(\alpha,\beta))$ then by (2.16) and (2.18) $\alpha\beta x \in K_2(A)$. Conversely, if $\alpha\beta x \in K_2(A)$, then from (2.18) and the Jacobi identity (I.4.1) in A^- we have $x \in N(A(\alpha,\beta))$. Therefore,

$$N(A(\alpha,\beta)) = \{x \in A : \alpha\beta x \in K_2(A)\}. \tag{2.20}$$

If $\alpha+\beta$ is not a zero divisor, then by (2.17) and (2.19) $K(A(\alpha,\beta)) = Z(A)$ and since $Z(A) \subseteq N(A(\alpha,\beta))$ by (2.18), we obtain

$$Z(A(\alpha,\beta)) = K(A(\alpha,\beta)) = Z(A). \tag{2.21}$$

If α and β are not zero divisors, then (2.20) gives the relation

$$N(A(\alpha,\beta)) = K_2(A). \tag{2.22}$$

We use these to show:

Proposition 2.10. *If A is a prime associative algebra over F and α,β are fixed elements in $Z(A)$ with $\alpha \neq 0$ or $\beta \neq 0$, then $Z(A(\alpha,\beta)) = Z(A)$.*

Proof. Recall from Lemma I.4.15 that if A is prime (so semiprime) then $K_2(A) = K(A) = Z(A)$. If $\alpha \neq 0$ or $\beta \neq 0$ then either $\alpha+\beta \neq 0$ or $\alpha \neq 0 \neq \beta$. In the former case, it follows from Lemma 2.9 and (2.21) that $K(A(\alpha,\beta)) = Z(A)$. If $\alpha \neq 0 \neq \beta$, then by Lemma I.4.15 and (2.22) we have: $N(A(\alpha,\beta)) = K_2(A) = Z(A)$. Therefore, if A is prime and $\alpha \neq 0$ or $\beta \neq 0$ in $Z(A)$, then $Z(A(\alpha,\beta)) = K(A(\alpha,\beta)) \cap N(A(\alpha,\beta)) = Z(A)$. □

To extend the relations above to a more general mutation algebra $A(p,q)$, we need the restrictions on A and p and q as given in Theorem 2.7.

Theorem 2.11. *Let A be an associative algebra over F with unit element 1 and p,q be fixed elements of A such that $\alpha p + \beta q$ is invertible in A for some $\alpha,\beta \in Z(A)$. Assume that $A(p,q)$ satisfies the equivalent conditions (i) – (viii) in Corollary 2.3 and Theorem 2.7. Then,*

(i) $K(A(p,q)) = \{x \in A : x(p+q) \in Z(A)\} = \{x \in A : (p+q)x \in Z(A)\}$.

(ii) $N(A(p,q)) = \{x \in A : xpq(\alpha p+\beta q)^{-1} \in K_2(A)\}$
$\qquad = \{x \in A : (\alpha p+\beta q)^{-1}pqx \in K_2(A)\}$.

(iii) $Z(A(p,q)) \supseteq Z(A)(\alpha p+\beta q)^{-1}$.

(iv) *If $p+q$ is not a zero divisor, then*

$$Z(A(p,q)) = K(A(p,q)) = Z(A)(\alpha p + \beta q)^{-1}.$$

(v) *If p and q are not zero divisors then*

$$N(A(p,q)) = K_2(A)(\alpha p + \beta q)^{-1} = (\alpha p + \beta q)^{-1}K_2(A).$$

(vi) *If A is prime then $Z(A(p,q)) = Z(A)(\alpha p + \beta q)^{-1}$.*

Proof. Recall first that the map $\phi : A(r,s) \to A(p,q)$ given by (2.9) is an isomorphism with inverse $\phi^{-1} : A(p,q) \to A(r,s) : x \to x(\alpha p + \beta q)$, where $r = (\alpha p + \beta q)^{-1}p$ and $s = (\alpha p + \beta q)^{-1}q$ are in $Z(A)$ by our assumption (Theorem 2.7). But, $r(\alpha p + \beta q) = (\alpha p + \beta q)r = p$, so $r = p(\alpha p + \beta q)^{-1}$, and similarly, $s = (\alpha p + \beta q)^{-1}q = q(\alpha p + \beta q)^{-1}$. We readily see that the map $\tilde{\phi}$ given by

$$\tilde{\phi} : A(p,q) \to A(r,s) : x \to (\alpha p + \beta q)x$$

is an isomorphism.

(i) For any $x \in A$, we have $x \in K(A(p,q))$ if and only if $\phi^{-1}(x) \in K(A(r,s))$. Since $r, s \in Z(A)$, by (2.19) $K(A(p,q)) = \{x \in A : (r+s)x(\alpha p + \beta q) \in Z(A)\} = \{x \in A : x(p+q) \in Z(A)\}$. Similarly, since $x \in K(A(p,q))$ if and only if $\tilde{\phi}(x) \in K(A(r,s))$, $K(A(p,q)) = \{x \in A : (p+q)x \in Z(A)\}$ and this proves part (i).

(ii) Using the maps $\phi^{-1}, \tilde{\phi}$ and (2.20), we have

$$\begin{aligned} N(A(p,q)) &= \{x \in A : rsx(\alpha p + \beta q) \in K_2(A)\} \\ &= \{x \in A : rs(\alpha p + \beta q)x \in K_2(A)\} \end{aligned}$$

which implies part (ii), since $rsx(\alpha p + \beta q) = xrs(\alpha p + \beta q) = xpq(\alpha p + \beta q)^{-1}$ and similarly, $rs(\alpha p + \beta q)x = pq(\alpha p + \beta q)^{-1}x$.

(iii) For $x, y \in A$, $[x(p+q), y] = [x(\alpha p + \beta q)(r+s), y] = (r+s)[x(\alpha p + \beta q), y] = (\alpha p + \beta q)^{-1}(p+q)[x(\alpha p + \beta q), y]$. Hence by part (i) we have

$$Z(A)(\alpha p + \beta q)^{-1} \subseteq K(A(p,q)), \tag{2.23}$$

and moreover, if $p+q$ is not a zero divisor, then since $x(p+q) \in Z(A)$ if and only if $x(\alpha p + \beta q) \in Z(A)$,

$$Z(A)(\alpha p + \beta q)^{-1} = K(A(p,q)). \tag{2.24}$$

For $x, y, z \in A$, we obtain

$$\begin{aligned} [[xpq(\alpha p + \beta q)^{-1}, y], z] &= [[xrs(\alpha p + \beta q), y], z] \\ &= rs[[x(\alpha p + \beta q), y], z] \\ &= (\alpha p + \beta q)^{-2}pq[[x(\alpha p + \beta q), y], z]. \end{aligned}$$

It follows from this and part (ii) that

$$Z(A)(\alpha p+\beta q)^{-1}\subseteq N(A(p,q)), \tag{2.25}$$

which combines with (2.23) to verify part (iii).

(iv) If $p+q$ is not a zero divisor, then (2.24) holds and hence by (2.25) $Z(A(p,q)) = K(A(p,q))$, showing part (iv).

(v) Assume that p and q are not zero divisors. From the relation following (2.24) and part (ii), we find that $x\in N(A(p,q))$ if and only if $x(\alpha p+\beta q)\in K_2(A)$. Hence, $N(A(p,q)) = K_2(A)(\alpha p+\beta q)^{-1}$. Using the second relation in part (ii), we similarly argue that $x\in N(A(p,q))$ if and only if $(\alpha p+\beta q)x\in K_2(A)$, so that $N(A(p,q)) = (\alpha p+\beta q)^{-1}K_2(A)$.

(vi) If A is prime, then by Proposition 2.10 $Z(A(r,s)) = Z(A)$ and hence $Z(A(p,q)) = \phi(Z(A(r,s))) = \phi(Z(A)) = Z(A)(\alpha p+\beta q)^{-1}$. □

If $A(p,q)$ has a unit element e, then by Theorem 2.8 the hypotheses of Theorem 2.11 are satisfied with $\alpha = -\beta = 1$ and $e = (p-q)^{-1}$. Thus, as an immediate consequence of Theorem 2.11, we have:

Corollary 2.12. *Let A be an associative algebra over F and let p,q be fixed elements of A. If $A(p,q)$ has a unit element e, then the following hold:*

(i) $K(A(p,q)) = \{x\in A : x(p+q)\in Z(A)\} = \{x\in A : (p+q)x\in Z(A)\}$

(ii) $N(A(p,q)) = \{x\in A : xpqe\in K_2(A)\} = \{x\in A : epqx\in K_2(A)\}$.

(iii) $Z(A(p,q))\supseteq Z(A)e$.

(iv) *If $p+q$ is not a zero divisor, then $Z(A(p,q)) = K(A(p,q)) = Z(A)e$.*

(v) *If p and q are not zero divisors, then $N(A(p,q)) = K_2(A)e = eK_2(A)$.*

(vi) *If A is prime then $Z(A(p,q)) = Z(A)e$.* □

3. Simple and prime mutation algebras

The primary objective of this section is to investigate conditions for $A(p,q)$ to be prime or simple for an associative algebra A. The results in this section are expanded and detailed versions of the results in [**EM1**] which also generalize several earlier works of [**Os2,My2,EGM1**]. We begin with:

Lemma 3.1. *Let A be an associative algebra over F and p,q be fixed elements of A. If $A(p,q)$ is prime or simple, then so is A.*

Proof. It is immediate that any ideal of A is an ideal of $A(p,q)$ and that if $A(p,q) * A(p,q) \neq 0$ then $A^2 \neq 0$. Hence, the simplicity of $A(p,q)$ implies the same of A. Assume $A(p,q)$ is prime, and let B_1 and B_2 be ideals of A such that $B_1B_2 = 0$. If we let $B = B_2B_1$ then $B \subseteq B_1 \cap B_2$ and $B * B \subseteq B_1B_2 = 0$, so $B = 0$ and $B_1 * B_2 \subseteq B_1B_2 + B_2B_1 = 0$. Thus, $B_1 = 0$ or $B_2 = 0$, showing that A is prime. □

As we show later, for certain simple or prime associative algebras A, it is easy to construct mutation algebras $A(p,q)$ that are not simple nor prime. Therefore, we focus on conditions for the fixed elements p and q under which the simplicity or primeness of A implies the same of $A(p,q)$. For this, the ideals $R^0(p,q)$ and $J(p)$ defined by (1.6) play a central role. We first need to develop some relationships between the ideals of $A(p,q)$ and of A.

Lemma 3.2. *Let A be a prime associative algebra over F and p,q be fixed elements of A.*

(i) *If $p+q \neq 0$, then the subspace C of A spanned by the elements $xpy + yqx$ for all $x,y \in A$ contains a nonzero ideal of A.*

(ii) *If $p-q \neq 0$, then $A(p,q) * A(p,q)$ contains a nonzero ideal of A.*

Proof. In either case, $p \neq 0$ or $q \neq 0$, and hence we may assume $p \neq 0$. Also, let $C = A(p,q) * A(p,q)$, so C is the subspace spanned by the elements $xpy - yqx$ for $x,y \in A$. For any $x,y,z \in A$, we have

$$xp(zqy) \pm (zqy)qx \in C,$$
$$(yqx)pz \pm zq(yqx) \in C$$

where the double sign "±" corresponds to the case $p \pm q \neq 0$, respectively. Subtracting the relations above, we obtain

$$(xpz)qy - yq(xpz) \in C \tag{3.1}$$

in both cases. Since $(xpz)py \pm yq(xpz) \in C$, adding this to (3.1) for $p+q \neq 0$ and subtracting from (3.1) for $p-q \neq 0$, one gets $(xpz)(p \pm q)y \in C$ and so $ApA(p \pm q)A \subseteq C$. Since A is prime and $p \neq 0 \neq p \pm q$, $ApA(p \pm q)A$ is a nonzero ideal of A in either case. □

Using the subspace C and the ideal $ApA(p+q)A(\subseteq C)$ of A as above, we prove:

Corollary 3.3. *Let A be a prime associative algebra over F and p,q be fixed elements of A with $p+q \neq 0$. If B is a nonzero ideal of $A(p,q)$, then either B contains a nonzero ideal of A or $(p-q)B(p-q) = 0$.*

Proof. Let C be the subspace in Lemma 3.2(i) and let $D = ApA(p+q)A$. Then, D is a nonzero ideal of A with $D \subseteq C$. For $x, z \in A$ and $u \in B$, by (2.1)

$$\begin{aligned}(x,u,z)^* &= upzqx - uqxpz + xpzqu - zqxpu \in B,\\ (zqx) * u &= zqxpu - uqzqx \in B,\\ u * (xpz) &= upxpz - xpzqu \in B.\end{aligned}$$

These three relations add to $u(p-q)(xpz + zqx) \in B$ and hence $B(p-q)C \subseteq B$. Similarly, we get $C(p-q)B \subseteq B$. Thus,

$$DA(p-q)B(p-q)AD \subseteq C(p-q)B(p-q)C \subseteq C(p-q)B \subseteq B.$$

If $E = DA(p-q)B(p-q)AD \neq 0$, then E is an ideal of A with $E \subseteq B$, as desired. If $E = 0$ then since $D \neq 0$, by the primeness of A, $A(p-q)B(p-q)A = 0$ and so $(p-q)B(p-q) = 0$. (If A is prime then $AaA = 0$ for $a \in A$ if and only if $a = 0$.) □

If $(p-q)B(p-q) = 0$ in Corollary 3.3 then we obtain additional relations:

Lemma 3.4. *Let A, p, q and B be the same as in Corollary 3.3 but with $p \neq \pm q$. If $(p-q)B(p-q) = 0$ then*

$$pB(p-q) = qB(p-q) = (p-q)Bp = (p-q)Bq = 0. \tag{3.2}$$

Proof. Let $D = ApA(p+q)A$. As in the proof of Corollary 3.3, $D(p-q)B \subseteq B$ and $B(p-q)D \subseteq B$. For $x \in A, b \in B$ and $d \in D$, we have

$$\begin{aligned}(p-q)xpb(p-q)d(p-q) &= (p-q)[xp(b(p-q)d)](p-q)\\ = (p-q)[x * (b(p-q)d)](p-q) &+ (p-q)[(b(p-q)d)qx](p-q).\end{aligned}$$

Since $b(p-q)d \in B, x * (b(p-q)d) \in B$ and hence $(p-q)[x * (b(p-q)d](p-q) = 0$. On the other hand, $(p-q)b(p-q) = 0$ and so $(p-q)xpb(p-q)d(p-q) = 0$. This gives the relation

$$A(p-q)ApB(p-q)D(p-q)A = 0. \tag{3.3}$$

Since $p - q \neq 0$, A is prime and $D \neq 0$ is an ideal of A, we have $A(p-q)A \neq 0$ and $D(p-q)A \neq 0$. Therefore, it follows from (3.3) and the primeness of A that $pB(p-q) = 0$. But, $qB(p-q) = pB(p-q) - (p-q)B(p-q) = 0$, and we similarly show that $(p-q)Bp = (p-q)Bq = 0$. □

If $p = q$, then the (p,p)–mutation $A(p,p)$ of A is a Lie algebra, that is, $A(p,p) = A(p)^-$ where $A(p)$ denotes the p–homotope of A (so associative; see (I.4.27)). If $A(p,p)$ is prime or simple, then by Lemma 3.1 $A(p)$ is still prime or simple. However, since the structure of $A(p,p)$ relies on that of the Lie algebra $A(p)^-$, the converse of the above behaves quite differently from the case of $p \neq q$. In fact, as we shall see in Section II.4, under some general conditions (e.g., A is finite–dimensional or artinian), it turns out that the center $Z(B)$ of a simple associative algebra B is not zero and is an ideal of B^- squaring to zero. Thus, for such a simple associative algebra, the (p,p)–mutation cannot be prime or simple. We first treat the case $p \neq q$.

Proposition 3.5. *Let A be a prime associative algebra over F and p,q be any fixed elements of A with $p \neq q$. If B is a nonzero ideal of $A(p,q)$, then either B contains a nonzero ideal of A or*

$$\Lambda(p,q)B = 0. \tag{3.4}$$

Proof. We first assume $p \neq -q$, and suppose that B contains no nonzero ideals of A. Hence, by Corollary 3.3 and Lemma 3.4 we have $(p-q)B(p-q) = 0$ and relation (3.2). For $x \in A$ and $b \in B$,

$$pbpx(p-q) = p(b*x)(p-q) + pxqb(p-q) = 0$$

since $b*x \in B$ and $pB(p-q) = qB(p-q) = 0$. Hence, $pBpA(p-q) = 0$, and since $A(p-q)A \neq 0$ by the primeness of A, we deduce $pBp = 0$. Similarly, $qbqx(p-q) = -q(b*x)(p-q) + qxpb(p-q) = 0$ and so $qBq = 0$. But, $pBq = (p-q)Bq + qBq = 0$ and $qBp = qB(p-q) + qBq = 0$, showing (3.4) for $p \neq -q$.

Assume now that $p = -q$. For $x \in A$ and $b \in B$,

$$b*(bpx - xpb) = bpbpx - xpbpb \in B. \tag{3.5}$$

Since $bpb = \frac{1}{2}(b*b) \in B$,

$$(bpb)*x = bpbpx + xpbpb \in B. \tag{3.6}$$

It follows from (3.5) and (3.6) that $bpbpx$ and $xpbpb$ are contained in B. Hence, for any $x, z \in A$ and $b \in B$,

$$xpbpbpz = (xpbpb)*z - (zpx)pbpb \in B,$$

which implies $ApbpbpA \subseteq B$. Consequently, by the primeness of A, either B contains a nonzero ideal of A or $pbpbp = 0$ for all $b \in B$. In the latter case, linearizing $pbpbp = 0$ in B gives the relation $pbpcp + pcpbp = 0$ for all $b, c \in B$, so $p(B*B)p = 0$. Using this, we have for $x \in A$ and $b \in B$

$$\begin{aligned} 0 = p(b*(b*x))p &= p[bp(bpx + xpb)]p + p[(bpx + xpb)pb]p \\ &= 2pbpxpbp \end{aligned}$$

since $(pbpbp)xp = px(pbpbp) = 0$. Hence, $pbpApbp = 0$, and since A is prime, $pbp = 0$ for all $b \in B$, so $pBp = 0$. Similarly, the remaining relations follow. □

We use Proposition 3.5 to prove our principal result for the case $p \neq q$:

Theorem 3.6. *Let A be a prime associative algebra over F and let p, q be fixed elements of A with $p \neq q$. Then, $A(p,q)/R^0(p,q)$ is prime.*

Proof. Suppose that there exist ideals I_1, I_2 of $A(p,q)$ such that $I_1 * I_2 \subseteq R^0(p,q)$ and $I_1, I_2 \not\subseteq R^0(p,q)$. Thus, by Proposition 3.5 A contains ideals J_1 and J_2 with $0 \neq J_i \subseteq I_i$ $(i = 1, 2)$. Since A is prime, $0 \neq J_1 \cap J_2 \equiv J$ but $J * J \subseteq R^0(p,q)$. For any $x, y \in J$, by (1.3) we have

$$x(p-q)y(p-q)x = 2\{x, \{x,y\}^*\}^* - \{\{x,x\}^*, y\}^* \in R^0(p,q).$$

Therefore, $A(p-q)x(p-q)J(p-q)x(p-q)A = 0$ for all $x \in J$, and since A is prime and $J \neq 0$, this gives $A(p-q)J(p-q)A = 0$. Thus, $p - q = 0$ by the primeness of A, a contradiction. This shows that $A(p,q)/R^0(p,q)$ is prime. □

As before, we denote $R^0(p) = R^0(p,0) = R^0(p,p)$ (see (1.6)). Note that if $A(p)$ denotes the p–homotope then $A(p) = A(p,0)$ and $A(p)^+ = A(\frac{p}{2}, -\frac{p}{2})$. The following results give necessary and sufficient conditions for $A(p,q)$ and $A(p)$ to be simple or prime.

Corollary 3.7. *Let A be an associative algebra over F and let p, q be fixed elements of A with $p \neq q$.*

(i) *$A(p,q)$ is prime (respectively simple) if and only if A is prime (respectively simple) and $R^0(p,q) = 0$.*

(ii) *If A is prime and $p \neq 0$, then $A(p)/R^0(p)$ and $A(p)^+/R^0(p)$ are prime.*

(iii) *$A(p)$ is prime (respectively simple) if and only if A is prime (respectively simple) and p is not a zero divisor in A.*

(iv) *A is prime if and only if A^+ is prime.*

Proof. (i) If A is prime and $R^0(p,q) = 0$, then it follows from Theorem 3.6 that $A(p,q)$ is prime. If A is simple and $R^0(p,q) = 0$, then $A(p,q) * A(p,q) \neq 0$ by the primeness of $A(p,q)$ and by Proposition 3.5 we find that $A(p,q)$ contains no proper ideals and hence is simple. For the converse, we note from the definition of $R^0(p,q)$ that

$$A(p,q) * (R^0(p,q) * R^0(p,q)) = (R^0(p,q) * R^0(p,q)) * A(p,q) = 0,$$

hence either $R^0(p,q) * R^0(p,q) = 0$, or $R^0(p,q) * R^0(p,q)$ is an ideal of $A(p,q)$ which annihilates $A(p,q)$. Therefore, if $A(p,q)$ is prime then it must be that $R^0(p,q) = 0$. This together with Lemma 3.1 proves the converse.

(ii) This follows from Theorem 3.6 by noting that $A(p)/R^0(p) = A(p,0)/R^0(p,0)$ and $A(p)^+/R^0(p) = A(\frac{p}{2}, -\frac{p}{2})/R^0(\frac{p}{2}, -\frac{p}{2})$.

(iii) Since p is not a zero divisor in A if and only if $R^0(p) = R^0(p,0) = 0$, the result follows from the first relation above and part (i).

(iv) If A has a unit element 1, then since $A^+ = A(\frac{1}{2}, -\frac{1}{2})$ and $R^0(\frac{1}{2}, -\frac{1}{2}) = 0$, by part (i) the result holds in this case. Otherwise, we adjoin a unit element 1 to A to obtain $\tilde{A} = F1 \oplus A$ and $\tilde{A}^+ = F1 \oplus A^+$. It is clear that every ideal of A (or A^+) is an ideal of $\tilde{A}$ (or $\tilde{A}^+$), and that every proper ideal J of $\tilde{A}$ (or $\tilde{A}^+$) has a nonzero intersection with A (or A^+). Otherwise, $J \cap A = 0$ would imply $\tilde{A} = J \oplus A$ and hence A contains 1. Therefore, A is prime if and only if $\tilde{A}$ is prime, and A^+ is prime if and only if $\tilde{A}^+$ is prime. Since the primeness is equivalent for $\tilde{A}$ and $\tilde{A}^+$ by the above, we have that A is prime if and only if A^+ is prime. □

Corollary 3.8. *Let A, p, q be the same as in Corollary 3.7 and let Γ be the centroid of A. Assume that there are elements $\alpha, \beta \in \Gamma$ such that $\alpha p + \beta q$ is not a zero divisor in A. Then, $A(p,q)$ is prime or simple if and only if A is prime or simple. In particular, if $A(p,q)$ has a unit element, then $A(p,q)$ is prime or simple if and only if A is prime or simple.*

Proof. If $\alpha p + \beta q$ is not a zero divisor for some $\alpha, \beta \in \Gamma$, then

$$R^0(p,q) \subseteq \{x \in A : (\alpha p + \beta q)x(\alpha p + \beta q) = 0\} = 0$$

and hence the assertion follows from Corollary 3.7. If $A(p,q)$ has a unit element, then by Theorem 2.8 $p - q$ is invertible in A and this is a special case with $\alpha = -\beta = I \in \Gamma$. □

The following example shows that the condition $R^0(p,q) = 0$ does not imply the existence of an element in $\Gamma p + \Gamma q$ which is not a zero divisor.

Example 3.9. Let $A = M_3(F)$ be the algebra of 3×3 matrices over a field F and choose

$$p = \begin{pmatrix} 0 & 0 & 0 \\ 0 & 0 & -1 \\ 0 & 1 & 0 \end{pmatrix}, \quad q = \begin{pmatrix} 0 & 0 & 1 \\ 0 & 0 & 0 \\ -1 & 0 & 0 \end{pmatrix}.$$

It is easy to check that $R^0(p,q) = 0$ and the matrix of the form

$$\alpha p + \beta q = \begin{pmatrix} 0 & 0 & \beta \\ 0 & 0 & -\alpha \\ -\beta & \alpha & 0 \end{pmatrix}$$

is a zero divisor for all $\alpha, \beta \in F$, since its rank is 0 or 2. In this case, $F = \Gamma$. □

However, for a generalized quaternion algebra Q (see Table I.1), the condition $R^0(p,q) = 0$ is equivalent to the existence of an invertible element in $\Gamma p + \Gamma q$:

Proposition 3.10. *Let Q be a generalized quaternion algebra Q over F and p, q be fixed elements of Q with $p \neq q$. Then, the following are equivalent:*

(i) *$Q(p,q)$ is simple.*

(ii) *One of p, q and $p+q$ is invertible in Q.*

(iii) *There exist scalars $\alpha, \beta \in F$ such that $\alpha p + \beta q$ is invertible in Q.*

Proof. Since Q is simple, in view of Corollary 3.8 it suffices to verify that (i) implies (ii). Thus, assume that $Q(p,q)$ is simple but p and q are not invertible in Q. Then, Q is not a division algebra over F and hence is isomorphic to the algebra $M_2(F)$ of 2×2 matrices over F (Theorem I.3.6). We identify $Q = Hom_F V$, the algebra of linear transformations on a two–dimensional vector space V over F. By Theorem 3.7, the simplicity of $Q(p,q)$ gives $R^0(p,q) = 0$. We contend that $\ker p \cap \ker q = 0$. If not, then there is a basis $\{v, w\}$ of V with $v \in \ker p \cap \ker q$ and the linear map $\varphi : V \to V$ given by $\varphi(v) = v, \varphi(w) = 0$ satisfies $p\varphi = q\varphi = 0$, so $\varphi \in R^0(p,q)$, a contradiction. Hence, $\ker p \cap \ker q = 0$. Since p and q are not invertible and $R^0(p,q) = 0$, we have $\ker p = Fv$ and $\ker q = Fw$, so $\{v, w\}$ is a basis of V. Moreover, $\{p(w), q(v)\}$ is a basis of V; for, otherwise, $p(w)$ and $q(v)$ are linearly dependent and $R^0(p,q)$ would contain the linear map $0 \neq \varphi : V \to V$ with $\varphi(p(w)) = \varphi(q(v)) = 0$. But then, $\{p(w), q(v)\} = \{(p+q)(v), (p+q)(w)\}$ is a basis of V and so $p+q$ is invertible in Q. This shows that (i) implies (ii). □

We next show that if $p \neq q$ and $A(p,q)$ is simple or prime then the centroids of $A(p,q)$ and A coincide. Let L_x^* and R_x^* denote the left and right multiplications in $A(p,q)$ by x. By (1.1) we have

$$L_x^* = L_x L_p - R_x R_q, \quad R_x^* = R_x R_p - L_x L_q. \tag{3.7}$$

It follows from (3.7) that the multiplication algebra $M(A(p,q))$ of $A(p,q)$ is contained in the multiplication algebra $M(A)$ of A (see Section I.1):

$$M(A(p,q)) \subseteq M(A). \tag{3.8}$$

Thus, if Γ and $\Gamma_{p,q}$ denote the centroids of A and of $A(p,q)$, respectively, then by (3.8)

$$\Gamma \subseteq \Gamma_{p,q}. \tag{3.9}$$

Proposition 3.11. *Let A be an associative algebra over F and p, q be fixed elements of A with $p \neq q$. If $A(p,q)$ is simple then $\Gamma = \Gamma_{p,q}$.*

Proof. Since $A(p,q)$ is simple, so is A by Lemma 3.1. By Theorem I.2.3 Γ is a field extending F and by (3.9) A and $A(p,q)$ are regarded as algebras over Γ. Hence, we may assume that $\Gamma = F$, that is, A is central simple over F (Theorem I.2.5). It suffices to show that $\Gamma_{p,q} = F$. If K is an extension field of F, then

$$A(p,q)_K = K \otimes_F A(p,q) = A_K(1 \otimes p, 1 \otimes q),$$
$$R^0(1 \otimes p, 1 \otimes q) = K \otimes_F R^0(p,q) = 0.$$

Since A_K is simple, it follows from this and Corollary 3.7 (i) that $A(p,q)_K$ is simple over K. Theorem I.2.5 now shows that $A(p,q)$ is central simple over F, so $\Gamma_{p,q} = F = \Gamma$. □

In fact, we can show that Proposition 3.11 is valid for any prime mutation algebra $A(p,q)$. For this, we need:

Lemma 3.12. *Let A be an associative algebra over F and p, q be fixed elements of A with $p \neq q$. For any element γ in the centroid $\Gamma_{p,q}$ of $A(p,q)$, we have:*

(i) *$[\gamma, L_A]A$ and $[\gamma, R_A]A$ are ideals of $A(p,q)$.*

(ii) *The subspace $S = \{x \in A : [\gamma, L_{x(p-q)}] = [\gamma, R_{(p-q)x}] = 0\}$ contains $xpy+yqx$ for all $x, y \in A$.*

Proof. (i) We often write $\gamma x = x^\gamma$ for $x \in A$. Denote $H = [\gamma, L_A]A$. For $x, y, z \in A$, we have

$$\begin{aligned} z * ([\gamma, L_x]y) &= L_z^*[\gamma, L_x]y = (L_z^*\gamma L_x - L_z^* L_x \gamma)y \\ &= (\gamma L_z^* L_x - L_z^* L_x \gamma)y = [\gamma, L_z^* L_x]y \\ &= [\gamma, L_{zpx}]y - [\gamma, R_z R_q L_x]y \end{aligned}$$

by (3.7). Hence, it suffices to show $[\gamma, R_z R_q L_x]y \in H$ for all $x, y, z \in A$. But

$$\begin{aligned} [\gamma, R_z R_q L_x]y &= (xyqz)^\gamma - xy^\gamma qz = (xyqz)^\gamma - x[zpy^\gamma - (zpy)^\gamma + (yqz)^\gamma] \\ &= (\gamma L_x)yqz - (L_x\gamma)yqz - (L_{xzp}\gamma)y + (\gamma L_{xzp})y \\ &\quad -(\gamma L_x)zpy + (L_x\gamma)zpy \\ &= -[\gamma, L_x](z * y) + [\gamma, L_{xzp}]y \in H, \end{aligned}$$

as desired. To show $R_z^*[\gamma, L_x]y \in H$, we simply note that a subspace of A is an ideal of $A(p,q)$ if and only if it is an ideal of $A(q,p)$, R_z^* is minus the left multiplication in $A(q,p)$ by z, and $\Gamma_{p,q} = \Gamma_{q,p}$. Using $y^\gamma pz - zqy^\gamma = (ypz)^\gamma - (zqy)^\gamma$, we similarly obtain

$$R_z^*[\gamma, R_x]y = [\gamma, R_{pxz}]y - [\gamma, R_{pzx}]y + [\gamma, R_x](y * z) \in [\gamma, R_A]A$$

for all $x, y, z \in H$. This completes the proof of part (i).

(ii) Using (2.1), we compute for $x, y, z \in A$

$$\begin{aligned} L_{(xpy+yqx)(p-q)}(z) &= -(x*z)*y + x*(z*y) + (xpy)*z + z*(yqx) \\ &= [L_x^*, R_y^*]z + L_{xpy}^* z + R_{yqx}^* z, \end{aligned}$$

and similarly,

$$R_{(p-q)(xpy+yqx)} = [R_y^*, L_x^*] + L_{yqx}^* + R_{xpy}^*.$$

Since $\gamma \in \Gamma_{p,q}$, it follows from these that $xpy + yqx \in S$ for all $x, y \in A$. □

Theorem 3.13. *Let A be an associative algebra over F and p, q be fixed elements of A with $p \neq q$. If $A(p,q)$ is prime then $\Gamma = \Gamma_{p,q}$.*

Proof. As in Lemma 3.12, denote $H = [\gamma, L_A]A$ and $J = [\gamma, R_A]A$ for each $\gamma \in \Gamma_{p,q}$. In view of (3.9), it suffices to verify that $H = J = 0$ for any $\gamma \in \Gamma_{p,q}$, i.e., $\gamma \in \Gamma$. Let S denote the subspace of A given in Lemma 3.12 (ii), so that

$$S = \{x \in A : (x(p-q)z)^\gamma = x(p-q)z^\gamma, (z(p-q)x)^\gamma = z^\gamma(p-q)x \text{ for all } z \in A\}.$$

We first show that S contains a nonzero ideal of A. If $p + q \neq 0$, then since A is prime by Theorem 3.7, it follows from Lemma 3.2(i) that S contains a nonzero ideal of A. Assume then that $p + q = 0$ or $p = -q$. Then,

$$S = \{x \in A : [\gamma, L_{xp}] = [\gamma, R_{px}] = 0\},$$

and for $x, y \in S$, $[\gamma, L_{xpyp}] = [\gamma, L_{xp}L_{yp}] = [\gamma, L_{xp}]L_{yp} + L_{xp}[\gamma, L_{yp}] = 0$ and similarly, $[\gamma, R_{pxpy}] = [\gamma, R_{py}R_{px}] = 0$. Hence, S is a subring of the p–homotope $A(p)$ of A. Since $xpy + yqx = xpy - ypx \in S$ for all $x, y \in A$ (Lemma3.12), $[A(p), A(p)] \subseteq S$. Since A is prime and $R^0(p,0) = R^0(p,-p) = 0$ by Corollary 3.7(i), $A(p) = A(p,0)$ is prime as well. Hence, it follows from Lemma I.4.16 that either S contains a nonzero ideal of $A(p) = A(p,0)$ or $S \subseteq Z(A(p))$. In the former, by Proposition 3.5 S contains a nonzero ideal of A since $R^0(p,-p) = 0$. In the latter, $[[A(p), A(p)], A(p)] = 0$ and by Lemma I.4.15 $A(p)$ is commutative. Thus, $A(p,q) = A(p,-p) = A(2p)^+ = A(2p) = A(2p,0)$, being a case where $p \neq -q$. Therefore, S contains a nonzero ideal I of A.

For $x, y \in A$ and $u \in I$, we have

$$(xu(p-q)y)^\gamma = ((xu)(p-q)y)^\gamma = xu(p-q)y^\gamma = x(u(p-q)y)^\gamma,$$

and hence, $(xy)^\gamma = xy^\gamma$ for all $x \in A$ and $y \in I(p-q)A$. Since A is prime and $p-q \neq 0$, $A(p-q)I(p-q)A \neq 0$ or $(p-q)I(p-q)A \neq 0$. Let v be a nonzero element of $(p-q)I(p-q)A \subseteq I \subseteq S$. Using these, for $x, y \in A$, we have $(xyv)^\gamma = (xy)^\gamma v$ since $v \in (p-q)I$ and $I \subseteq S$, and $(xyv)^\gamma = x(yv)^\gamma$ since $yv \in A(p-q)I(p-q)A \subseteq I(p-q)A$.

Also, one has $(yv)^\gamma = y^\gamma v$ since $v \in (p-q)I$ and $I \subseteq S$. We use these relations to derive

$$([\gamma, L_x]y)v = [(xy)^\gamma - xy^\gamma]v = (xyv)^\gamma - x(yv)^\gamma = (xyv)^\gamma - (xyv)^\gamma = 0,$$

and so $Hv = 0$.

Suppose $H \neq 0$. Then, by Lemma 3.12, Proposition 3.5 and Corollary 3.7, H contains a nonzero ideal D of A. Hence, $DAvA \subseteq DvA \subseteq HvA = 0$ since $Hv = 0$. By the primeness of A, it must be that $v = 0$, a contradiction. Therefore, we have shown that $H = 0$. Similarly, for $x, y \in A$ and $u \in I$, we find $(x(p-q)uy)^\gamma = (x(p-q)u)^\gamma y$ and so $(xy)^\gamma = x^\gamma y$ for all $x \in A(p-q)I$ and $y \in A$. Hence, as before, there is a nonzero $v \in A(p-q)I(p-q)$. As above, we argue that

$$v([\gamma, R_x]y) = v((yx)^\gamma - y^\gamma x) = (vyx)^\gamma - (vy)^\gamma x = 0$$

for all $x, y \in A$. This gives $vJ = 0$, and consequently, $J = 0$. Therefore, we have $\Gamma = \Gamma_{p,q}$. □

We now treat the case $p = q$. Recall from (1.6) that $J(p)$ is an ideal of the Lie algebra $A(p,p) = A(p)^-$ and $J(p) \supseteq R^0(p)$. The following results are analogous to Theorem 3.6 and Corollary 3.7.

Theorem 3.14. *Let A be an associative algebra over F and p be a fixed element of A.*

(i) *If A is prime then $A(p,p)/J(p)$ is zero or prime.*

(ii) *$A(p,p)$ is prime if and only if A is prime and $J(p) = 0$.*

(iii) *If A is simple, then $(A(p,p)/J(p)) * (A(p,p)/J(p))$ is zero or simple.*

(iv) *$A(p,p)$ is simple if and only if A is simple, $J(p) = 0$ and $A(p,p) * A(p,p) = A(p,p)$.*

Proof. (i) If $p = 0$ then clearly $A(p,p)/J(p) = 0$. Thus, assume $p \neq 0$. It follows from Theorem 3.6 that $B \equiv A(p)/R^0(p) = A(p,0)/R^0(p,0)$ is a prime associative algebra over F. Hence, by Theorem I.4.21 $B^-/Z(B)$ is zero or prime, and the result follows since $B^- = A(p,p)/R^0(p)$, $Z(B) = J(p)/R^0(p)$ by (1.6) and

$$A(p,p)/J(p) \cong (A(p,p)/R^0(p))/(J(p)/R^0(p)) = B^-/Z(B). \tag{3.10}$$

(ii) Since $J(p) * A(p,p) \subseteq R^0(p) = R^0(p,0)$ and $A(p,p) * (R^0(p,p) * R^0(p,p)) = 0$, $J(p)$ is a solvable ideal of the Lie algebra $A(p,p)$. Hence, if $A(p,p)$ is prime, then $J(p) = 0$ and by Lemma 3.1 A is prime also. The converse follows from part (i).

(iii) By (3.10) we have

$$\begin{aligned}(A(p,p)/J(p)) * (A(p,p)/J(p)) \\ \cong [B^-/Z(B), B^-/Z(B)] \cong [B,B]/(Z(B)\cap[B,B]),\end{aligned}$$

and the result follows from Theorem I.4.19.

(iv) If $A(p,p)$ is simple, then A is simple by Lemma 3.1, $J(p)=0$ by part (ii) and $A(p,p) * A(p,p) = A(p,p)$. The converse is immediate from part (iii). □

As we show in Section II.4, $J(p)$ is always not zero for simple artinian associative algebras (Theorem 4.7) and hence Theorem 3.14 (iv) applies only for nonartinian algebras.

4. Mutations of simple associative algebras

Having determined conditions for $A(p,q)$ to be prime or simple in Section II.3, this section is devoted to the determination of the structure of $A(p,q)$ for a simple artinian associative algebra A over F. If A is a simple artinian associative algebra, then since $R(A)=0$, the ideal $R^0(p,q)$ of $A(p,q)$ is nilpotent with $R^0(p,q)^3=0$ (Proposition 1.5(iii)). Moreover, if $p\neq q$ then $R^0(p,q)$ is shown to be the radical $R(A(p,q))$ of $A(p,q)$ (see Section II.5). On the other hand, if $p=q$ then $A(p,p)$ is a Lie algebra and $R^0(p,p)$ behaves quite differently. In any case, $A(p,q)$ decomposes as $A(p,q)=R^0(p,q)\oplus S$ for a subalgebra S of $A(p,q)$. Therefore, the primary task in this section is to determine the subalgebra S.

Our discussion is essentially based on the work in [**EM1**] and generalizes the known results for the structure of mutations of finite–dimensional associative algebras. The structure of simple or semisimple artinian associative algebras plays a central role for our investigation (see Theorem 1.1 and Proposition 1.2).

Let A be an associative algebra over F. For a subset H of A, let $l(H)$ and $r(H)$ denote the left and right annihilators of H in A, respectively, that is,

$$l(H)=\{x\in A : xH=0\},\ r(H)=\{x\in A : Hx=0\}.$$

Clearly, $l(H)$ is a left ideal of A and $r(H)$ is a right ideal of A. If $H=\{a_1,\dots,a_n\}$, then we write $l(H)=l(a_1,\dots,a_n)$. We first describe the ideal $R^0(p,q)$ in terms of left and right annihilators.

Lemma 4.1. *Let A be a semisimple artinian associative algebra over F and p,q be fixed elements of A. Then, $R^0(p,q)=l(p,q)+r(p,q)$.*

Proof. Clearly, $l(p,q)+r(p,q) \subseteq R^0(p,q)$. On the other hand, by Proposition 1.2 there exist idempotents e and f in A such that $Ap + Aq = Ae$ and $pA + qA = fA$. Hence, $x \in R^0(p,q)$ if and only if $(Ap + Aq)x(pA + qA) = AexfA = 0$ if and only if $exf = 0$, since A has a unit element 1. Therefore, if $x \in R^0(p,q)$ then $x = (x - xf) + xf$ with $x - xf \in l(f) = l(p,q)$ and $xf \in r(e) = r(p,q)$. This implies $R^0(p,q) \subseteq l(p,q) + r(p,q)$. □

We use Lemma 4.1 to decompose $A(p,q)$ as $A(p,q) = R^0(p,q) \oplus S$ for some subalgebra S of $A(p,q)$.

Theorem 4.2. *Let A be a simple artinian associative algebra over F and p,q be fixed elements of A.*

(i) *$R^0(p,q)$ is nilpotent in $A(p,q)$ of index ≤ 3 and there exists a subalgebra S of $A(p,q)$ such that*

$$A(p,q) = R^0(p,q) \oplus S. \tag{4.1}$$

(ii) *If $p \neq q$, then S is simple, $A(p,q) * A(p,q) = A(p,q)$ and $R^0(p,q)$ is the unique maximal ideal of $A(p,q)$ which contains all proper ideals of $A(p,q)$.*

Proof. (i) As noted earlier, since $R(A) = 0$, by Proposition 1.5 (iii) $R^0(p,q)^3 = 0$ in $A(p,q)$. By Lemma 4.1, $R^0(p,q) = l(p,q) + r(p,q) = l(f) + r(e)$ where e and f are idempotents in A such that $Ap + Aq = Ae$ and $pA + qA = fA$. Since $A = Af + A(1-f)$ and $Af \cap l(f) = 0$, we have $l(f) = A(1-f)$, and similarly, $r(e) = (1-e)A$. Therefore,

$$A = eA + (1-e)A = eAf + eA(1-f) + (1-e)A \subseteq eAf + l(f) + r(e)$$

and so $A = R^0(p,q) + eAf$. Let $S = eAf$. For $x, y \in S$, we have $x * y = xpy - yqx \in SAS = eAfAeAf \subseteq eAf = S$, and hence S is a subalgebra of $A(p,q)$. For $x \in S \cap R^0(p,q)$, we write $x = a + b$ with $a \in l(f)$ and $b \in r(e)$. But, $x \in S = eAf$ and so

$$x = exf = e(af) + e(bf) = 0.$$

This gives the decomposition (4.1).

(ii) Since $p \neq q$, by Lemma 3.2(ii) $A(p,q) * A(p,q)$ contains a nonzero ideal of A, and since A is simple, $A(p,q) * A(p,q) = A(p,q)$. Therefore, $A(p,q) \neq R^0(p,q)$, since $R^0(p,q)$ is nilpotent. If I is a proper ideal of $A(p,q)$, then since A is simple, by Proposition 3.5 I satisfies (3.4) and so $I \subseteq R^0(p,q)$. Thus, $R^0(p,q)$ is the unique maximal ideal of $A(p,q)$. From $A(p,q) * A(p,q) = A(p,q)$, we obtain $S * S = S$. If S_0 is a proper ideal of S, then $R^0(p,q) + S_0$ is a proper ideal of $A(p,q)$ which must be contained in $R^0(p,q)$, a contradiction. Thus, S is simple. □

If $p = q$, then each of the conditions holding for $p \neq q$ in Theorem 4.2 is not true in general.

Example 4.3. Let $A = M_2(F)$ and let $p = \begin{pmatrix} 1 & 0 \\ -1 & 0 \end{pmatrix}$. Then, $A(p,p) = A(p)^-$, and $R^0(p,p)$ is given by

$$R^0(p,p) = \left\{ \begin{pmatrix} \alpha & \alpha \\ \beta & \gamma \end{pmatrix} : \alpha, \beta, \gamma \in F \right\}.$$

We have the decomposition $A(p,p) = R^0(p,p) \oplus S$ for a one–dimensional subalgebra with $S * S = 0$. Moreover, $A(p,p)^{(1)} = A(p,p) * A(p,p) \subseteq R^0(p,p)$ and hence $A(p,p)$ is a solvable Lie algebra with $A(p,p)^{(3)} = 0$. Note that the ideal FI_2 of $A(p,p)$ is not contained in $R^0(p,p)$. □

More general examples can be given following our main structure theorem for $A(p,q)$. Let D be a division (associative) algebra over F. Recall from Section II.1 that $M_n(D)$ is identified with the associative algebra $Hom_D V$ of D–linear maps $: V \to V$ for a vector space V over D of dimension n where each element in $M_n(D)$ acts on V from the left and scalars in D act on V from the right. Let $\{v_1, \ldots, v_n\}$ be a basis of V over D. For $x \in M_n(D)$, the matrix representation $x = (x_{ij})$ of x relative to this basis is given by

$$xv_j = \sum_{i=1}^{n} v_i x_{ij}, \quad j = 1, \ldots, n. \tag{4.2}$$

For $1 \leq r, s \leq n$, let $K = \{i_1, \ldots, i_r\}$ and $L = \{j_1, \ldots, j_s\}$ be fixed subsets of $\{1, \ldots, n\}$. If $x \in M_n(D)$, then let $\hat{x}$ denote the $n \times n$ matrix obtained from x by replacing all rows with indices not in K and all columns with indices not in L by zeros. Let

$$\hat{M}_n(D) = \{\hat{x} : x \in M_n(D)\}. \tag{4.3}$$

Using the matrix product, it is easy to see that $\hat{M}_n(D)$ is a subalgebra of $M_n(D)$ and is isomorphic to the linear space $M_{r \times s}(D)$ of $r \times s$ matrices over D. Moreover, we have

$$\hat{M}_n(D) M_n(D) \hat{M}_n(D) \subseteq \hat{M}_n(D).$$

In particular, for fixed elements $p, q \in M_n(D)$, the (p,q)–mutation $\hat{M}_n(D)(p,q)$ is well defined and is a subalgebra of $M_n(D)(p,q)$. In fact, if $\hat{x}, \hat{y} \in \hat{M}_n(D)$ and $a \in M_n(D)$, then we see that those row j's of a with $j \notin L$ and column i's of a with $i \notin K$ play no roles in performing the matrix product $(\hat{x}a)\hat{y} = \hat{x}(a\hat{y})$. Thus, for $a \in M_n(D)$, let $\tilde{a}$ denote the $s \times r$ matrix formed from a by discarding those row j's with $j \notin L$ and column i's with $i \notin K$. Similarly, if $x \in M_n(D)$, then denote by x' the $r \times s$ matrix obtained from x by discarding row i's with $i \notin K$ and column j's with $j \notin L$. Then, it is immediate that the map $\hat{M}_n(D) \to M_{r \times s}(D) : \hat{x} \to x'$ gives an algebra isomorphism of the subalgebra $\hat{M}_n(D)(p,q)$ to the twisted mutation algebra $M_{r \times s}(D)(\tilde{p}, \tilde{q})$ (Definition 1.2). We summarize these in:

Lemma 4.4. *For $1 \leq r, s \leq n$, let $K = \{i_1, \dots, i_r\}$ and $L = \{j_1, \dots, j_s\}$ be fixed subsets of $\{1, \dots, n\}$ and let p, q be fixed elements of $M_n(D)$. Then, the subalgebra $\hat{M}_n(D)(p,q)$ of $M_n(D)(p,q)$ is isomorphic to the twisted mutation algebra $M_{r\times s}(D)(\tilde{p}, \tilde{q})$ where $\hat{M}_n(D)$ is given by (4.3) and $\tilde{p}, \tilde{q}$ are the $s \times r$ matrices obtained from p, q as above.* □

For a positive integer r, let D^r denote the vector space over D of all row vectors $x = (\alpha_1, \dots, \alpha_r)$, $\alpha_1, \dots, \alpha_r \in D$, and the corresponding column vector is denoted by x^t using the transpose. With these preparations, we now prove the following structure theorem for mutations of a simple artinian associative algebra.

Theorem 4.5. *Let A be a simple artinian associative algebra over F, so that $A = M_n(D)$ for some division algebra D over F and positive integer n, and let p, q be fixed elements of A. Then, there exists a subalgebra S of $A(p,q)$ such that $A(p,q) = R^0(p,q) \oplus S$ and S is isomorphic to the twisted mutation algebra $M_{r\times s}(D)(u,v)$ for some $1 \leq r, s \leq n$ and $u, v \in M_{s\times r}(D)$ which satisfy the conditions*

$$\{x \in D^r : ux^t = vx^t = 0\} = \{y \in D^s : yu = yv = 0\} = 0. \tag{4.4}$$

Moreover, if $p \neq q$ then $u \neq v$ and S is simple. Conversely, any twisted mutation algebra $M_{r\times s}(D)(u,v)$ for $u, v \in M_{s\times r}(D)$ satisfying (4.4) is isomorphic to a subalgebra S of $A(p,q)$ with $S \cong A(p,q)/R^0(p,q)$ for some simple artinian associative algebra A and fixed elements $p, q \in A$.

Proof. Assume that A is a simple artinian associative algebra over F, and let p, q be fixed elements of A. Thus, we can let $A = M_n(D) = Hom_D V$ for some division algebra D over F where V is a vector space over D of dimension n. If $r = \dim_D(\ker p \cap \ker q)$, then we can choose a basis $\{v_1, \dots, v_n\}$ of V such that $\{v_{r+1}, \dots, v_n\}$ is a basis of $\ker p \cap \ker q$. Then, by (4.2) the matrices of p and q relative to this basis are given by

$$p = \begin{pmatrix} p_{11} & \cdots & p_{1r} & 0 & \cdots & 0 \\ p_{21} & \cdots & p_{2r} & 0 & \cdots & 0 \\ \vdots & \ddots & \vdots & \vdots & \ddots & \vdots \\ p_{n1} & \cdots & p_{nr} & 0 & \cdots & 0 \end{pmatrix},$$

(4.5)

$$q = \begin{pmatrix} q_{11} & \cdots & q_{1r} & 0 & \cdots & 0 \\ q_{21} & \cdots & q_{2r} & 0 & \cdots & 0 \\ \vdots & \ddots & \vdots & \vdots & \ddots & \vdots \\ q_{n1} & \cdots & q_{nr} & 0 & \cdots & 0 \end{pmatrix}.$$

Since $r(p,q) = \{x \in A : px = qx = 0\} = \{x \in A : xV \subseteq \ker p \cap \ker q\}, r(p,q)$

consists of all matrices of the form

$$\begin{pmatrix} 0 & \cdots & 0 \\ \vdots & \ddots & \vdots \\ 0 & \cdots & 0 \\ a_{r+1,1} & \cdots & a_{r+1,n} \\ \vdots & \ddots & \vdots \\ a_{n1} & \cdots & a_{nn} \end{pmatrix}. \tag{4.6}$$

For $x \in A$, let $x = (x_{ij})$ be the matrix representation of x relative to the basis $\{v_1, \ldots, v_n\}$. Equating $xp = xq = 0$ and using (4.5), we find that $x \in l(p,q)$ if and only if each row $(x_{i1} \ldots x_{in})$ of x is a solution to the system of linear equations in $\alpha_1, \ldots, \alpha_n$ given by

$$(\alpha_1, \ldots, \alpha_n) \begin{pmatrix} p_{11} & \cdots & p_{1r} & q_{11} & \cdots & q_{1r} \\ p_{21} & \cdots & p_{2r} & q_{21} & \cdots & q_{2r} \\ \vdots & \ddots & \vdots & \vdots & \ddots & \vdots \\ p_{n1} & \cdots & p_{nr} & q_{n1} & \cdots & q_{nr} \end{pmatrix} = 0. \tag{4.7}$$

Let $d(p,q)$ denote the $n \times 2r$ matrix on the left side of (4.7), and let s be the row rank (= column rank) of $d(p,q)$. There exists a subset $L = \{j_1, \ldots, j_s\}$ of $\{1, \ldots, n\}$ such that any solution to the system (4.7) is uniquely expressed as

$$\begin{aligned} \alpha_{j_1} &= \textstyle\sum_{k \notin L} \alpha_k d_1^k \\ &\vdots \\ \alpha_{j_s} &= \textstyle\sum_{k \notin L} \alpha_k d_s^k \end{aligned} \tag{4.8}$$

for $d_l^k \in D$, $k \notin L$ and $l = 1, \ldots, s$.

Therefore, since $R^0(p,q) = r(p,q) + l(p,q)$ by Lemma 4.1, we have that $x \in R^0(p,q)$ if and only if x is a sum of a matrix of the form (4.6) and a matrix with columns $j_1, \ldots, j_s$ determined by (4.8). Thus,

$$A(p,q) = R^0(p,q) \oplus \hat{M}_n(D)$$

where $\hat{M}_n(D)$ is the subalgebra of $M_n(D)$ given by (4.3) associated with the subsets $K = \{1, \ldots, r\}$ and $L = \{j_1, \ldots, j_s\}$. For $x = x_0 + \hat{x}, y = y_0 + \hat{y} \in A$ with $x_0, y_0 \in R^0(p,q)$ and $\hat{x}, \hat{y} \in \hat{M}_n(D)$,

$$x * y = x_0 * y_0 + x_0 * \hat{y} + \hat{x} * y_0 + \hat{x} * \hat{y}.$$

Since $R^0(p,q)$ is an ideal of $A(p,q)$ and $\hat{M}_n(D)$ is a subalgebra of $A(p,q)$, we have the decomposition

$$A(p,q) = R^0(p,q) \oplus \hat{M}_n(D)(p,q)$$

as an algebra. In view of Lemma 4.4, $\hat{M}_n(D)(p,q)$ is isomorphic to the twisted mutation algebra $M_{r\times s}(D)(\tilde{p},\tilde{q})$ for some $\tilde{p},\tilde{q} \in M_{s\times r}(D)$.

Since s is the row rank of $d(p,q)$, each row k with $k \notin L$ is a linear combination of row j's with $j \in L$. This can be also seen from (4.8) by putting $\alpha_k = 1$ for $k \notin L$ and $\alpha_h = 0$ for $h \notin L$ with $h \neq k$. Hence, if $p \neq q$ then $\tilde{p} \neq \tilde{q}$ by the construction of $\tilde{p},\tilde{q}$. In this case, the simplicity of $M_{r\times s}(D)(\tilde{p},\tilde{q})$ follows from Theorem 4.2. Let $x = (\alpha_1,\ldots,\alpha_r) \in D^r$ be such that $\tilde{p}x^t = \tilde{q}x^t = 0$. Then, by (4.2) and (4.5) $v_1\alpha_1 + \ldots + v_r\alpha_r \in \ker p \cap \ker q$ and by the choice of $\{v_1,\ldots,v_n\}$ we have $x = 0$. If there is a nonzero $y = (\beta_1,\ldots,\beta_s) \in D^s$ such that $y\tilde{p} = y\tilde{q} = 0$, then $\alpha_{j_h} = \beta_h$ for $h = 1,\ldots,s$, and $\alpha_k = 0$ for $k \notin L$ would give a solution to the system (4.7) which is not represented by (4.8). This proves relations (4.4).

Conversely, assume that the conditions (4.4) hold for some $u, v \in M_{s\times r}(D)$ and some division algebra D over F. If $n = \max\{r,s\}$, then let $A = M_n(D)$. Let $p = (p_{ij})$ and $q = (q_{ij})$ be the matrices in $M_n(D)$ defined by

$$p_{ij} = u_{ij} \text{ for } 1 \leq i \leq s, 1 \leq j \leq r \text{ and } p_{ij} = 0, \text{ otherwise;}$$
$$q_{ij} = v_{ij} \text{ for } 1 \leq i \leq s, 1 \leq j \leq r \text{ and } q_{ij} = 0, \text{ otherwise.}$$

It follows easily from (4.4) that $l(p,q)$ and $r(p,q)$ respectively consist of those matrices in $M_n(D)$ with the first r rows and first s columns equal to zero. Hence, by Theorem 4.2 $A(p,q) = R^0(p,q) \oplus \hat{M}_n(D)(p,q)$ where $\hat{M}_n(D)$ is given by (4.3) associated with the subsets $K = \{1,\ldots,r\}$ and $L = \{1,\ldots,s\}$, and by Lemma 4.4 $\hat{M}_n(D)(p,q) \cong M_{r\times s}(u,v)$. This completes the proof. □

Note in Theorem 4.5 that $S \cong A(p,q)/R^0(p,q) \cong M_{r\times s}(D)(\tilde{p},\tilde{q})$. In particular, if $r = s$ then $A(p,q)/R^0(p,q)$ is isomorphic to $M_r(D)(\tilde{p},\tilde{q})$ which is again a mutation of a simple associative algebra. But, not every twisted mutation is isomorphic to a mutation of an associative algebra.

Example 4.6. Let $A = M_2(F)$ and let $p = \begin{pmatrix} 1 & 0 \\ 0 & 0 \end{pmatrix}$, $q = \begin{pmatrix} 0 & 0 \\ 1 & 0 \end{pmatrix}$, so that $r = 1$ and $s = 2$ with $\tilde{p} = \begin{pmatrix} 1 \\ 0 \end{pmatrix}$ and $\tilde{q} = \begin{pmatrix} 0 \\ 1 \end{pmatrix}$. Then, $l(p,q) = 0$ and

$$r(p,q) = \left\{ \begin{pmatrix} 0 & 0 \\ \alpha & \beta \end{pmatrix} : \alpha, \beta \in F \right\},$$

and hence $R^0(p,q) = r(p,q)$. It follows from Theorem 4.5 that

$$A(p,q)/R^0(p,q) \cong M_{1\times 2}(F)(\tilde{p},\tilde{q})$$

and $A(p,q)/R^0(p,q)$ is a simple algebra over F of dimension 2. If this twisted mutation is isomorphic to a mutation $C(a,b)$ of an associative algebra C over F, then by Lemma 3.1 C is simple and hence is commutative. This implies that $C(a,b)$ is commutative, but $(-1,1)=(1,0)*(0,1)\neq(0,1)*(1,0)=(0,0)$ in $M_{1\times 2}(F)(\tilde{p},\tilde{q})$, a contradiction. □

We note that if $p=q$ then $A(p,p)=A(p)^-$ is a Lie algebra over F. In this case, Theroem 4.5 has an exceptional feature in that the subalgebra $S\cong A(p,p)/R^0(p,p)$ is not necessarily simple and $R^0(p,p)$ is not maximal. In fact, $R^0(p,p)=R^0(p)$ is contained in the ideal $J(p)$ of $A(p,p)$ defined by (1.6).

If $R(A)=0$ then $R^0(p,q)=R(p,q)$, given by Definition 1.3. However, the difference is that $R(p,q)$ is used for the artinian case in our investigation (Section II.5) whereas $R^0(p,q)$ is essentially utilized for prime or simple associative algebras including the artinian case (Sections II.3 and II.4). A similar observation applies to $J(p)$ although if $R(A)=0$ then $J(p)=J(p,p)$, given by Definition 1.3.

The following result treats the special case of $p=q$ in Theorem 4.5.

Theorem 4.7. *Let A be a simple artinian associative algebra over F, so that $A\cong M_n(D)$ for some division algebra D over F, and let p be a nonzero element of A. Let $A=R^0(p,p)\oplus S$ be the decomposition given in Theorem 4.5.*

(i) *The subalgebra S is isomorphic to $M_r(D)^-$ where r is the rank of p.*

(ii) *$J(p)$ is the unique maximal solvable ideal of $A(p,p)$ such that $J(p)=R^0(p,p)\oplus Z(S)$ and $J(p)/R^0(p,p)=Z(A(p,p)/R^0(p,p))\cong Z(S)$.*

(iii) *Either $A(p,p)=J(p)$ or $B(p)\equiv A(p,p)/J(p)$ is a Lie algebra such that $[B(p),B(p)]$ is simple.*

Proof. If $p=q$ in Theorem 4.5, then since by (4.5) r equals the column rank of p and by (4.7) s equals the row rank of p, it must be that $r=s$. Hence, S is isomorphic to $M_r(D)(u,u)$ for some $u\in M_r(D)$. It follows from (4.4) that $l(u)=r(u)=0$, i.e., u is invertible in $M_r(D)$. Therefore, the map

$$\varphi : M_r(D)(u)\to M_r(D) : x\to xu$$

is an algebra isomorphism which is also an isomorphism of $M_r(D)(u,u)=M_r(D)(u)^-$ to $M_r(D)^-$. This proves part (i).

By the definition of $J(p)$, we find that $x\in J(p)$ if and only if $x+R^0(p,p)\in Z(A(p,p)/R^0(p,p))\cong Z(S)$, hence $J(p)/R^0(p,p)=Z(A(p,p)/R^0(p,p))$. Clearly, $R^0(p,p)+Z(S)\subseteq J(p)$, and if $x\in J(p)$ and $x=x_0+s$ for $x_0\in R^0(p,p)$, $s\in S$, then $[x,S]\subseteq R^0(p,p)$ and $[s,S]=0$ or $s\in Z(S)$. This implies $J(p)=R^0(p,p)\oplus Z(S)$, and hence $J(p)$ is a solvable ideal of $A(p,p)$ with $J(p)^{(3)}=0$.

If $r=1$ and D is a field, then $S\cong M_r(D)^-$ is a zero algebra, i.e., $S*S=0$ and by the definition of $J(p)$ we have $J(p)=A(p,p)$, which is the first case of part (iii).

Assume then that $r > 1$ or D is not a field, so $M_r(D)$ is not commutative. It follows from Theorem I.4.17 that any ideal of $M_r(D)^-$ either is contained in the center $Z(M_r(D)) = Z(D)I_r$ (I_r is the identity matrix) or contains the derived algebra $[M_r(D), M_r(D)]$. Moreover, by Theorem I.4.19 the Lie algebra

$$[M_r(D), M_r(D)]/Z(M_r(D)) \cap [M_r(D), M_r(D)] \tag{4.9}$$

is simple. In particular, $[M_r(D), M_r(D))]$ is not solvable. This shows that $Z(M_r(D))$ is the unique maximal solvable ideal of $M_r(D)^-$. Therefore, if I is a solvable ideal of $A(p,p)$, then $(I + R^0(p,p))/R^0(p,p) \cong I/I \cap R^0(p,p)$ is solvable and so $(I + R^0(p,p))/R^0(p,p) \subseteq Z(A(p,p)/R^0(p,p)) \cong Z(M_r(D))$. This implies that $I \subseteq J(p)$ by the above and hence $J(p)$ is the unique maximal solvable ideal of $A(p,p)$. This completes the proof of part (ii).

Finally, assume that $J(p) \neq A(p,p)$, and let $B(p) = A(p,p)/J(p)$. Then, for $S = M_r(D)^-$,

$$[B(p), B(p)] = ([S,S] + J(p))/J(p) \cong [S,S]/J(p) \cap [S,S] = [S,S]/Z(S) \cap [S,S]$$

is simple by (4.9). This proves part (iii) □

In general, $J(p)$ need not be nilpotent, as shown by Example 4.3, where $J(p) = A(p,p)$ is solvable but not nilpotent, since

$$\begin{pmatrix} 0 & 0 \\ 1 & 0 \end{pmatrix} * \begin{pmatrix} 1 & 0 \\ -1 & 0 \end{pmatrix} = \begin{pmatrix} 0 & 0 \\ 1 & 0 \end{pmatrix}.$$

The following example shows that $J(p)$ need not be complemented by a subalgebra of $A(p,p)$.

Example 4.8. Let F be a field of characteristic $n > 2$ and let $A = M_n(F)$. If we choose p to be the identity matrix I_n, then $A(p,p) = A^-$, $R^0(p,p) = 0$ and $J(p) = Z(A) = FI_n$ by Theorem 4.7. From Proposition 1.6(ii) we have $Z(A) \subseteq [A,A]$. If there exists a subalgebra S of A^- such that $A = Z(A) \oplus S$, then $Z(A) \subseteq [A,A] \subseteq [Z(A),S] + [S,S] \subseteq S$, a contradiction. □

However, for the characteristic zero case, we have:

Corollary 4.9. *Let A be a finite-dimensional simple associative algebra over a field F of characteristic 0 and let p be a nonzero element of A. If $J(p) \neq A(p,p)$ then there is a simple subalgebra T of $A(p,p)$ such that $A(p,p) = J(p) \oplus T$.*

Proof. We can let $A = M_n(D)$ for a finite-dimensional division algebra D over F. By Theorem 4.7(i) there is a subalgebra S of $A(p,p)$ such that $A(p,p) =$

$R^0(p,p) \oplus S$ and $S \cong M_r(D)^-$. Since the characteristic is zero, from Proposition 1.6(i) we have:

$$M_r(D) = Z(M_r(D)) \oplus [M_r(D), M_r(D)] \tag{4.10}$$

where $[M_r(D), M_r(D)]$ is a simple Lie algebra. Since $J(p) = R^0(p,p) \oplus Z(S)$ by Theorem 4.7(ii), from (4.10) we obtain

$$A(p,p) = R^0(p,p) \oplus S = J(p) \oplus T$$

where $T = [S,S]^*$, as desired. □

We close this section with some results on division mutation algebras. We show that if A is an artinian associative algebra and if $A(p,q)$ is a division algebra (Definition I.3.3), then A is also a division algebra. However, the converse of this is not in general true, as shown for a generalized quaternion algebra Q. In fact, even if Q is a division algebra, $Q(p,q)$ cannot be a division algebra for $p,q \in Q$ with $n(p) = n(q)$, where $n(x)$ denotes the norm of $x \in Q$ (Definition I.3.2).

Theorem 4.10. *Let A be an artinian associative algebra over F and let p,q be fixed elements of A. If $A(p,q)$ is a division algebra then so is A.*

Proof. Note first that any division algebra B is simple. In fact, if B has a proper ideal I, then the equation $ax = b$ with $0 \neq a \in I$ and $b \notin I$ would have a solution in B. Assume that $A(p,q)$ is a division algebra. Since $A(p,q)$ is simple, by Lemma 3.1 A is simple artinian. Hence, by Theorems 4.5 and 4.7 $p \neq q$ and $R^0(p,q) = 0$. It follows from Lemma 4.1 that $l(p,q) = r(p,q) = 0$. If $0 \neq x \in l(p)$ and $0 \neq y \in l(q)$, then $x * y = (xp)y - (yq)x = 0$, a contradiction. Therefore, $l(p) = 0$ or $l(q) = 0$, so that either p or q is invertible in A.

Suppose that p is invertible in A. Then, the map

$$A(p,q) \to A(1, p^{-1}q) : x \to xp$$

is an algebra isomorphism. Thus, we may assume $p = 1$. If A is not a division algebra then we can let $A = M_n(D)$ for some division algebra D over F and some $n > 1$. Let e_{ij} denote the matrix unit in A for $i,j = 1,\ldots,n$. Then, $qe_{11} = \alpha_1 e_{11} + \ldots + \alpha_n e_{n1}$ for some $\alpha_1, \ldots, \alpha_n \in D$. If we choose a $0 \neq (\beta_1, \ldots, \beta_n) \in D^n$ with $\beta_1\alpha_1 + \ldots + \beta_n\alpha_n = 0$, then

$$\begin{aligned} e_{11} * (\beta_1 e_{n1} + \ldots + \beta_n e_{nn}) &= e_{11}(\beta_1 e_{n1} + \ldots + \beta_n e_{nn}) \\ &\quad -(\beta_1 e_{n1} + \ldots + \beta_n e_{nn}) q e_{11} \\ &= -(\beta_1\alpha_1 + \ldots + \beta_n\alpha_n) e_{n1} = 0, \end{aligned}$$

a contradiction. Therefore, $n = 1$ and $A = D$ is a division algebra. If q is invertible, then the result follows similarly. □

For a generalized quaternion algebra Q, the converse of Theorem 4.10 holds with a restriction on p and q.

Theorem 4.11. *Let Q be a generalized quaternion algebra over F and let p, q be fixed elements of Q. Then, $Q(p,q)$ is a division algebra if and only if Q is a division algebra and $n(p) \neq n(q)$.*

Proof. Recall that Q is a division algebra if and only if $n(x) \neq 0$ for $x \neq 0$ in Q (Lemma I.3.5). Assume that Q is a division algebra and $n(p) \neq n(q)$. If $Q(p,q)$ is not a division algebra, then $x * y = 0$ for some nonzero elements $x, y \in Q$ and so $xpy = yqx$. But, $n(x)n(p)n(y) = n(y)n(q)n(x)$, and since $n(x) \neq 0$ and $n(y) \neq 0$, $n(p) = n(q)$, a contradiction. Conversely, suppose that $Q(p,q)$ is a division algebra. Thus, by Theorem 4.10 Q is a division algebra. Assume $n(p) = n(q)$. Since $pQ = Q$ and $Qq = Q$, there is an $x \neq 0$ in Q such that $t(px) = t(xq) = 0$ where t denotes the trace in Q (Theorem I.3.11). Then,

$$\begin{aligned}(\bar{p}+\bar{q})px - xq(\bar{p}+\bar{q}) &= n(p)x + \bar{q}px - xq\bar{p} - n(q)x \\ &= \bar{q}px - xq\bar{p} = \bar{q}px - t(xq)\bar{p} + \bar{q}\bar{x}\bar{p} \\ &= \bar{q}px + \bar{q}\bar{x}\bar{p} = \bar{q}(px + \overline{px}) = \bar{q}t(px) = 0.\end{aligned}$$

Hence, if $p \neq -q$, then $(\bar{p}+\bar{q}) * x = 0$ and $Q(p,q)$ is not a division algebra, a contradiction. If $p = -q$, then $Q(p,q) = Q(2p)^+$ which is either a zero algebra (for $p = 0$) or isomorphic to Q^+ (for $p \neq 0$). In any case, $Q(p,q)$ is not a division algebra. This completes the proof. □

5. Radicals of mutation algebras

In this section we investigate the structure of the radical $R(A(p,q))$ and its relationships with the ideals $R^0(p,q)$, $J(p,q)$ and $R(A)$ for an artinian associative algebra A. Recall from Proposition 1.5 that if A is an artinian associative algebra over F, then

$$R(A) \subseteq R(p,q) \subseteq J(p,q) \subseteq R(A(p,q)) \tag{5.1}$$

where $R(A)$ is the maximal nilpotent ideal of A, $R(p,q)$ is a nilpotent ideal of $A(p,q)$ and $J(p,q)$ is a solvable ideal of $A(p,q)$ with $J(p,q)^{(3)} \subseteq R(A)$. It turns out that $J(p,q)$ plays the role of the solvable radical of $A(p,q)$ and that the structure of $A(p,q)/J(p,q)$ is determined as the direct sum of twisted mutation algebras and certain Lie algebras. If A is artinian, then since $R(A) = J(A)$, the radical of A (Lemma 1.3), $A/R(A)$ is semisimple artinian and is the direct sum

$$A/R(A) = A_1 \oplus \ldots \oplus A_n \tag{5.2}$$

of simple artinian algebras $A_1, \ldots, A_n$ (Theorem 1.1).

In view of (5.1) and (5.2), it is natural to determine the structure of $R(A(p,q))$ relative to $R(A)$, namely, $R(A(p,q))/R(A)$. If A is semisimple, i.e., $R(A) = 0$, then $R^0(p,q) = R(p,q)$. For the simple artinian case, we have:

Lemma 5.1. *Let A be a simple artinian associative algebra over F and p, q be fixed elements of A. Then, $R(p,q) = J(p,q)$ if and only if $p \neq q$.*

Proof. If $p \neq q$, then by definition, $J(p,q) \supseteq R(p,q)$, but by Theorem 4.2(ii) $J(p,q)$ is a proper ideal of $A(p,q)$, so $J(p,q) = R(p,q)$. If $R(p,q) = J(p,q)$, then since $J(p) = J(p,p)$ (Definition 1.3), it follows from Theorem 4.7 that $p \neq q$. □

Let A denote an artinian associative algebra over F, so that $A/R(A)$ is decomposed as in (5.2). For fixed elements $p, q \in A$, let

$$\begin{aligned} \bar{p} &\equiv p + R(A) = p_1 + \ldots + p_n, \\ \bar{q} &\equiv q + R(A) = q_1 + \ldots + q_n \end{aligned} \tag{5.3}$$

with $p_i, q_i \in A_i, i = 1, \ldots, n$. Reordering the A_i, we can let

$$p_i \neq q_i \quad (i = 1, \ldots, r), \quad p_j = q_j \quad (j = r+1, \ldots, n). \tag{5.4}$$

Let $R_i(p_i, q_i)$ and $J_i(p_i, q_i)$ denote the ideals of $A_i(p_i, q_i)$ defined by (1.9) and (1.10), respectively. Since $R(A_i) = 0$, $R_i(p_i, q_i) = R_i^0(p_i, q_i)$.

Lemma 5.2. *Let A be an artinian associative algebra over F and let p, q be fixed elements of A with decompositions (5.3). Then,*

$$A(p,q)/R(A) = (A/R(A))(\bar{p}, \bar{q}) = A_1(p_1, q_1) \oplus \ldots \oplus A_n(p_n, q_n), \tag{5.5}$$

$$R(p,q)/R(A) = R_1(p_1, q_1) \oplus \ldots \oplus R_n(p_n, q_n), \tag{5.6}$$

$$J(p,q)/R(A) = J_1(p_1, q_1) \oplus \ldots \oplus J_n(p_n, q_n). \tag{5.7}$$

Proof. For $x \in A$, denote $\bar{x} = x + R(A) = \sum_{i=1}^n x_i$ with $x_i \in A_i$, $i = 1, \ldots, n$. The product $*$ in $(A/R(A))(\bar{p}, \bar{q})$ is given by

$$\bar{x} * \bar{y} = \bar{x}\bar{p}\bar{y} - \bar{y}\bar{q}\bar{x} = x * y + R(A) = \sum_{i=1}^n x_i * y_i,$$

which implies (5.5). Let $\Lambda(p,q)$ be given by (1.5). We note from (1.9) that $x \in R(p,q)$ if and only if $\Lambda(p,q)x \in R(A)$ if and only if

$$0 = \overline{\Lambda(p,q)x} = \Lambda(\bar{p}, \bar{q})\bar{x} = \sum_{i=1}^n \Lambda(p_i, q_i)x_i$$

if and only if $\Lambda(p_i, q_i)x_i = 0$ for all $i = 1, \ldots, n$, or equivalently, $\bar{x} \in \sum_{i=1}^n R_i(p_i, q_i)$. This gives (5.6). It follows from (1.10), (5.1), and (5.5) that $x \in J(p,q)$ if and only if $\bar{x} * \bar{y}, \bar{y} * \bar{x} \in R(p,q)/R(A) = R_1(p_1, q_1) \oplus \ldots \oplus R_n(p_n, q_n)$ for all $y \in A$, or equivalently, $x_i \in J_i(p_i, q_i)$ for all $i = 1, \ldots, n$, that is, $\bar{x} \in J_1(p_1, q_1) \oplus \ldots \oplus J_n(p_n, q_n)$. □

Theorem 5.3. *Let A be an artinian associative algebra over F and p, q be fixed elements of A with decompositions (5.3) and (5.4).*

(i) *$J(p,q)$ is the solvable radical of $A(p,q)$ which contain all solvable ideals of $A(p,q)$.*

(ii) *$A(p,q)/J(p,q)$ is the direct sum,*

$$A(p,q)/J(p,q) = S_1 \oplus \ldots \oplus S_r \oplus S_{r+1} \oplus \ldots \oplus S_n$$

of ideals S_i where S_i for $i = 1, \ldots, r$ is a simple algebra isomorphic to a twisted mutation algebra over F given by Definition 1.2 and S_j for $j = r+1, \ldots, n$ is a Lie algebra over F isomorphic to $M_m(D)^-/Z(M_m(D))$.

Proof. (i) Note first that $R_i(p_i, q_i) = J_i(p_i, q_i)$ for $i = 1, \ldots, r$ by Lemma 5.1 and $J_j(p_j, q_j)/R_j(p_j, q_j) = Z(A_j(p_j, q_j)) \neq 0$ for $j = r+1, \ldots, n$ by Theorem 4.7. Thus, in any case, it follows from Theorems 4.2 and 4.7 that each $J_i(p_i, q_i)$ is the solvable radical of $A_i(p_i, q_i)$ such that $A_i(p_i, q_i)/J_i(p_i, q_i)$ contains no nonzero solvable ideals. Using this and decompositions (5.5) and (5.7), we find that $J(p,q)$ contains all solvable ideals of $A(p,q)$.

(ii) We note from (5.5) and (5.7) that

$$\begin{aligned} A(p,q)/J(p,q) &\cong (A(p,q)/R(A))/(J(p,q)/R(A)) \\ &\cong A_1(p_1,q_1)/J_1(p_1,q_1) \oplus \ldots \oplus A_n(p_n,q_n)/J_n(p_n,q_n). \end{aligned}$$

For $i = 1, \ldots, r$, i.e, $p_i \neq q_i$, by Theorem 4.5 $A_i(p_i, q_i)/J_i(p_i, q_i)$ is isomorphic to a twisted mutation algebra. If $j = r+1, \ldots, n$, then the result follows from Theorem 4.7, since

$$\begin{aligned} A_j(p_j,q_j)/J_j(p_j,q_j) &\cong (A_j(p_j,q_j)/R_j(p_j,q_j))/(J_j(p_j,q_j)/R_j(p_j,q_j)) \\ &\cong M_m(D)^-/Z(M_m(D)). \end{aligned}$$

□

For an ideal I of A, define the left ideal $l_I(p,q)$ and right ideal $r_I(p,q)$ of A by

$$l_I(p,q) = \{x \in A : xp, xq \in I\}, \quad r_I(p,q) = \{x \in A : px, qx \in I\}.$$

If $I = 0$, then $l_I(p,q) = l(p,q)$ and $r_I(p,q) = r(p,q)$. Generalizing Lemma 4.1, we have:

Lemma 5.4. *Let A, p and q be the same as in Theorem 5.3. Then,*

$$R(p,q) = l_{R(A)}(p,q) + r_{R(A)}(p,q).$$

Proof. Since $A/R(A)$ is semisimple artinian, $x \in R(p,q)$ if and only if $\bar{x} \in R^0(\bar{p}, \bar{q}) = l(\bar{p}, \bar{q}) + r(\bar{p}, \bar{q})$ by Lemma 4.1. But, $l_{R(A)}(p,q) = \{x \in A : \bar{x} \in l(\bar{p}, \bar{q})\}$ and $r_{R(A)}(p,q) = \{x \in A : \bar{x} \in r(\bar{p}, \bar{q})\}$. □

Proposition 5.5. *Let A be an artinian associative algebra over F and p, q be fixed elements of A. Then, $J(p,q) = R(A)$ if and only if $R(p,q) = R(A)$ and there are no maximal ideals M of A such that $A^2 \not\subseteq M$ and $p - q \in M$.*

Proof. Assume $J(p,q) = R(A)$. Since $R(A) \subseteq R(p,q) \subseteq J(p,q)$ by (5.1), $R(p,q) = R(A)$. Let M be any maximal ideal of A with $A^2 \not\subseteq M$. Clearly, $R(A) \subseteq M$, and so $M/R(A)$ is a maximal ideal of $A/R(A)$. It follows from (5.2) that $M/R(A) = A_1 \oplus \ldots \oplus A_{i-1} \oplus A_{i+1} \oplus \ldots \oplus A_n$ for some $1 \leq i \leq n$. Since $J(p,q) = R(A)$, by (5.7) $J_i(p_i, q_i) = 0$. Since by Theorem 4.7 $J_i(p_i, q_i) \neq 0$ for $p_i = q_i$, it must be that $p_i \neq q_i$. This implies $p - q \notin M$, since if $p - q \in M$ then $p_i - q_i \in M/R(A)$, so $p_i - q_i = 0$.

Conversely, assume that $R(p,q) = R(A)$ and that there are no maximal ideals M of A such that $A^2 \not\subseteq M$ and $p - q \in M$. Consider the ideals M_i of A given by $M_i/R(A) = A_1 \oplus \ldots \oplus A_{i-1} \oplus A_{i+1} \oplus \ldots \oplus A_n (i = 1, \ldots, n)$ in the decomposition (5.2). Clearly, the M_i are maximal and do not contain A^2. Thus, $p - q \notin M_i$, and as argued above, $p_i \neq q_i$ for all $i = 1, \ldots, n$. But by Lemma 5.1 $R_i(p_i, q_i) = J_i(p_i, q_i)$ for $i = 1, \ldots, n$, and by (5.6) and (5.7) $J(p,q) = R(p,q)$. Since $R(p,q) = R(A)$, we have $J(p,q) = R(A)$. □

If $A(p,q)$ has a unit element, then we have a stronger result:

Corollary 5.6. *Let A, p and q be the same as in Proposition 5.5. If $A(p,q)$ has a unit element, then $R(A) = R(p,q) = J(p,q) = R(A(p,q))$.*

Proof. If $A(p,q)$ has a unit element, then by Theorem 2.8 $p - q$ is invertible in A and so cannot be contained in any maximal ideal of A. If $xp, xq \in R(A)$ for $x \in A$, then $x(p-q) \in R(A)$ and $x \in R(A)$, since $p - q$ is invertible. Hence, $l_{R(A)}(p,q) \subseteq R(A)$, and similarly, $r_{R(A)}(p,q) \subseteq R(A)$, so $R(p,q) = R(A)$ by Lemma 5.4. Consequently, we obtain $J(p,q) = R(A)$ from Proposition 5.5. Since $p - q + R(A) = \sum(p_i - q_i)$ by (5.3), $(p-q)^{-1} + R(A) = \sum(p_i - q_i)^{-1}$ and $p_i - q_i$ is invertible in A_i for $i = 1, \ldots, n$. Hence, by Corollary 3.8, each $A_i(p_i, q_i)$ is simple and so $A(p,q)/R(A)$ is semisimple. This implies $R(A(p,q)) \subseteq R(A)$ (see the last part of the proof of Theorem I.1.7). Thus, by (5.1) we have the desired relation. □

Under a weaker condition on p and q, we can show:

Theorem 5.7. *Let A be an artinian associative algebra over F and p, q be fixed elements of A with decompositions (5.3). Then, $R(p,q) = J(p,q)$ if and only if $p - q \notin M$ for any maximal ideal M of A with $A^2 \not\subseteq M$. In this case, $R(p,q) = J(p,q) = R(A(p,q))$.*

Proof. In view of Lemma 5.1, (5.6) and (5.7), $R(p,q) = J(p,q)$ if and only if $p_i \neq q_i$ for all $i = 1, \ldots, n$. But, the maximal ideals M_i of A with $A^2 \not\subseteq M_i$ are given by $M_i/R(A) = A_1 \oplus \ldots \oplus A_{i-1} \oplus A_{i+1} \oplus \ldots \oplus A_n, i = 1, \ldots, n$. Thus, the condition of $p_i \neq q_i$ for all $i = 1, \ldots, n$ is equivalent to the condition that $p - q \notin M$ for any maximal ideal M of A with $A^2 \not\subseteq M$. But, by (5.5) and (5.6), we have

$$A(p,q)/R(p,q) \cong (A(p,q)/R(A))/(R(p,q)/R(A)) \cong \sum_{i=1}^{n} \oplus A_i(p_i,q_i)/R_i(p_i,q_i)$$

where each $A_i(p_i, q_i)$ is simple by Theorem 4.2, since $p_i \neq q_i$ for all $i = 1, \ldots, n$. Hence, $A(p,q)/R(p,q)$ is semisimple, and as before, $R(A(p,q)) \subseteq R(p,q)$. This gives $R(p,q) = R(A(p,q))$ by (5.1), as desired. □

If $p - q$ is invertible in A, then $p - q \notin M$ for any maximal ideal M of A. Thus, we have:

Corollary 5.8. *Let A, p and q be the same as in Theorem 5.7. If $p - q$ is invertible in A, then $R(p,q) = J(p,q) = R(A(p,q))$.* □

Therefore, under the condition in Theorem 5.7 or Corollary 5.8, the radical $R(A(p,q))$ is the solvable (nilpotent) radical of $A(p,q)$ (Proposition 1.5 and Theorem 5.3).

In the remainder of this section, we determine the structure of $R(A(p,q))$ relative to $R(A)$ for a finite–dimensional associative algebra A over F. Assume that fixed elements p, q of A are decomposed as in (5.3) and (5.4), so that $p_i \neq q_i$ for $i = 1, \ldots, r$ and $p_j = q_j$ for $j = r+1, \ldots, n$. We first show:

Lemma 5.9. *Let B be any algebra over F and let $R(B)$ denote the radical of B. If I is an ideal of B contained in $R(B)$, then $R(B/I) = R(B)/I$.*

Proof. Let M be a maximal ideal of B with $B^2 \not\subseteq M$. Then, $I \subseteq M$ since $I \subseteq R(B) \subseteq M$, and M/I is a maximal ideal of B/I with $(B/I)^2 \not\subseteq M/I$ since if $(B/I)^2 \subseteq M/I$ then $(B^2 + I)/I \subseteq M/I$, so $B^2 \subseteq M$, a contradiction. Conversely, it is easy to see that any maximal ideal $\bar{M}$ of B/I with $(B/I)^2 \not\subseteq \bar{M}$ is of the form M/I for some maximal ideal M of B with $B^2 \not\subseteq M$. If we let $\{M_\alpha\}$ be the collection of all such maximal ideals of B, then $R(B/I) = \bigcap_\alpha (M_\alpha/I) = (\bigcap_\alpha M_\alpha)/I = R(B)/I$. □

Therefore, if A is an artinian associative algebra over F and if $A/R(A)$ is decomposed as in (5.2), then since $R(A) \subseteq R(A(p,q))$, by Lemma 5.9, Proposition I.1.5 and (5.5) we obtain

$$\begin{aligned} R(A(p,q)/R(A)) &= R(A(p,q))/R(A) \\ &= R(A_1(p_1,q_1)) \oplus \ldots \oplus R(A_n(p_n,q_n)). \end{aligned} \tag{5.8}$$

If $p_i \neq q_i$, then by Lemma 5.1 and Theorem 5.7 $J_i(p_i, q_i) = R(A_i(p_i, q_i))$. Thus, it remains to determine $R(A(p,p))$ for a finite–dimensional simple associative algebra A over F. For this, we first note:

Lemma 5.10. *Let A be a simple noncommutative associative algebra over F. Then, $[A/Z(A), A/Z(A)]$ is a simple Lie algebra, and there are two possibilities:*

(i) $A^-/Z(A) = [A/Z(A), A/Z(A)]$ *and* $R(A^-/Z(A)) = 0$,

(ii) $0 \neq [A/Z(A), A/Z(A)] \subset A^-/Z(A)$ *and* $R(A^-/Z(A)) = A^-/Z(A)$.

Proof. It follows from Theorem I.4.19 that

$$\begin{aligned} [A/Z(A), A/Z(A)] &= ([A,A] + Z(A))/Z(A) \\ &\cong [A,A]/(Z(A) \cap [A,A]) \end{aligned} \tag{5.9}$$

is a simple Lie algebra. Moreover, Theorem I.4.19 implies that any nonzero ideal of $A^-/Z(A)$ contains $[A/Z(A), A/Z(A)]$. Therefore, if $A^-/Z(A) = [A/Z(A), A/Z(A)]$ then by Definition I.1.4 $R(A^-/Z(A)) = 0$. Case (ii) is similarly obtained. □

Proposition 5.11. *Let A be a finite–dimensional simple noncommutative associative algebra over F and $p \neq 0$ be a fixed element of A. Let $A(p,p) = R(p,p) \oplus A$ be the decomposition given in Theorem 4.7 where $R(p,p) = R^0(p,p)$ and $S = M_n(D)^-$ for some division algebra D over F. Let mn be the degree of $M_n(D)$ over F (Definition 1.4).*

(i) *If the characteristic of F does not divide mn, then $R(A(p,p)) = J(p,p)$.*

(ii) *If the characteristic of F divides mn, then $R(A(p,p)) = A(p,p)$.*

Proof. (i) By Proposition 1.6(i), $S = M_n(D)^- = Z(S) \oplus [S,S]$, and $S/Z(S) = [S/Z(A), S/Z(A)] \cong [S,S]$ by (5.9). Hence, we are in case (i) of Lemma 5.10, and so $R(S/Z(S)) = 0$. But,

$$\begin{aligned} A(p,p)/J(p,p) &= (J(p,p) + S)/J(p,p) \\ &\cong S/J(p,p) \cap S \cong S/Z(S) \end{aligned} \tag{5.10}$$

and by Lemma 5.9 $R(A(p,p)/J(p,p)) = R(A(p,p))/J(p,p) \cong R(S/Z(S)) = 0$, so $R(A(p,p)) = J(p,p)$.

(ii) From Proposition 1.6(ii) we have $0 \neq [S/Z(S), S/Z(S)] \subset S/Z(S)$, which is in case (ii) of Lemma 5.10. Hence, $R(S/Z(A)) = S/Z(S)$, but as in (5.10) $A(p,p)/J(p,p) \cong S/Z(S)$, implying $R(A(p,p)) = A(p,p)$. □

We note that if A is commutative then $A(p,p)$ is a zero algebra and hence $J(p,p) = R(A(p,p)) = A(p,p)$ (Definition I.1.4). Let A be a finite–dimensional associative algebra over F with decomposition (5.2) and let p, q be fixed elements of A with decompositions (5.3) and (5.4). Thus, $p_j = q_j$ for $j = r+1, \ldots, n$ and

$$A_j(p_j, p_j)/J_j(p_j, p_j) \cong M_{m_j}(D_j)^-/Z(M_{m_j}(D_j))$$

for some division algebra D_j over F, $j = r+1, \ldots, n$. Let $s, r+1 \leq s \leq n$, be such that the degree of $M_{m_j}(D_j)$ is not divisible by the characteristic of F for $j = r+1, \ldots, s$ and is divisible by the characteristic of F for $j = s+1, \ldots, n$. With these preparations we prove the following structure theorem for $R(A(p,q))$ relative to $R(A)$.

Theorem 5.12. *Let A be a finite-dimensional associative algebra over F and p, q be any fixed elements of A with decompositions as above. Then, the radical $R(A(p,q))$ of $A(p,q)$ is determined by*

$$\begin{aligned} &R(A(p,q))/R(A) \\ &\quad = J_1(p_1, q_1) \oplus \ldots \oplus J_s(p_s, q_s) \oplus A_{s+1}(p_{s+1}, p_{s+1}) \oplus \ldots \oplus A_n(p_n, p_n) \end{aligned}$$

where s is defined as above with $r+1 \leq s \leq n$. In particular, if the characteristic of F is zero, then $R(A(p,q)) = J(p,q)$.

Proof. We use the decomposition (5.8) for $R(A(p,q))/R(A)$. For $i = 1, \ldots, r$, since $p_i \neq q_i$, from Lemma 5.1 and Theorem 5.7 we have $R(A_i(p_i, q_i)) = J_i(p_i, q_i)$. If $j = r+1, \ldots, s$, then by Proposition 5.11(i) $R(A_j(p_j, q_j)) = J_j(p_j, q_j)$ while by Proposition 5.11(ii) $R(A_j(p_j, q_j)) = A_j(p_j, q_j)$ for $j = s+1, \ldots, n$. If the characteristic of F is 0, then we are in case (i) of Proposition 5.11 for $j = r+1, \ldots, n$ and hence $R(A(p,q)) = J(p,q)$ by (5.7). □

CHAPTER III

MUTATIONS OF ALTERNATIVE ALGEBRAS

1. Introduction

The primary concern in this chapter is to develop a structure theory for mutations of alternative algebras which is parallel to the structure theory for mutations of associative algebras. The framework which organizes this chapter is therefore very similar to that of Chapter II. Since a simple alternative algebra is either associative or a Cayley–Dickson algebra over its center (Theorem I.3.17), this chapter is in large devoted to the structure of mutations of Cayley–Dickson algebras.

Let A denote an alternative algebra with multiplication xy over a field F of characteristic $\neq 2$. For a fixed element $a \in A$, the (left) *a–homotope* $A(a)$ of A is the algebra over F with multiplication

$$x \cdot y = (xa)y \tag{1.1}$$

defined on A. Recall from Definition I.4.4 that if p and q are fixed elements of A then the (left) (p,q)*–mutation* $A(p,q)$ of A is defined to be the algebra with multiplication

$$x * y = (xp)y - (yq)x \tag{1.2}$$

for $x, y \in A$. The right (p,q)–mutation of A is the left $(-q,-p)$–mutation of the opposite algebra A^{op} (see the remark following Definition I.4.4). Thus, the results for a left mutation of A hold also for a right mutation of A. Henceforth, all mutations of A are assumed to be left mutations.

The commutator $[x,y]^*$ and anticommutator $\{x,y\}^*$ in $A(p,q)$ are given by

$$[x,y]^* = x * y - y * x = (x(p+q))y - (y(p+q))x, \tag{1.3}$$

$$\{x,y\}^* = \frac{1}{2}(x * y + y * x) = \frac{1}{2}[(x(p-q))y + (y(p-q))x] \tag{1.4}$$

and, as for the associative case, we obtain

$$A(p,q)^- = A(p+q)^-, \tag{1.5}$$

$$A(p,q)^+ = A(p-q)^+. \tag{1.6}$$

Thus, $A(p,q)$ is Malcev–admissible as well as Jordan–admissible (Proposition I.4.14 and (I.5.2)).

The (p,q)–mutation of an alternative algebra was first introduced in [**My3**], and some basic results for mutations of an associative algebra were extended to the alternative case [**MSh1,2**]. This chapter contains expanded and more detailed versions of the work in [**Mo1,3**]. Analogous to the associative case (Section II.1), in this section we present some basic definitions and results which will be used for later discussions.

As for associative algebras, an alternative algebra over a field F will be called *left (or right) artinian* if it satisfies the descending chain condition on left (or right) ideals (Definition II.1.1). Recall that the radical $R(B)$ of any algebra B is the intersection of all maximal ideals M of B with $B^2 \not\subseteq M$ (Definition I.1.4). Henceforth, a left (or right) artinian algebra will be simply called artinian. For an alternative algebra A, there exists a radical for A analogous to the Jacobson radical in the associative theory. Using the same notation as in the associative case, we denote this radical by $J(A)$ (commonly called the Zhevlakov radical of A in the alternative theory [**ZSSS1**, p.210]). The radical $J(A)$ satisfies virtually all characterizations that hold for the Jacobson radical of an associative algebra. For example, $J(A)$ contains all one–sided quasi–regular ideals of A, and if A is artinian then $J(A)$ is the unique maximal nilpotent ideal of A [**ZSSS1**, Chapters 10–12]. If $J(A) = 0$ then A is called *semisimple*. The following Wedderburn–Artin–Zhevlakov theorem is an extension of Theorem II.1.1 to artinian alternative algebras [**ZSSS1**, p.250].

Theorem 1.1. *Let A be an artinian alternative algebra over F.*

(i) *A is simple if and only if it is isomorphic either to a complete matrix algebra $M_n(D)$ over a division algebra D over F, or to a Cayley–Dickson algebra over an extension field K of F (K is in this case the center $Z(A)$).*

(ii) *A is semisimple if and only if it is isomorphic to a finite direct sum of complete matrix algebras over division algebras (over F) and Cayley–Dickson algebras over some extension fields of F.* □

Using Theorem 1.1 and the same argument as in the associative case (Lemma II.1.3), the following can be easily proven.

Proposition 1.2. *Let A be an alternative algebra over a field F and p, q be fixed elements of A. Let N be any nilpotent ideal of A or of $A(p,q)$.*

(i) *N is contained in every maximal ideal M of A or of $A(p,q)$ such that $A^2 \not\subseteq M$ or $A(p,q) * A(p,q) \not\subseteq M$.*

(ii) *If A is artinian, then $J(A) = R(A) \subseteq R(A(p,q))$.* □

As for associative algebras (Section II.5), we have $R(A) \neq R(A(p,q))$ in general. In fact, for a Cayley–Dickson algebra C (so $R(C) = 0$), it occurs that $\dim R(C(p,q)) = 8, 7, 6$ or 4. On the other hand, it can be shown that if A is artinian and $A(p,q)$ has a unit element then $R(A) = R(A(p,q))$ (Theorem 6.6(ii)). In general, $R(A)$ is far apart from the radical $R(A(p,q))$. It is desirable to have an ideal of $A(p,q)$ between $R(A)$ and $R(A(p,q))$ that serves as the solvable radical of $A(p,q)$. For this we follow the approach employed for mutations of associative algebras.

Let L_a and R_a denote the left and right multiplications in A by a. For fixed elements $p, q \in A$, put

$$\Lambda(p,q) = \{L_pR_p, L_pR_q, L_qR_p, L_qR_q\}.$$

Definition 1.1. For an alternative algebra A over F and fixed elements $p, q \in A$, define the subsets $R^0(p,q), R(p,q)$ and $J(p,q)$ of A by

$$\begin{aligned} R^0(p,q) &= \{x \in A : \Lambda(p,q)x = 0\}, \\ R(p,q) &= \{x \in A : \Lambda(p,q)x \subseteq R(A)\}, \\ J(p,q) &= \{x \in A : x * A \subseteq R(p,q) \text{ and } A * x \subseteq R(p,q)\}. \end{aligned}$$

□

We note that

$$R^0(p,q) = \{x \in A : pxp = p(xq) = q(xp) = qxq = 0\}, \tag{1.7}$$

and that $R^0(p,q) = R(p,q)$ if $R(A) = 0$. Unlike the associative case, it is not immediate that $R^0(p,q)$ and $R(p,q)$ are ideals of $A(p,q)$. If A is a composition algebra over F, then $R^0(p,q)$ will be shown to be an ideal of $A(p,q)$ (Theorem 2.7), and consequently $R(p,q)$ is an ideal of $A(p,q)$ for any artinian alternative algebra A (Proposition 6.3). Thus, if A is artinian then the relation $R(A) \subseteq R(p,q) \subseteq J(p,q)$ follows immediately from Definition 1.1. In Section III.6 we show that if A is artinian then $J(p,q)$ is the solvable radical of $A(p,q)$ which contains all solvable ideals of $A(p,q)$, and $A(p,q)/J(p,q)$ is isomorphic to a finite direct sum of simple algebras and certain semisimple Lie algebras. Those simple summands that do not arise from mutations of an associative algebra are of the form $C(p,q)/J(p,q)$ for a Cayley–Dickson algebra C. Accordingly, the main effort is focused on the structure of mutations of Cayley–Dickson algebras.

This chapter is organized as follows. In Section III.2 we prove some basic results about composition algebras and its mutations, which are instrumental for the main results in this chapter. It is shown in Section III.3 that virtually all results about identities for mutations of associative algebras are also valid for mutations of alternative algebras. Section III.4 is devoted to the structure of mutations of Cayley–Dickson algebras C. For $p \neq q$, $R^0(p,q)$ is the unique maximal ideal of $C(p,q)$ and is nilpotent with $R^0(p,q)^{*4} = 0$, i.e., all products of any four elements in $R^0(p,q)$ are zero. In this case, the algebra $C(p,q)/R^0(p,q)$ is completely determined and is a simple algebra of dimension $1, 2, 4$ or 8. If $p = q$, then it turns out that $C(p,p)$ is either a solvable Malcev algebra with $C(p,q)^{(3)} = 0$, or isomorphic to the Malcev algebra C^-. However, unlike the associative case (Theorem II.4.5), it is an open problem whether $C(p,q) = R^0(p,q) \oplus S$ for some subalgebra S of $C(p,q)$ when

$\dim R^0(p,q) = 4$. In Section III.5 we show that an alternative algebra A over F of characteristic $\neq 2, 3$ is prime (or simple) if and only if $A(p,q)$ $(p \neq q)$ is prime (or simple) and $R^0(p,q) = 0$, and that if $A(p,q)$ is prime then the centroids of A and $A(p,q)$ coincide (cf. Section II.3). Finally, in Section III.6 we investigate relationships among $R(A), R(p,q), J(p,q)$ and $R(A(p,q))$, and determine the structure of $A(p,q)/J(p,q)$ and $R(A(p,q))/R(A)$ for any artinian alternative algebra A.

2. Basic results

Let A denote an alternative algebra with multiplication xy over a field F of characteristic $\neq 2$. Recall from Section I.3 the Moufang identities for A:

$$(xax)y = x(a(xy)), \tag{2.1}$$
$$y(xax) = ((yx)a)x, \tag{2.2}$$
$$(xy)(ax) = x(ya)x. \tag{2.3}$$

For an idempotent $e \in A$, let

$$A_{ij} = \{x_{ij} \in A : ex_{ij} = ix_{ij},\ x_{ij}e = jx_{ij}\},\ i,j = 0,1. \tag{2.4}$$

Then, A has a Peirce decomposition $A = A_{00} + A_{01} + A_{10} + A_{11}$ relative to e. Using the Moufang identities, it is easy to verify the relations [**Sc3**, p.32]:

$$A_{ij}A_{jk} \subseteq A_{ik},\ A_{ij}A_{ij} \subseteq A_{ji},\ i,j,k = 0,1, \tag{2.5}$$

where all other products are zero. We note from (2.4) that $A = A_{11}$ if and only if e is the unit element in A.

Proposition 2.1. *Let A be an alternative algebra over F and let $N(A)$ denote the nucleus of A (Definition I.2.2).*

(i) *If A^+ has a unit element e then e is also the unit element of A.*

(ii) *If A has a unit element and $u \in N(A)$ is invertible in A with inverse u^{-1} then $u^{-1} \in N(A)$.*

(iii) *For a fixed $a \in A$, the a-homotope $A(a)$ defined by (1.1) has a unit element e if and only if A has a unit element 1 and a is invertible. In this case, $e = a^{-1}$.*

Proof. (i) Since $e = e \circ e = \frac{1}{2}(e^2 + e^2) = e^2$, e is an idempotent in A, and $ex + xe = 2x$ for all $x \in A$. If A_{ij} denotes the Peirce space relative to e given by (2.4), then $2x = ex + xe = 0$ for $x \in A_{00}$ and $2x = ex + xe = x$ for $x \in A_{01} \cup A_{10}$. Hence, $A_{00} = A_{01} = A_{10} = 0$ and $A = A_{11}$, so e is the unit element in A.

(ii) Recall first that if $x \in A$ is invertible in A then $(yx)x^{-1} = x^{-1}(xy) = y$ for all $y \in A$ [**ZSSS1**, p.204]. Thus, for any $x, y \in A$, $u(u^{-1}, x, y) = u[(u^{-1}x)y] - u[u^{-1}(xy)] = -(u, u^{-1}x, y) + [u(u^{-1}x)]y - u[u^{-1}(xy)] = 0 + xy - xy = 0$. Thus, $(u^{-1}, x, y) = 0$ for all $x, y \in A$, since $u \in N(A)$, giving $u^{-1} \in N(A)$.

(iii) Assume $A(a)$ has a unit element e. Hence, for all $x \in A$, $e \cdot x = (ea)x = x = (xa)e = x \cdot e$, and in particular, ea is a left unit element in A, so ea is an idempotent in A. This implies that the Peirce spaces A_{00} and A_{01} relative to ea are zero and $A = A_{10} + A_{11}$. From $(ea)(ae) = e \cdot (ae) = ae$ and $(ea)(ae) = ((ea)a)e = (ea) \cdot e = ea$, we obtain $ae = ea$. But, $(ea)a = e \cdot a = a$, $a(ea) = (ae)a = (ea)a = a$ and $(ea)e = e = e(ae) = e(ea)$, hence $e, a \in A_{11}$. For any $x \in A_{10}$, $x = x \cdot e = (xa)e = 0$, since $xa \in A_{10}A_{11} = 0$ by (2.5). Thus, $A = A_{11}$, and $ea = ae$ is the unit element 1 of A, so a is invertible and $e = a^{-1}$. Conversely, if A has a unit element 1 and a is invertible in A, then $a^{-1} \cdot x = (a^{-1}a)x = x$ and $x \cdot a^{-1} = (xa)a^{-1} = x(aa^{-1}) = x$ [**ZSSS1**, p.204], hence a^{-1} is the unit element in $A(a)$. □

If $a \in A$ is invertible in A, then $A(a)$ is called the (left) *a–isotope* of A. Unlike the associative case, the a–isotope $A(a)$ is not in general isomorphic to A [**Mc2**]. However, they are isomorphic for a composition algebra. Let A now be a composition algebra with unit element 1 over F and let n denote the norm of A. Thus, A satisfies the quadratic equation

$$x^2 - t(x)x + n(x)1 = 0 \tag{2.6}$$

for all $x \in A$ where $t : A \to F$ is the trace of A. We recall from Section I.3 some basic facts about composition algebras. The linear map : $A \to A : x \mapsto \bar{x} = t(x)1 - x$ is an (standard) involution on A and satisfies the relations

$$t(x) = t(\bar{x}) \ , \quad x\bar{x} = \bar{x}x = n(x)1 \tag{2.7}$$

for all $x \in A$. Let (x, y) denote the bilinear form on A given by

$$(x, y) = \frac{1}{2}[n(x + y) - n(x) - n(y)]. \tag{2.8}$$

Then, (,) is symmetric and nondegenerate. Since $2(x, y) = n(x+y) - n(x) - n(y) = (x + y)(\bar{x} + \bar{y}) - x\bar{x} - y\bar{y} = x\bar{y} + y\bar{x} = x\bar{y} + \overline{x\bar{y}} = t(x\bar{y})$ by (2.7), we have

$$(x, y) = \frac{1}{2}(x\bar{y} + y\bar{x}) = \frac{1}{2}t(x\bar{y}) \tag{2.9}$$

for $x, y \in A$.

Proposition 2.2. *Let A be a composition algebra over F and a be a fixed invertible element of A. Then, $A(a)$ is isomorphic to A.*

Proof. Since a is invertible, $n(a) \neq 0$ (in fact, by (2.7) $a^{-1} = n(a)^{-1}\bar{a}$). If we define $n_a : A \to F$ by $n_a(x) = n(a)n(x) = n(ax)$ for $x \in A$, then the quadratic form n_a is similar to n, since $A \to A \ : \ x \mapsto ax$ is a linear isomorphism. For $x, y \in A$, $n_a(x \cdot y) = n_a((xa)y) = n(a)n(x)n(a)n(y) = n_a(x)n_a(y)$. Thus, $A(a)$ is a composition algebra with norm n_a and with unit element a^{-1} by Proposition 2.1(iii). The result now follows from Theorem I.3.7. □

For any elements x, y in a composition algebra A over F, we have $t(xy)x = (xy + \overline{xy})x = xyx + (\bar{y}\bar{x})x = xyx + \bar{y}(\bar{x}x) = xyx + n(x)\bar{y}$ and hence

$$xyx = t(xy)x - n(x)\bar{y}, \tag{2.10}$$

which is linearized to the identity

$$x(yz) + z(yx) = (xy)z + (zy)x = t(xy)z + t(zy)x - 2(x,z)\bar{y}. \tag{2.11}$$

Since $x(\bar{x}z) = (x\bar{x})z = n(x)z = (x,x)z$ for $x, z \in A$, linearizing this we get $x(\bar{y}z) + y(\bar{x}z) = 2(x,y)z$ which by (2.9) implies

$$x(\bar{y}z) + y(\bar{x}z) = 2(x,y)z = t(x\bar{y})z, \tag{2.12}$$

$$(z\bar{y})x + (z\bar{x})y = 2(x,y)z = t(x\bar{y})z \tag{2.13}$$

for all $x, y, z \in A$. Identities (2.10)–(2.13) will be frequently used throughout this chapter.

A composition algebra A over F is a division algebra over F if and only if $n(x) \neq 0$ for all $x \neq 0$ in A (Lemma I.3.5). Otherwise, there is an element $x \neq 0$ in A with $n(x) = 0$, and A is the unique split composition algebra over F of dimension $2, 4$ or 8 which is isomorphic to $F \oplus F$, $M_2(F)$ or to the split Cayley–Dickson algebra $C(F)$ given by (I.3.12) (Theorem I.3.6). If $A = F \oplus F$ and $x = (\alpha, \beta) \in A$ then $t(x) = \alpha + \beta$, $n(x) = \alpha\beta$ and the bilinear form associated with n is given by $(x,y) = \frac{1}{2}(\alpha\beta' + \alpha'\beta)$ for $y = (\alpha', \beta')$. For $A = M_2(F)$, we have $x^2 - (\operatorname{trace} x)x + (\det x)1 = 0$ for all $x \in A$, hence $t(x) = \operatorname{trace} x$ and $n(x) = \det x$. $C(F)$ can be realized by a Zorn vector matrix algebra over F. Thus, each element of $C(F)$ has the form $x = \begin{pmatrix} \alpha & a \\ b & \beta \end{pmatrix}$ where $a, b \in F^3$ and $\alpha, \beta \in F$. If we let $a_1 = (1,0,0)$, $a_2 = (0,1,0)$ and $a_3 = (0,0,1)$ then the matrices

$$e_1 = \begin{pmatrix} 1 & 0 \\ 0 & 0 \end{pmatrix}, e_2 = \begin{pmatrix} 0 & 0 \\ 0 & 1 \end{pmatrix}, u_i = \begin{pmatrix} 0 & -a_i \\ 0 & 0 \end{pmatrix}, v_i = \begin{pmatrix} 0 & 0 \\ a_i & 0 \end{pmatrix}, \tag{2.14}$$

$i = 1, 2, 3$, form a (standard) basis of $C(F)$ with multiplication given by

$$\begin{gathered} e_i^2 = e_i \ (i = 1,2), \ e_1u_i = u_i = u_ie_2, \ e_2v_i = v_i = v_ie_1, \\ u_iu_j = \epsilon_{ijk}v_k, \ v_iv_j = \epsilon_{ijk}u_k, \ u_iv_j = -\delta_{ij}e_1, \ v_iu_j = -\delta_{ij}e_2 \end{gathered} \tag{2.15}$$

for $i, j, k = 1, 2, 3$ (see [**My4**, p. 284] and (I.3.12)). In this case, we have $t(x) = \alpha + \beta$ and $n(x) = \alpha\beta - (a, b)$ with the inner product (a, b) in F^3. Hence, the associated bilinear form (x, y) is given by

$$(x,y) = \frac{1}{2}[\alpha\beta' + \alpha'\beta - (a,b') - (a',b)] \tag{2.16}$$

for $y = \begin{pmatrix} \alpha' & a' \\ b' & \beta' \end{pmatrix}$. Let

$$U = \sum_{i=1}^{3} Fu_i, \; V = \sum_{i=1}^{3} Fv_i. \tag{2.17}$$

Recall that a subspace S of A is called totally isotropic if $(S, S) = 0$ and that the dimension of a maximal totally isotropic subspace of A is the Witt index of $(\, , \,)$ in A which cannot exceed $\frac{1}{2} \dim A$. If A is a split composition algebra over F then it follows easily from the foregoing remarks that the Witt index of $(\, , \,)$ is $\frac{1}{2} \dim A$. In fact, for $C(F)$ we obtain from (2.14), (2.16) and (2.17)

$$(e_1, e_2) = \frac{1}{2}, \; (e_i, e_i) = (e_i, U) = (e_i, V) = (U, U) = (V, V) = 0 \tag{2.18}$$

for $i = 1, 2$. Hence, $Fe_i + U$ and $Fe_i + V$ are maximal totally isotropic subspaces of A. The structure of the (p, q)–mutation of a Cayley–Dickson algebra over F depends in large on whether or not $Fp + Fq$ is totally isotropic. The following is well known.

Lemma 2.3. *A composition algebra A over F is split if and only if A contains an idempotent $e \neq 0, 1$. For any idempotent $e \neq 0, 1$ in $C(F)$, $e_1 = e$ and $e_2 = 1 - e$ can be completed to a standard basis of $C(F)$ with multiplication given by (2.15).*

Proof. We only prove the first part, and a proof of the second part can be found, for example, in [**Sc3**] or [**ZSSS1**]. If $e \neq 0, 1$ is an idempotent of A then $e(1 - e) = 0$ and $n(e) = 0$ or $n(1 - e) = 0$, so A is split. Assume A is split. If we let $N = \{a \in A : n(a) = 0\}$, then N contains nonzero elements. Linearizing (2.6) and using (2.8) we obtain $xy + yx - t(x)y - t(y)x + 2(x, y) = 0$ for all $x, y \in A$, which implies

$$(x,y) = \frac{1}{2}\left[t(x)t(y) - t(xy)\right] \tag{2.19}$$

for $x, y \in A$, since $t(xy) = t(yx)$ and $t(1) = 2$. If $t(N) = 0$, then since $n(aA) = 0$ and so $aA \subseteq N$ for all $a \in N$, we have $(N, A) = 0$ by (2.19). This is impossible, since $(\, , \,)$ is nondegenerate. Thus, there is an $a \in N$ with $t(a) \neq 0$, and $t(a)^{-1}a \neq 0, 1$ is an idempotent of A. □

For a subset S of an algebra B over F, let $l(S)$ denote the left annihilator of S in B:

$$l(S) = \{x \in B : xS = 0\}.$$

Lemma 2.4. *Let C be a split Cayley–Dickson algebra over F, and let e_1, e_2 be orthogonal idempotents in C with $e_1 + e_2 = 1$. Let U and V be the subspaces of C given by (2.17).*

(i) *If $uu' = 0$ for $u, u' \in U$ then u and u' are linearly dependent.*

(ii) *If $vv' = 0$ for $v, v' \in V$ then v and v' are linearly dependent.*

(iii) *If $S = Fe_1 + Fu$ for $0 \neq u \in U$ then $l(S) = \bar{S} = Fe_2 + Fu$.*

(iv) *If $S = Fu + Fv$ for nonzero elements $u \in U$ and $v \in V$ such that $uv = vu = 0$, then $l(S) = Fu + Fv = S = \bar{S}$.*

Proof. (i) Let $u = \sum_{i=1}^{3} \alpha_i u_i$ and $u' = \sum_{i=1}^{3} \beta_i u_i$ be elements of U such that $uu' = 0$. We may assume $u \neq 0$, say $\alpha_1 \neq 0$. It follows from (2.15) that $\alpha_1\beta_2 - \alpha_2\beta_1 = \alpha_2\beta_3 - \alpha_3\beta_2 = \alpha_3\beta_1 - \alpha_1\beta_3 = 0$. Hence, $\beta_3 = \alpha_1^{-1}\beta_1\alpha_3$, $\beta_2 = \alpha_1^{-1}\beta_1\alpha_2$ and $\beta_1 = \alpha_1^{-1}\beta_1\alpha_1$. This gives $u' = \alpha_1^{-1}\beta_1 u$.

(ii) is similar to (i).

(iii) Since $e_2, u \in l(S)$ by (2.15), $\bar{S} = Fe_2 + Fu \subseteq l(S)$. If $x = \alpha e_1 + \beta e_2 + u' + v'$ is in $l(S)$ for $\alpha, \beta \in F$, $u' \in U$ and $v' \in V$ then $0 = xe_1 = \alpha e_1 + v'$, so $\alpha = 0$ and $v' = 0$. By part (i), $0 = xu = u'u$ implies $u' \in Fu$, since $u \neq 0$, and hence $l(S) = Fe_2 + Fu = \bar{S}$.

(iv) Since $u^2 = v^2 = uv = vu = 0$, $Fu + Fv \subseteq l(S)$. If $x = \alpha e_1 + \beta e_2 + u' + v'$ is in $l(S)$, then $0 = xu = \alpha u + u'u + v'u \in Fe_2 + U + V$ by (2.15) and hence $\alpha = 0$ and $u'u = v'u = 0$. By part (i), this implies $u' \in Fu$ and $u'v = 0$. Hence, $0 = xv = (\beta e_2 + u' + v')v = \beta v + v'v$, and so $\beta = 0$, $v'v = 0$. But, by part (ii) $v' \in Fv$, so $x = u' + v' \in Fu + Fv$, showing $l(S) = Fu + Fv = S = \bar{S}$. □

Proposition 2.5. *Let C be a split Cayley–Dickson algebra over F and S be a subspace of C of dimension 2 such that $\bar{S}S = 0$. Then, $l(S) = \bar{S}$.*

Proof. If $\bar{F}$ denotes the algebraic closure of F, then $\overline{\bar{F} \otimes_F S} = \bar{F} \otimes_F \bar{S}$ and $l(\bar{F} \otimes_F S) = \bar{F} \otimes_F l(S)$. Thus, we may assume that F is algebraically closed.

Assume first that S contains an idempotent $e \neq 0, 1$. By Lemma 2.3 $e_1 = e$ and $e_2 = 1 - e$ are completed to a basis of C given by (2.15). Let U and V be the subspaces of C given by (2.17). Since $0 = \bar{e}_1 S = e_2 S$, $S \subseteq Fe_1 + U$, and so $S = Fe_1 + Fu$ for some $0 \neq u \in U$, since $\dim S = 2$. Thus, by Lemma 2.4(iii) $l(S) = \bar{S}$.

Assume then that S does not contain an idempotent $\neq 0, 1$. Then, $(1, S) = t(S) = 0$; otherwise, for an $x \in S$ with $t(x) \neq 0$, $t(x)^{-1}x$ would be an idempotent $\neq 0, 1$ since $n(x) = x\bar{x} = 0$ and so $x^2 - t(x)x = 0$. If we let $C_0 = \{x \in C : t(x) = 0\}$, then $\dim(S^\perp \cap C_0) = \dim S^\perp + \dim C_0 - \dim(S^\perp + C_0) \geq 6 + 7 - 8 = 5$. Thus, $S^\perp \cap C_0$ is not totally isotropic and there is an $a \in S^\perp \cap C_0$ with $n(a) = (a, a) \neq 0$.

Since F is algebraically closed, we may assume $n(a) = -1$. If we let $e_1 = \frac{1}{2}(1-a)$ and $e_2 = \frac{1}{2}(1+a)$ then $e_i^2 = e_i$, $e_ie_j = 0$ for $i \neq j$ and $e_1 + e_2 = 1$. We now complete $\{e_1, e_2\}$ to a standard basis of C with U and V as before. Since $a \in S^\perp$, $(e_i, S) = 0$ for $i = 1, 2$. It follows from this and (2.18) that $S \subseteq U + V$. Since $S\bar{S} = 0$ and $\dim S = 2$, by Lemma 2.4(i),(ii) it can not occur that $S \subseteq U$ or $S \subseteq V$. Hence, there exists an $x \neq 0$ in S such that $x = u + v$ for some $0 \neq u \in U$ and $0 \neq v \in V$. Since $\bar{S}S = 0$, $uv = vu = 0$. For any $y = u' + v' \in S$, $xy = 0$ gives $uu' = vv' = 0$, and by Lemma 2.4 $u' \in Fu$ and $v' \in Fv$. Therefore, $S = Fu + Fv$ and so $l(S) = \bar{S}$ by Lemma 2.4(iv). □

In the remainder of this section we present some basic properties about the subspace $R^0(p, q)$ of A defined by (1.7) for a composition algebra A. We begin with:

Lemma 2.6. *Let A be a composition algebra over F and let p, q be fixed elements of A. If $Fp + Fq$ is not totally isotropic then $R^0(p, q) = 0$.*

Proof. Since $Fp + Fq$ is not totally isotropic, there is an $a \in Fp + Fq$ with $n(a) = (a, a) \neq 0$, so a is invertible in A. Thus, by (1.7)

$$R^0(p, q) \subseteq \{x \in A : axa = 0\} = 0.$$

□

Theorem 2.7. *For any fixed elements p, q of a composition algebra A over F, $R^0(p, q)$ is an ideal of $A(p, q)$.*

Proof. If $Fp + Fq$ is not totally isotropic then the result is immediate from Lemma 2.6. Thus, we assume that $Fp + Fq$ is totally isotropic. Recall first from Theorem I.3.11 the relation

$$t((xy)z) = t(x(yz)) \tag{2.20}$$

for all $x, y, z \in A$. Using (2.10), (2.20) and the Moufang identity (2.2), for $x \in A$ and $y \in R^0(p, q)$ we deduce

$$\begin{aligned} p(x * y)p &= t\left((x * y)p\right)p - n(p)\overline{x * y} \\ &= t\left(((xp)y - (yq)x)p\right)p, \qquad (n(p) = 0) \\ &= \left(t(x(pyp)) - t((xp)(yq))\right)p \\ &= -t\left(x(p(yq))\right)p = 0. \end{aligned}$$

Similarly, we obtain $p(y * x)p = q(x * y)q = q(y * x)q = 0$. Now,

$$\begin{aligned} q\left((x * y)p\right) &= q\left(((xp)y - (yq)x)p\right) \\ &= q\left(x(pyp)\right) - q\left(((yq)x)p\right) \\ &= -q\left(((yq)x)p\right) \end{aligned}$$

by (2.2). Since $t(((yq)x)p) = t((p(yq))x) = 0$ by (2.20) and $a = t(a)1 - \bar{a}$, $q(((yq)x)p) = -q(\bar{p}(\bar{x}(\bar{q}\bar{y}))) = p(\bar{q}(\bar{x}(\bar{q}\bar{y})))$ using (2.12) and $(p,q) = 0$. Hence, by (2.1) $q(((yq)x)p) = p((\bar{q}\bar{x}\bar{q})\bar{y})$. Since $t(y(qxq)) = t(((yq)x)q) = t(q((yq)x)) = t((qyq)x) = 0$ by (2.20) or (2.2), we get

$$p\left((\bar{q}\bar{x}\bar{q})\bar{y}\right) = -p\left(y(qxq)\right) = p\left(y(t(xq)q)\right) = t(xq)p(yq) = 0,$$

using (2.10). Thus, from the above, $q((x*y)p) = 0$ and similarly, $p((y*x)q) = 0$. Finally, we compute

$$\begin{aligned} p\left((x*y)q\right) &= p\left(((xp)y - (yq)x)q\right) \\ &= p\left((xp)y)q\right) - p\left(y(qxq)\right), && \text{by (2.2)} \\ &= p\left((xp)y)q\right) - t(xq)p(yq), && \text{by (2.10)} \\ &= p\left((xp)y)q\right). \end{aligned}$$

Since $t(((xp)y)q) = t(x(p(yq))) = 0$ by (2.20), we derive

$$p\left((x*y)q\right) = -p\left(\bar{q}(\bar{y}(\bar{p}\bar{x}))\right) = q\left(\bar{p}(\bar{y}(\bar{p}\bar{x}))\right) = q\left((\overline{pyp})\bar{x}\right) = 0,$$

using (2.12) for the second equality. Similarly, we obtain $q((y*x)p) = 0$, and hence $R^0(p,q)$ is an ideal of $A(p,q)$. □

If A is a composition algebra of dimension ≤ 4 then it is associative and Theorem 2.7 is immediate from the definition of $R^0(p,q)$. Hence, the proof of Theorem 2.7 is essentially concerned with Cayley–Dickson algebras.

Lemma 2.8. *Let A and p, q be the same as in Theorem 2.7. If $Fp + Fq$ is totally isotropic then for any $x \in A$, $y \in R^0(p,q)$ and $a \in Fp + Fq$ we have*

$$t(ya) = 0, \tag{2.21}$$
$$(xa)y = -(x\bar{y})\bar{a}, \tag{2.22}$$
$$(x\bar{a})\bar{y} = -(xy)a, \tag{2.23}$$
$$y(ax) = -\bar{a}(\bar{y}x), \tag{2.24}$$
$$\bar{y}(\bar{a}x) = -a(xy). \tag{2.25}$$

Proof. Since $y \in R^0(p,q)$ and $Fp + Fq$ is totally isotropic, $n(a) = 0$ and by (2.10) $0 = aya = t(ya)a - n(a)\bar{y} = t(ya)a$. This gives (2.21). It follows from (2.13) and (2.21) that $(xa)y + (x\bar{y})\bar{a} = t(ya)x = 0$ and $(x\bar{a})\bar{y} + (xy)a = t(\bar{y}\bar{a})x = t(ay)x = t(ya)x = 0$, implying (2.22) and (2.23). Relations (2.24) and (2.25) follow similarly from (2.12) and (2.21). □

Proposition 2.9. *Let A be a composition algebra over F and let p, q be fixed elements of A with $Fp + Fq$ totally isotropic. If $J = F\bar{p} + F\bar{q}$, then*

(i) $J \subseteq R^0(p,q)$,

(ii) $J * R^0(p,q) = R^0(p,q) * J = 0$,

(iii) $(R^0(p,q))^{*3} \subseteq J$.

Proof. (i) Since $Fp + Fq$ is totally isotropic, $p\bar{p}p = n(p)p = 0 = p(\bar{p}q) = (p\bar{p})q = n(p)q$. Also, from (2.10) $q\bar{p}q = t(q\bar{p})q - n(q)p = 2(q,p)q - n(q)p = 0$ using (2.9), and so $\bar{p} \in R^0(p,q)$. Similarly, $\bar{q} \in R^0(p,q)$ and hence $J \subseteq R^0(p,q)$.

(ii) If $x \in R^0(p,q)$, then $\bar{p} * x = (\bar{p}p)x - (xq)\bar{p} = -(xq)\bar{p} = (\bar{q}\bar{x})\bar{p} = \overline{p(xq)} = 0$ by (2.21). Similarly, we have $\bar{q} * x = x * \bar{p} = x * \bar{q} = 0$.

(iii) We write $x \equiv y$ if $x - y \in J$. Let $x, y, z \in R^0(p,q)$. By (2.10) and (2.21), $x * x = x(p-q)x = t(x(p-q))x - n(x)(\overline{p-q}) = -n(x)(\overline{p-q}) \in J$. Thus, $x * x \equiv 0$ and so, $x * y + y * x \equiv 0$ or $2x * y \equiv [x,y]^*$, which implies $4(x*y)*z \equiv [[x,y]^*, z]^* \equiv 4z * (y * x)$. Hence, it suffices to verify $[[x,y]^*, z]^* \equiv 0$. Using (1.3) we first expand $[[x,y]^*, z]^*$ and simplify its terms as follows:

$$\begin{aligned}
[[x,y]^*,z]^* &= \left(((x(p+q))y - (y(p+q))x\right)(p+q)z \\
&\quad -(z(p+q))\left((x(p+q))y - (y(p+q))x\right), \\
((x(p+q))y)(p+q) &= x\left((p+q)y(p+q)\right) = 0, \quad \text{by (2.2)}, \\
((y(p+q))x)(p+q) &= y\left((p+q)x(p+q)\right) = 0, \quad \text{by (2.2)}, \\
(z(p+q))((x(p+q))y) &= \left((\overline{p+q})\bar{z}\right)\left((x\bar{y})(\overline{p+q})\right), \quad \text{by (2.21) and (2.23)}, \\
&= (\overline{p+q})(\bar{z}(x\bar{y}))(\overline{p+q}), \quad \text{by (2.3)} \\
&= t\left((\bar{z}(x\bar{y}))(\overline{p+q})\right)(\overline{p+q}) - n(p+q)(\bar{z}(x\bar{y})), \\
&\qquad \text{by (2.10)} \\
&= t\left((\bar{z}(x\bar{y}))(\overline{p+q})\right)(\overline{p+q}) \equiv 0.
\end{aligned}$$

Similarly, it follows that $(z(p+q))((y(p+q))x) \equiv 0$. Thus, $[[x,y]^*, z]^* \equiv 0$ for all $x, y, z \in R^0(p,q)$. □

Recall from Proposition II.1.5(iii) that if A is associative then for any fixed $p, q \in A$, $R^0(p,q)^{*3} = 0$. On the other hand, it is shown in Section III.4 that if C is a Cayley–Dickson algebra over F, then $R^0(p,q)^{*4} = 0$ for any fixed $p, q \in C$ but $R^0(p,q)^{*3} \neq 0$ for some $p, q \in C$ with $Fp + Fq$ totally isotropic (Theorem 4.9).

3. Identities

Following the results developed in Section II.2 for mutations of an associative algebra, in this section we investigate conditions for mutations of an alternative algebra to satisfy certain well known identities. Two primary identities considered are third power–associativity and flexibility. The main effort focuses on conditions that ensure the equivalence of the two identities above, power–associativity and noncommutative Jordan identities. It turns out that many of the results in Section II.2 are directly extended to mutations of an alternative algebra. The discussion in this section is also an extended and generalized version of the results in [**My3**] and [**MSh1**] where the fixed elements p, q are taken from the nucleus of an alternative algebra.

Let A denote an alternative algebra over F of characteristic $\neq 2$ and let p, q be fixed elements of A. For the associator $(x, y, z)^* = (x * y) * z - x * (y * z)$ in $A(p, q)$, we first compute from (1.2)

$$\begin{aligned}(x,y,x)^* &= (((xp)y-(yq)x)p)\,x-(xq)\,((xp)y-(yq)x)\\ &\quad -(xp)\,((yp)x-(xq)y)+(((yp)x-(xq)y)q)\,x\\ &= (((xp)y)p)\,x-(((yq)x)p)\,x-(xq)((xp)y)+(xq)((yq)x)\\ &\quad -(xp)((yp)x)+(xp)((xq)y)+(((yp)x)q)\,x-(((xq)y)q)\,x\end{aligned}$$

for $x, y \in A$. On the other hand, by (2.2) and (2.3) we obtain

$$\begin{aligned}(((xp)y)p)\,x &= x(pyp)x=(xp)((yp)x),\\ (((xq)y)q)\,x &= x(qyq)x=(xq)((yq)x),\end{aligned}$$

which imply

$$\begin{aligned}(x,y,x)^* &= (xp)((xq)y)+(((yp)x)q)x-(xq)((xp)y)-(((yq)x)p)x\\ &= (xp)((xq)y)+(yp)(xqx)-(xq)((xp)y)-(yq)(xpx)\end{aligned}\tag{3.1}$$

for all $x, y \in A$. Therefore, by (2.3)

$$(x,x,x)^* = 2x\,(p(xq)-q(xp))\,x. \tag{3.2}$$

In the following, we give a condition for $A(p, q)$ to be third power–associative or flexible, which is analogous to the associative case (Theorems II.2.1 and II.2.2).

Theorem 3.1. *Let A be an alternative algebra over F and let p, q be fixed elements of A.*

(i) *$A(p, q)$ is third power–associative if and only if*

$$x(p(xq)-q(xp))x=0 \tag{3.3}$$

for all $x \in A$.

(ii) *$A(p,q)$ is flexible if and only if*

$$x(p(yq)-q(yp))x=0 \tag{3.4}$$

for all $x,y\in A$.

Proof. (i) This follows immediately form (3.2).

(ii) Assume that $A(p,q)$ is flexible, so $A(p,q)$ is third power–associative. By (2.3), (3.3) becomes $0=(xp)((xq)x)-(xq)((xp)x)$, and linearizing this via $\lambda x+y$, the coefficient of λ^2 gives $(yp)((xq)x)+(xp)((yq)x)+(xp)((xq)y)-(yq)((xp)x)-(xq)((yp)x)-(xq)((xp)y)=0$, or equivalently, by (3.1)

$$(x,y,x)^*+(xp)((yq)x)-(xq)((yp)x)=0$$

for all $x,y\in A$. Since $(xp)((yq)x)=x(p(yq))x$ and $(xq)((yp)x)=x(q(yp))x$ by (2.3), the last identity becomes

$$(x,y,x)^*+x(p(yq)-q(yp))x=0 \tag{3.5}$$

for all $x,y\in A$. It follows from (3.5) that flexibility in $A(p,q)$ is equivalent to (3.4). □

As for mutations of associative algebras, third power–associativity in $A(p,q)$ implies power–associativity in $A(p,q)$. For the proof of this we need:

Lemma 3.2. *If $A(p,q)$ is third power–associative then for any $x\in A$ and any natural number n, we have*

$$[[(x(p-q))^n x]\,p]\,x = (xp)((x(p-q))^n x), \tag{3.6}$$
$$[[(x(p-q))^n x]\,q]\,x = (xq)((x(p-q))^n x). \tag{3.7}$$

Proof. Since the proofs of (3.6) and (3.7) are identical, we verify only (3.6). We proceed by induction on n. Let $r=p-q$. By (2.2) and (2.3),

$$(((xr)x)p)x=(xr)(xpx)=x(r(xp))x. \tag{3.8}$$

But, since $A(p,q)$ is third power–associative, by (3.3)

$$\begin{aligned} x(r(xp))x &= x(pxp)x-x(q(xp))x = xpxpx-x(p(xq))x \\ &= x(p(xr))x=(xp)(xrx), \end{aligned}$$

using (2.3). It follows from this and (3.8) that (3.6) holds for $n=1$. For $n>1$, by Artin's theorem (Theorem I.3.1; xr and xpx generate an associative subalgebra of

A), we have

$$\begin{aligned}
[[(xr)^n x]\, p]\, x &= (xr)^n(xpx), \quad \text{by (2.2)} \\
&= (xr)^{n-1}((xr)(xpx)) \\
&= (xr)^{n-1}((xp)(xrx)), \quad \text{by (3.8) and (3.6) with } n=1 \\
&= (xr)^{n-1}[((xpx)r)x], \quad \text{by (2.2)} \\
&= (xr)^{n-2}[(xr)[((xpx)r)x]], \quad \text{by Artin's theorem} \\
&= (xr)^{n-2}[x(r(xpx)r)x], \quad \text{by (2.3)} \\
&= (xr)^{n-2}[x((r(xp))(xr))x], \quad \text{by (2.3)} \\
&= (xr)^{n-2}[[x(r(xp))](xrx)], \quad \text{by (2.3)} \\
&= (xr)^{n-2}[(xrx)p(xrx)], \quad \text{by (2.1)} \\
&= \big([(xr)^{n-2}(xrx)]p\big)\,(xrx), \quad \text{by (2.2)} \\
&= \Big(\big((((xr)^{n-1}x)p)x\big)\, r\Big)\, x, \quad \text{by Artin's theorem and (2.2)} \\
&= \Big(\big[(xp)((xr)^{n-1}x)\big]\, r\Big)\, x, \quad \text{by induction hypothesis.}
\end{aligned}$$

Making repeated use of (2.2), it follows easily that

$$\begin{aligned}
z((xr)^m x) &= z(xrx\cdots xrx) \\
&= ((\cdots(((zx)r)x)\cdots)r)x
\end{aligned} \tag{3.9}$$

for all $x, r, z \in A$ and any positive integer m. Using (3.9) with $z = xp$, the last expression in the above becomes

$$\begin{aligned}
\Big(\big[(xp)((xr)^{n-1}x)\big]\, r\Big)\, x &= (([(\cdots(((xp)x)\cdots)r)x])r)x \\
&= (xp)((xr)^n x),
\end{aligned}$$

which completes the proof of (3.6). □

Theorem 3.3. *Let A be an alternative algebra over F and let p, q be fixed elements in A. Then, $A(p,q)$ is third power–associative if and only if it is power–associative.*

Proof. Assume that $A(p,q)$ is third power–associative. For a positive integer $k > 1$, define $x^{*k} = x^{*(k-1)} * x$ with $x^{*1} = x$. We first show that the identity

$$x^{*n} = (xr)^{n-1}x \tag{3.10}$$

holds for all $x \in A$ and integers $n > 1$, where we set $r = p - q$ as before. If $n = 1$, then (3.10) is obvious. Assume that (3.10) holds for $n > 1$. Then,

$$\begin{aligned}
x^{*(n+1)} &= x^{*n} * x = (x^{*n}p)x - (xq)x^{*n} \\
&= (((xr)^{n-1}x)p)x - (xq)((xr)^{n-1}x) \\
&= (((xr)^{n-1}x)p)x - (((xr)^{n-1}x)q)x, \quad \text{by (3.7)} \\
&= (((xr)^{n-1}x)r)x = (xr)^n x,
\end{aligned}$$

using Artin's theorem. This verifies (3.10).

For $m, n > 0$, we use (3.10) to compute

$$\begin{aligned}
(x^{*m}p)x^{*n} &= (((xr)^{m-1}x)p)((xr)^{n-1}x) \\
&= ((\cdots(([((xr)^{m-1}x)p]x)r)\cdots)r)x, \quad \text{by (3.9)} \\
&= ((\cdots(([(xp)((xr)^{m-1}x)])r)\cdots)r)x, \quad \text{by (3.6)} \\
&= ((\cdots((((xpx)r)x)r)\cdots)r)x, \quad \text{by (3.9)} \\
&= (xp)((xr)^{m+n-1}x), \quad \text{by (3.9)} \\
&= (((xr)^{m+n-1}x)p)x, \quad \text{by (3.6)} \\
&= (x^{*(m+n-1)}p)x, \quad \text{by (3.10)}.
\end{aligned}$$

Similarly, we obtain $(x^{*m}q)x^{*n} = (x^{*(m+n-1)}q)x$. Therefore, for any $m, n > 0$,

$$\begin{aligned}
x^{*m} * x^{*n} &= (x^{*m}p)x^{*n} - (x^{*n}q)x^{*m} \\
&= (x^{*(m+n-1)}p)x - (x^{*(m+n-1)}q)x \\
&= (x^{*(m+n-1)}r)x \\
&= (((xr)^{m+n-2}x)r)x, \quad \text{by (3.10)} \\
&= (xr)^{m+n-1}x = x^{*(m+n)}.
\end{aligned}$$

This shows that $A(p, q)$ is power–associative. □

Using Theorems 3.1 and 3.3, we extend Corollary II.2.3 to mutations of an alternative algebra.

Theorem 3.4. *Let A and p, q be the same as in Theorem 3.3. Consider the following conditions:*

(i) *$p(xq) = q(xp)$ for all $x \in A$,*

(ii) *$A(p, q)$ is flexible,*

(iii) *$A(p, q)$ is a noncommutative Jordan algebra,*

(iv) *$A(p, q)$ is third power–associative,*

(v) *$A(p, q)$ is power–associative.*

Then, we have the implications: (i)⇒(ii)⇔(iii)⇒(iv)⇔(v). Moreover, if A contains an element which is not a zero divisor in A then the conditions (i)–(v) are all equivalent.

Proof. As for the associative case (Corollary II.2.3), by Theorem 3.1(ii) condition (i) implies (ii), while the equivalence of (ii) and (iii) follows from Proposition I.5.1, since $A(p, q)$ is Jordan–admissible by (1.6). The equivalence of (iv) and (v) is a consequence of Theorem 3.3. For the last part, it suffices to show (iv)⇒(i). Let

z be an element of A that is not a zero divisor of A. Assume that $A(p,q)$ is third power–associative. By Theorem 3.1(i) A satisfies the identity $x(p(xq)-q(xp))x = x[(p(xq)-q(xp))x] = 0$. Linearizing the latter with $x+\lambda z$ ($\lambda \in F$), the coefficient of λ^2 gives $x([p(zq)]z)+z([p(xq)]z)+z([p(zq)]x) = x([q(zp)]z)+z([q(xp)]z)+z([q(zp)]x)$ for all $x \in A$. Since z is not a zero divisor in A, from $z(p(zq)-q(zp))z = 0$ we get $p(zq)-q(zp) = 0$. It follows from this and the relation above that $p(xq) = q(xp)$ for all $x \in A$. □

In particular, if A has a unit element 1 then the conditions (i)—(v) above are equivalent. Conditions for $A(p,q)$ to have a unit element can be given exactly the same as for the associative case (Theorem II.2.8).

Theorem 3.5. *Let A be an alternative algebra over F and let p,q be fixed elements of A. Then, $A(p,q)$ has a unit element e if and only if A has a unit element 1, $p-q$ is invertible in A with inverse $e=(p-q)^{-1}$ and the equivalent conditions (i)-(v) in Theorem 3.4 hold for $A(p,q)$.*

Proof. Assume that $A(p,q)$ has a unit element e. Since e is a unit element of $A(p,q)^+ = A(p-q)^+$, it follows from Proposition 2.1(i),(iii) that A has a unit element 1 and $p-q$ is invertible in A with $e=(p-q)^{-1}$. Now, for any $x \in A$, we have

$$\begin{aligned}(p-q)x(p-q) &= e^{-1}(x*e)e^{-1}\\ &= e^{-1}((xp)e)e^{-1} - e^{-1}((eq)x)e^{-1}\\ &= (e^{-1}(xp)(ee^{-1}) - (e^{-1}(eq))(xe^{-1}), \quad \text{by (2.3)}\\ &= e^{-1}(xp) - q(xe^{-1}) = (p-q)(xp) - q(x(p-q)),\end{aligned}$$

which implies $(p-q)(xq) = q(x(p-q))$ and so $p(xq) = q(xp)$. Thus, condition (i) in Theorem 3.4 holds for $A(p,q)$.

Conversely, suppose that A contains a unit element 1, $p-q$ is invertible in A with $e=(p-q)^{-1}$, and $p(xq) = q(xp)$ holds for all $x \in A$. Then, for any $x \in A$,

$$\begin{aligned}e^{-1}(e*x)e^{-1} &= e^{-1}((ep)x - (xq)e)e^{-1}\\ &= (e^{-1}(ep))(xe^{-1}) - (e^{-1}(xq))(ee^{-1}), \quad \text{by (2.3)}\\ &= p(xe^{-1}) - e^{-1}(xq) = p(xp) - p(xq) - e^{-1}(xq)\\ &= p(xp) - q(xp) - e^{-1}(xq) = e^{-1}xe^{-1},\end{aligned}$$

which implies $e*x = x$ for all $x \in A$. We similarly argue that $x*e = x$ for all $x \in A$, showing that e is a unit element of $A(p,q)$. □

Extending Definition II.2.1 for the associative case, we put:

Definition 3.1. An algebra over F is called *generalized quasialternative* if it is isomorphic to a mutation algebra $A(p,q)$ of an alternative algebra A with $p, q \in Z(A)$ (cf. Section I.5). □

The following result generalizes Theorem II.2.7 for the associative case to mutations of alternative algebras.

Theorem 3.6. *Let A be an alternative algebra over a field F of characteristic $\neq 2, 3$ and let p, q be fixed elements in A. If there exists elements $\alpha, \beta \in Z(A)$ such that $\alpha p + \beta q$ is invertible in the nucleus $N(A)$ of A, then the conditions (i)—(v) in Theorem 3.4 are equivalent to each of the following:*

(vi) *$(\alpha p + \beta q)^{-1} p$ and $(\alpha p + \beta q)^{-1} q$ are contained in $Z(A)$.*

(vii) *$A(p,q)$ is isomorphic to $A(r,s)$ for some $r, s \in Z(A)$ with $\alpha r + \beta s = 1$.*

(viii) *$A(p,q)$ is a generalized quasialternative algebra.*

Proof. The proof is the same as for the associative case (Theorem II.2.7). Note first that $(\alpha p + \beta q)^{-1} \in N(A)$, by Proposition 2.1(ii). Thus, as for the associative case, we find that if one of the conditions (i)–(v) holds, then $r = (\alpha p + \beta q)^{-1} p$ and $s = (\alpha p + \beta q)^{-1} q$ are contained in the commutative center $K(A)$. But, since the characteristic of F is not 3, $K(A) = Z(A)$ [**ZSSS1**], and hence r and s are in $Z(A)$. The remainder of the proof is exactly the same as for the proof of Theorem II.2.7. □

4. Mutations of Cayley–Dickson algebras

Let A be an artinian alternative algebra over a field F of characteristic $\neq 2$ and p, q be fixed elements of A. By Proposition 1.2(ii) $J(A) = R(A)$, and Wedderburn–Artin–Zhevlakov theorem implies that $A/R(A)$ is a direct sum of simple artinian associative algebras and Cayley–Dickson algebras over some extension fields of F. As for the associative case (Section II.5), in Section III.6 it is shown that $A(p,q)/R(A)$ is isomorphic to the direct sum of mutations $A_i(p_i, q_i)$ of the simple summands A_i $(i = 1, \ldots, n)$ of $A/R(A)$, and that $J(p,q)$ is the solvable radical of $A(p,q)$ and $A(p,q)/J(p,q)$ is the direct sum of $A_i(p_i,q_i)/J_i(p_i,q_i)$ $(i = 1, \ldots, n)$ where $J_i(p_i, q_i)$ is the solvable radical of $A_i(p_i, q_i)$. Therefore, the structure of $A(p,q)$ that is not covered by the associative case arises from mutations of a Cayley–Dickson algebra C over a field. This section is devoted to the structure of $C(p,q)$.

Recall from Theorem 2.7 that $R^0(p,q)$ is an ideal of $C(p,q)$. The structure of $C(p,q)$ depends in large on $R^0(p,q)$ and the subspace $Fp + Fq$. We begin with:

Lemma 4.1. *Let C be a Cayley–Dickson algebra over F and p, q be fixed elements of C. Then, $R^0(p,q)$ is nilpotent such that $R^0(p,q)^{*4} = 0$.*

Proof. If $Fp + Fq$ is not totally isotropic, then by Lemma 2.6, $R^0(p,q) = 0$ and the result is obvious. Assume that $Fp+Fq$ is totally isotropic. It follows from Proposition 2.9 that $R^0(p,q)^{*3} \subseteq J = F\bar{p} + F\bar{q}$ and $R^0(p,q) * J = J * R^0(p,q) = 0$, hence $R^0(p,q)^{*3} * R^0(p,q) = R^0(p,q) * R^0(p,q)^{*3} = 0$. From the proof of Proposition 2.9(iii), we note that $2a * b - [a,b]^* \in J$ for $a, b \in R^0(p,q)$, and hence

$$4(a * b) * (c * d) = [a,b]^* * [c,d]^* \tag{4.1}$$

for all $a, b, c, d \in R^0(p,q)$. But,

$$\begin{aligned} ((a(p+q))b)p &= -((a(p+q))\bar{p})\bar{b} = (((\overline{p+q})\bar{a})\bar{p})\bar{b}, \text{ by (2.22), (2.21)} \\ &= (\overline{p(a(p+q))})\bar{b} = 0, \end{aligned}$$

and similarly, $((a(p+q))b)q = 0$. Expanding the right side of (4.1) and using (1.2) and (1.3), we find that the left side of (4.1) is a sum of elements of the form

$$(((a(p+q))b)(p \text{ or } q))\,(c(p+q)d)$$

for $a, b, c, d \in R^0(p,q)$, which must vanish. Thus, $R^0(p,q)^{*2} * R^0(p,q)^{*2} = 0$ and so $R^0(p,q)^{*4} = 0$. □

We note from Theorem II.4.2 that if A is a simple artinian associative algebra over F then $R^0(p,q)^{*3} = 0$. In contrast, for any Cayley–Dickson algebra C over F there exist elements $p, q \in C$ such that $R^0(p,q)^{*4} = 0 \neq R^0(p,q)^{*3}$ (Theorem 4.9). The following result is useful for our discussion.

Lemma 4.2. *Let C be a Cayley–Dickson algebra over F and p, q be fixed elements of C. If I is a nontrivial ideal of $C(p,q)$, then*

(i) $t(I(p-q)) = n(p-q) = 0$,

(ii) $(p-q)I(p-q) = 0$,

(iii) $(p+q)(x(p-q)) = (p-q)(x(p+q))$ *for all* $x \in I$,

(iv) $p(xq) = q(xp)$ *for all* $x \in I$.

Proof. (i) Let $r = p - q$. For any $x \in I$ and $y \in C$, we have $x * y + y * x \in I$, and by (1.4) and (2.11)

$$\begin{aligned} x * y + y * x &= (xr)y + (yr)x \\ &= t(xr)y + t(yr)x - 2(x,y)\bar{r}. \end{aligned}$$

Hence, $t(xr)y - 2(x,y)\bar{r} \in I$ for all $x \in I$ and $y \in C$. Suppose that $t(xr) = \alpha \neq 0$ for some $x \in I$, so that $\alpha y - 2(x,y)\bar{r} \in I$ for all $y \in C$. It follows from this that I

contains $(Fx)^\perp = \{y \in C : (x,y) = 0\}$. Since $\dim(Fx)^\perp = 7$ and I is nontrivial, we have $I = (Fx)^\perp$. Thus, $(x,y) \neq 0$ for some $x, y \in I$, and consequently, $2(x,y)\bar{r} \in I$, or $\bar{r} \in I$. This implies that $\alpha y - 2(x,y)\bar{r} \in I$ for all $y \in C$, and so $I = C$, a contradiction. Therefore, $t(Ir) = 0$ and by (2.10)

$$rxr = n(r)\bar{x} = r\bar{r}\bar{x} \tag{4.2}$$

for all $x \in I$. If $n(r) \neq 0$, then r is invertible in C and by (4.2) $xr = \bar{r}\bar{x} = \overline{xr}$. Hence, $xr = \frac{1}{2}t(xr)1 = 0$ for all $x \in I$, and since r is invertible, we have $I = 0$, a contradiction. Thus, $n(r) = 0$.

(ii) Since $n(r) = n(p-q) = 0$, this follows from (3.2).

(iii) Let $s = p + q$. For $x \in I$ and $y \in C$, by (1.3) we have $x * y - y * x = (xs)y - (ys)x \in I$. Thus, by part (i)

$$t\left(((xs)y - (ys)x)r\right) = 0 \tag{4.3}$$

for all $x \in I$. But, by (2.20) $t(((xs)y)r) = t(((r(xs))y) = t(y(r(xs)))$, and hence form (4.3) we have

$$t(y(s(xr) - r(xs))) = 0$$

for all $y \in C$. The result now follows from the nondegeneracy of the trace (see (2.9)).

(iv) This follows from expanding the relation in (iii). □

Corollary 4.3. *Let C be a Cayley–Dickson algebra over F and p, q be fixed elements of C with $p - q$ invertible. Then, $C(p,q)$ is simple.* □

We show that the simplicity of $C(p,q)$ also follows if $p+q$ is invertible and $p \neq q$.

Lemma 4.4. *Let C be a Cayley–Dickson algebra over F. If p and q are fixed elements of C such that $p+q$ is invertible in C and $p \neq q$, then $C(p,q)$ is simple.*

Proof. Let $r = p - q$ and $s = p + q$. Since s is invertible, by Propositions 2.2 and 2.1 $C(s)$ is a Cayley–Dickson algebra over F isomorphic to C and has the unit element s^{-1}. The bilinear form associated with the norm n_s in $C(s)$ is given by (xs, ys), and the trace t_s in $C(s)$ has the form $t_s(x) = (s^{-1}s, xs) = (1, xs) = t(xs)$ for $x \in C(s)$.

Suppose that $C(p,q)$ has a nontrivial ideal I. Then, I is an ideal of $C(p,q)^- = C(s)^-$, and $C(s)^- = Z(C(s)) \oplus [C(s), C(s)]$ (see Section I.4), where $Z(C(s)) = Fs^{-1}$. From Theorem I.4.17 we find that $Z(C(s))$ and $[C(s), C(s)] = \{x \in C : t_s(x) = t(xs) = 0\}$ are the only nontrivial ideals of $C(s)^-$. Consequently, either $I = Fs^{-1}$ or $I = [C(s), C(s)]$. If $I = Fs^{-1}$, then $s^{-1} * s^{-1} = s^{-1}rs^{-1} \in I = Fs^{-1}$ and $s^{-1}r \in F1$, hence $r \in Fs$. This implies that r is invertible in C, contrary to Corollary 4.3. For the latter case, from Lemma 4.2(i) we have $I = \{x \in C : t(xs) = t(xr) = 0\}$. Since $\dim I = 7$ and the bilinear form $\langle x, y\rangle \equiv t(xy)$ is nondegenerate, r and s must be linearly dependent. Hence, $p - q = r \in Fs$, and since $p \neq q$, $p - q$ is invertible in C, a contradiction. Therefore, $C(p, q)$ is simple. □

We use the results above to prove the following characterization of the simplicity of $C(p, q)$.

Theorem 4.5. *Let C be a Cayley–Dickson algebra over F and p, q be fixed elements of C. The following are equivalent:*

(i) *$C(p, q)$ is simple,*

(ii) *$p \neq q$ and $R^0(p, q) = 0$,*

(iii) *$p \neq q$ and $Fp + Fq$ is not totally isotropic.*

Proof. (i)⇒(ii). If $p = q = 0$ then $C(p, p)$ is a zero algebra, i.e., $C(p, p) * C(p, p) = 0$. If $p = q \neq 0$, then $C(p, p) = C(p)^-$ and $\bar{p} * x = (\bar{p}p)x - (xp)\bar{p} = n(p)x - n(p)x = 0$ for all $x \in C$, so $F\bar{p}$ is a nonzero ideal of $C(p, p)$. Thus, if $C(p, q)$ is simple, then $p \neq q$. Since $R^0(p, q)$ is an ideal of $C(p, q)$ by Theorem 2.7, it suffices to show $R^0(p, q) \neq C$. Note first that

$$R^0(p, q) \subseteq \{x \in C : (p - q)x(p - q) = 0\}. \tag{4.4}$$

If $(p-q)$ is invertible in C, then the right side of (4.4) is zero. Otherwise, $n(p-q) = 0$, and by (2.10) the right side of (4.4) is contained in $\{x \in C : t(x(p - q)) = 0\}$, being of dimension 7. Hence, $R^0(p, q) \neq C$.

(ii)⇒(iii). If $Fp+Fq$ were totally isotropic, then by Proposition 2.9(i) we would have $0 \neq F\bar{p} + F\bar{q} \subseteq R^0(p, q)$.

(iii)⇒(i). As before, let $r = p - q$ and $s = p + q$. In view of Corollary 4.3 and Lemma 4.4, it suffices to treat the case $n(r) = n(s) = 0$, so that $(r, s) \neq 0$ since $Fp + Fq$ is not totally isotropic. Suppose $C(p, q)$ has a nontrivial ideal I. We assert that

$$Ir = 0 \quad \text{and} \quad I = F\bar{r} \tag{4.5}$$

hold. Let y be any element of I. From Lemma 4.2(iii) $r(ys) = s(yr)$, and hence $\bar{r}(s(yr)) = 0$ since $\bar{r}r = n(r) = 0$. But, by Lemma 4.2(i) $t(yr) = 0$, hence $yr = -\bar{r}\bar{y}$ and, using (2.1), (2.10) and (2.9) we obtain

$$0 = \bar{r}(s(\bar{r}\bar{y})) = (\bar{r}s\bar{r})\bar{y} = t(s\bar{r})\bar{r}\bar{y} = -2(s, r)yr.$$

Since $(s, r) \neq 0$, this gives $Ir = 0$. Hence, by Lemma 4.2(iii) $r(Is) = 0$ and using (2.11) we have

$$0 = r(ys) + s(yr) = t(ys)r + t(yr)s - 2(s, r)\bar{y} = t(ys)r - 2(s, r)\bar{y}$$

since $t(yr) = 0$. But, since $(s, r) \neq 0$, $y \in F\bar{r}$ and so $I = F\bar{r}$, as asserted.

For any $x \in C$, by (4.5) $x * \bar{r} - \bar{r} * x = (xs)\bar{r} - (\bar{r}s)x \in I$, and since $n(r) = 0$, by (4.5) again we get

$$0 = ((xs)\bar{r} - (\bar{r}s)x)r = -((\bar{r}s)x)r. \tag{4.6}$$

Since $\dim I = 1$, there exists an element $a \in C$ such that $(a, I) = 0$ but $n(a) \neq 0$. It follows from (4.6) and (2.13) that

$$(\bar{r}s\bar{r})\bar{a} = (\bar{r}s\bar{r})\bar{a} + ((\bar{r}s)a)r = 2(a, \bar{r})\bar{r}s = 0,$$

which implies $\bar{r}s\bar{r} = 0$. Thus, by (2.10) and (2.9) $0 = \bar{r}s\bar{r} = t(\bar{r}s)\bar{r} - n(\bar{r})s = 2(r, s)\bar{r}$, and since $(r, s) \neq 0$ and $r \neq 0$, this is impossible. Therefore, (iii) implies (i). □

Theorem 4.6. *Let C be a Cayley–Dickson algebra over F and p, q be fixed elements of C with $p \neq q$. Then, $R^0(p, q)$ is the unique maximal ideal of $C(p, q)$ such that $R^0(p, q) \subseteq \{x \in C : t(xp) = t(xq) = 0\}$.*

Proof. If $Fp + Fq$ is not totally isotropic then by Theorem 4.5 $C(p, q)$ is simple and $R^0(p, q) = 0$. Thus, we assume that $Fp + Fq$ is totally isotropic. Let I be a nontrivial ideal of $C(p, q)$. By Lemma 4.2(i), for any $x \in I$, $t(xp) = t(xq)$, and hence

$$pxp = t(xp)p, \tag{4.7}$$

$$2p(xq) = 2q(xp) = p(xq) + q(xp) = t(xp)q + t(xq)p = t(xp)(p + q), \tag{4.8}$$

$$qxq = t(xq)q = t(xp)q \tag{4.9}$$

where we have used (2.10) for (4.7) and (4.9), and Lemma 4.2(iv) and (2.11) for (4.8). Suppose there exists an element $x \in I$ with $t(xp) \neq 0$. Then, by (4.8) $0 = 2\bar{p}(p(xq)) = t(xp)(\bar{p}p + \bar{p}q) = t(xp)\bar{p}q$, and so $\bar{p}q = 0$. Hence, using (2.12) and (2.9) we obtain $p(xq) = p(xq) + \bar{x}(\bar{p}q) = 2(p, \bar{x})q = t(xp)q$, which by (4.8) implies $t(xp)q = p(xq) = \frac{1}{2}t(xp)(p + q)$. Therefore, $p = q$, a contradiction, and we have $t(Ip) = 0$. It now follows from (4.7)–(4.9) that $I \subseteq R^0(p, q)$ and $t(xp) = t(xq) = 0$ for all $x \in R^0(p, q)$. □

It follows from Theorems 4.5 and 4.6 that if $p \neq q$ then $R^0(p, q)$ is a nontrivial ideal of $C(p, q)$ if and only if $Fp + Fq$ is totally isotropic.

Corollary 4.7. *Let C and $p, q \in C$ be the same as in Theorem 4.6. Then, $R^0(p, q)$ is a nontrivial ideal of $C(p, q)$ if and only if $Fp + Fq$ is totally isotropic.* □

The structure of $C(p, q)$ and $R^0(p, q)$ for $p \neq q$ is quite different from the case $p = q$. In the latter, $C(p, p) = C(p)^-$ is a Malcev algebra (Theorem I.4.9), and the structure of $C(p,p)$ is easily described (Theorem 4.16). It turns out that $C(p,p)$ for $p \neq 0$ is isomorphic either to C^- or to a solvable Malcev algebra.

For the case $p \neq q$, the structure of $C(p,q)$ is more subtle. If $C(p,q)$ is not simple (so $R^0(p,q) \neq 0$), then below we determine completely the structure of $C(p,q)/R^0(p,q)$ as well as of $R^0(p,q)$. In contrast, it is an open problem to describe the isomorphism classes of simple mutations $C(p,q)$ in terms of the structure of C. In view of Theorem 4.5, $C(p,q)$ is simple for any division Cayley–Dickson algebra C. In fact, Theorem II.4.11 can be extended to show that $C(p,q)$ is a division algebra if and only if C is a division algebra and $n(p) \neq n(q)$ (Theorem 5.8). In Chapter IV, $\operatorname{Aut} C(p,q)$ and $\operatorname{Der} C(p,q)$ are determined for $C(p,q)$ with $p+q$ invertible (so $C(p,q)$ is simple). It turns out that if $C(p,q)$ is a division algebra, then $\operatorname{Der} C(p,q) \cong$ type G_2 (non–split) or $su(3,K)$ for some quadratic extension field K of F (Proposition IV.6.10). In particular, if $C(p,q)$ is a real division algebra (so C is the Cayley algebra), then $\operatorname{Der} C(p,q) \cong$ compact G_2 or $su(3)$ (Corollary IV.6.11). In the former, $C(p,q)$ is completely determined whereas the latter remains as an open problem.

In the remainder of this section we focus on the structure of $C(p,q)$ and $R^0(p,q)$ when $C(p,q)$ is not simple or equivalently $R^0(p,q) \neq 0$. We first treat the case $p \neq q$. Specifically, we determine the structure of $C(p,q)$ with respect to $R^0(p,q)$, that is, the structure of $C(p,q)/R^0(p,q)$ and $R^0(p,q)$. Since $Fp + Fq$ is totally isotropic in this case, by Theorem I.3.6 C is the split Cayley–Dickson algebra $C(F)$, and we are in one of the following three situations.

Lemma 4.8. *Let C be a Cayley–Dickson algebra over F and let p, q be fixed elements of C such that $p \neq 0$ or $q \neq 0$ and $Fp + Fq$ is totally isotropic. Then, one of the following possibilities holds.*

(i) *p and q are linearly dependent and $R^0(p,q)$ is given by*

$$R^0(p,q) = \{x \in C : t(xp) = t(xq) = 0\}, \tag{4.10}$$

with $\dim R^0(p,q) = 7$.

(ii) *p and q are linearly independent with $\bar{p}q = 0$, and*

$$R^0(p,q) = \{x \in C : t(xp) = t(xq) = 0\}, \tag{4.11}$$

with $\dim R^0(p,q) = 6$.

(iii) *p and q are linearly independent with $\bar{p}q \neq 0$ and*

$$R^0(p,q) = \{x \in C : (\bar{p}q)x = 0\} = (\bar{p}q)C, \tag{4.12}$$

with $\dim R^0(p,q) = 4$.

Proof. (i) If $p \neq 0$, then $q \in Fp$ and $R^0(p,q) = \{x \in C : pxp = 0\}$. Since $n(p) = 0$, by (2.10) $pxp = 0$ if and only if $t(xp) = t(xq) = 0$. This gives (4.10) with $\dim R^0(p,q) = 7$.

(ii) By Theorem 4.6 $R^0(p,q)$ is contained in the right side of (4.11). If $t(xp) = t(xq) = 0$, then by (2.12) $p(xq) = -\bar{x}(\bar{p}q) = 0$ and $q(xp) = -\bar{x}(\bar{q}p) = -\bar{x}(\overline{\bar{p}q}) = 0$. Also, by (2.10) $pxp = t(xp)p = t(xq)q = qxq = 0$, since $n(p) = n(q) = 0$, hence $x \in R^0(p,q)$. This gives (4.11), and since p and q are linearly independent, $\dim R^0(p,q) = 6$.

(iii) For $x \in R^0(p,q)$, by Theorem 4.6 $t(xq) = 0$ and so by (2.13) $(\bar{p}q)x + (\bar{p}\bar{x})\bar{q} = t(xq)\bar{p} = 0$, which implies $(\bar{p}q)x = 0$ since $(\bar{p}\bar{x})\bar{q} = \overline{q(xp)} = 0$. Let $(\bar{p}q)x = 0$ for $x \in C$. Then, $0 = ((\bar{p}q)x)q = \bar{p}(qxq) = t(xq)\bar{p}q$ by (2.10), so $t(xq) = 0$, and since $t(\bar{p}q) = 2(p,q) = 0$, $\bar{p}q = -\bar{q}p$. Hence, $0 = ((\bar{q}p)x)p = \bar{q}(pxp) = t(xp)\bar{q}p$ and so $t(xp) = 0$. Consequently, we have $pxp = qxq = 0$. Also, by (2.12) $p(xq) = -\bar{x}(\bar{p}q) = \bar{x}(\bar{q}p) = \overline{(\bar{p}q)x} = 0$ and $q(xp) = -\bar{x}(\bar{q}p) = -\overline{(\bar{p}q)x} = 0$. Hence, $x \in R^0(p,q)$ and $R^0(p,q) = \{x \in C : (\bar{p}q)x = 0\} = \ker L_{\bar{p}q}$, where L_a $(a \in C)$ denotes the left multiplication in C by a.

Since $\bar{p}q = -\overline{\bar{p}q} = -\bar{q}p$, $(\bar{p}q)((\bar{p}q)C) = (\bar{p}q)^2C = -(\bar{p}q)(\overline{\bar{p}q})C = -n(\bar{p}q)C = -n(\bar{p})n(q)C = 0$, hence $(\bar{p}q)C \subseteq \ker L_{\bar{p}q} = R^0(p,q)$. If $x \in R^0(p,q)$, then $(\bar{p}q)x = 0$ and so x is a zero divisor. Thus, $R^0(p,q)$ is totally isotropic, so $\dim R^0(p,q) \leq 4$. On the other hand, from $R^0(p,q) = \ker L_{\bar{p}q}$, $(\bar{p}q)C = \operatorname{Im} L_{\bar{p}q}$ and $\dim\ker L_{\bar{p}q} + \dim \operatorname{Im} L_{\bar{p}q} = 8$, we have $\dim(\bar{p}q)C = \dim R^0(p,q) = 4$ and $(\bar{p}q)C = R^0(p,q)$. □

We remark in Lemma 4.8 that the condition $p \neq 0$ or $q \neq 0$ is superfluous for cases (ii) and (iii) while (4.10) holds for $p = q \neq 0$ also. We analyze the structure of $C(p,q)/R^0(p,q)$ and $R^0(p,q)$, according to the three cases in Lemma 4.8. We begin with case (i).

Theorem 4.9. *Let C be a Cayley–Dickson algebra over F and let p, q be fixed elements of C such that p and q are linearly dependent with $p \neq q$ and $Fp + Fq$ is totally isotropic.*

(i) *$C(p,q)/R^0(p,q)$ is isomorphic to F, and there exists a subalgebra S of $C(p,q)$ such that $S \cong F$ and $C(p,q) = R^0(p,q) \oplus S$.*

(ii) *If $p = -q$, then $R^0(p,q)^{*3} = 0$.*

(iii) *If $p \neq -q$, then $R^0(p,q)^{*3} \neq 0 = R^0(p,q)^{*4}$.*

Proof. (i) Recall from Proposition 2.9 that $J \equiv F\bar{p} + F\bar{q} \subseteq R^0(p,q)$. We find an element $a \in C$ with $t(a(p-q)) \neq 0$, and since p and q are linearly dependent, $t(ap) \neq 0$, so $a \notin R^0(p,q)$ by (4.10). By (2.10) we have $a * a = a(p-q)a = t(a(p-q))a - n(a)(\overline{p-q})$, and since $\overline{p-q} \in R^0(p,q)$, if we let $\hat{a} = a + R^0(p,q)$, then $\hat{a} * \hat{a} = \alpha\hat{a}$ with $0 \neq \alpha \in F$. From this and $\dim R^0(p,q) = 7$, it follows that $C(p,q)/R^0(p,q) \cong F$. Since C is the split Cayley–Dickson algebra over F, it contains a basis consisting of elements x with $n(x) = 0$ (see (2.18)). Thus, we can choose an $a \in C$ with $a \notin R^0(p,q)$ and $n(a) = 0$, but $a * a = t(a(p-q))a$ with $t(a(p-q)) \neq 0$, so $S \equiv Fa \cong F$ and $C(p,q) = R^0(p,q) \oplus S$.

(ii) If $p = -q$, then for $x, y \in R^0(p,q)$, by (2.11) and (4.10)

$$x * y = (xp)y + (yp)x = t(xp)y + t(yp)x - 2(x,y)\bar{p} = -2(x,y)\bar{p} \in J.$$

The result now follows from Proposition 2.9(ii).

(iii) Let $q = \alpha p$ with $-1 \neq \alpha \in F$ and let $x, y, z \in R^0(p,q)$. By (1.3) we obtain

$$[[x,y]^*, z]^* = (1+\alpha)^2 \left[(((xp)y - (yp)x)p)z - (zp)((xp)y - (yp)x)\right]. \tag{4.13}$$

Since $((xp)y)p = x(pyp) = 0 = y(pxp) = ((yp)x)p$ by (2.2), (4.13) becomes

$$\begin{aligned}
[[x,y]^*, z]^* &= -(1+\alpha)^2 (zp)((xp)y - (yp)x) \\
&= -(1+\alpha)^2 (\bar{p}\bar{z})((x\bar{y})\bar{p} - (y\bar{x})\bar{p}), \text{ by (2.22) and (4.10)} \\
&= -(1+\alpha)^2 \bar{p}\,(\bar{z}(x\bar{y}) - \bar{z}(y\bar{x}))\,\bar{p}, \text{ by (2.3)} \\
&= -(1+\alpha)^2 2\bar{p}(\bar{z}(x\bar{y}))\bar{p} + (1+\alpha)^2 t(x\bar{y})\bar{p}\bar{z}\bar{p}, \\
&= -2(1+\alpha)^2 t\,(((y\bar{x})z)p)\,\bar{p}, \text{ by (2.10)}
\end{aligned}$$

since $\bar{p}\bar{z}\bar{p} = 0$ and $t(x\bar{y}) = x\bar{y} + y\bar{x}$. Thus, to show $R^0(p,q)^{*3} \neq 0$, by (4.10) it suffices to find elements $x, y, z \in R^0(p,q)$ with $(y\bar{x})z \notin R^0(p,q)$.

Since $\dim R^0(p,q) = 7$, there is an element $y \in R^0(p,q)$ with $n(y) \neq 0$. If there is an $x \in R^0(p,q)$ with $\bar{x} \notin R^0(p,q)$, then the relation $(y\bar{x})y = t(\bar{x}y)y - n(y)\bar{x}$ (by (2.10)) implies $(y\bar{x})y \notin R^0(p,q)$, as desired. Otherwise, we have $\bar{x} \in R^0(p,q)$ for all $x \in R^0(p,q)$, and so $x + \bar{x} = t(x)1 \in R^0(p,q)$ for any $x \in R^0(p,q)$. Since $1 \notin Fp + Fq$ and $\dim R^0(p,q) = 7$, there is an $x \in R^0(p,q)$ with $t(x) \neq 0$, so $1 \in R^0(p,q)$. Thus, letting $x = 1$, it suffices to choose $y, z \in R^0(p,q)$ with $yz \notin R^0(p,q)$. But this is possible since the maximum dimension of nontrivial subalgebras of C is 6 [**E2**,**R1**]. Hence, $R^0(p,q)^{*3} \neq 0$, but by Lemma 4.1 $R^0(p,q)^{*4} = 0$. □

The following result determines the structure of $C(p,q)/R^0(p,q)$ and $R^0(p,q)$ for case (ii) of Lemma 4.8.

Theorem 4.10. *Let C be a Cayley–Dickson algebra over F and let p, q be fixed elements of C such that p and q are linearly independent with $\bar{p}q = 0$ and $Fp + Fq$ is totally isotropic. Then,*

(i) *$C(p,q)/R^0(p,q)$ is isomorphic to the twisted mutation algebra*

$$M_{1\times 2}(F)\left(\begin{pmatrix} 0 \\ -1 \end{pmatrix}, \begin{pmatrix} -1 \\ 0 \end{pmatrix}\right)$$

with multiplication given by

$$u * u = u, \ u * v = 0, \ v * u = -u + v, \ v * v = -v, \tag{4.14}$$

where $u = (1,0)$ *and* $v = (0,1)$. *Moreover,* $C(p,q)$ *contains a subalgebra* T *such that* $C(p,q) = R^0(p,q) \oplus T$ *and* $T \cong C(p,q)/R^0(p,q)$.

(ii) $R^0(p,q)^{*3} = 0 \neq R^0(p,q)^{*2}$.

Proof. (i) By the nondegeneracy of $\langle x, y\rangle \equiv t(xy)$, we find elements $a, b \in C$ such that

$$t(aq) = t(bp) = 0, \ t(ap) = t(bq) = 1. \tag{4.15}$$

Let $\hat{a} = a + R^0(p,q)$ and $\hat{b} = b + R^0(p,q)$. Then, $\{\hat{a}, \hat{b}\}$ is a basis of $C(p,q)/R^0(p,q)$, since $\dim R^0(p,q) = 6$ by (4.11) and since if $\hat{a} = \alpha\hat{b}$ for $\alpha \in F$ then $a - \alpha b \in R^0(p,q)$, so $t(ap - \alpha bp) = 0$, a contradiction by (4.15). Hence, for any $x, y \in C$ we have $\hat{x} = t(xp)\hat{a} + t(xq)\hat{b}$ by (4.15) and

$$\hat{x} * \hat{y} = \widehat{x * y} = t((x * y)p)\hat{a} + t((x * y)q)\hat{b}. \tag{4.16}$$

To expand $t((x * y)p)$ and $t((x * y)q)$, we compute

$$t(((xp)y)p) = t(x(pyp)) = t(xp)t(yp),$$
$$t(((xp)y)q) = t((q(xp))y) = -t(\bar{x}(\bar{q}p)y) + t(qx)t(py) = t(xq)t(yp),$$

where we have used (2.2), (2.10) for the first relation and (2.12) with $\bar{q}p = 0$ for the second. Similarly, we get $t(((yq)x)p) = t(xq)t(yp)$ and $t(((yq)x)q) = t(xq)t(yq)$. Thus,

$$t((x * y)p) = t(x(p - q))t(yp), \ t((x * y)q) = t(xq)t(y(p - q))$$

and (4.16) becomes

$$\hat{x} * \hat{y} = t(x(p - q))t(yp)\hat{a} + t(xq)t(y(p - q))\hat{b}.$$

It follows from this and (4.15) that $\{\hat{a}, \hat{b}\}$ has the same multiplication as (4.14).

For the second part, it suffices to show that there is a subalgebra T of $C(p,q)$ such that $C(p,q) = R^0(p,q) \oplus T$. If we let $S = Fp + Fq$, then $\bar{S}S = 0$ since $\bar{p}q = 0$. Hence, the proof of Proposition 2.5 applies for S, since C is isomorphic to the split Cayley–Dickson algebra $C(F)$. Let $\{e_1, e_2, u_i, v_i : i = 1, 2, 3\}$ be a standard basis of C given by (2.15). If S contains an idempotent $\neq 0, 1$, then from the proof of Proposition 2.5 we can let $S = Fe_1 + Fu_1$. Otherwise, we can choose a standard basis of C such that $S = Fu_1 + Fv_2$. In the first case, $R^0(p,q) = \{x \in C : t(xS) = (x, \bar{S}) = 0\}$ is the subspace spanned by u_i $(i = 1, 2, 3)$, v_2, v_3 and e_2 (see (2.18)), and we let $T = Fe_1 + Fv_1$. If $S = Fu_1 + Fv_2$, then $R^0(p,q)$ is spanned by $e_1, e_2, u_1, u_3, v_2, v_3$ and we choose $T = Fu_2 + Fv_1$. It is easy to see from (2.15) that $(TS)T \subseteq T$ and hence $T * T \subseteq T$. Therefore, T is a subalgebra of $C(p,q)$ such that $C(p,q) = R^0(p,q) \oplus T$.

(ii) Since $\bar{S}S = 0$ as above, by Proposition 2.5 $\bar{S} = l(S)$. But, from Proposition 2.9(i) we have $l(S) = \bar{S} = F\bar{p} + F\bar{q} = J \subseteq R^0(p,q)$. Let x, y be any elements of $R^0(p,q)$. From the proof of Proposition 2.9(iii) $2x * y - [x,y]^* \in J = l(S)$, and $[x,y]^* = (x(p+q))y - (y(p+q))x$. To show $[x,y]^* \in l(S) = J$, we compute

$$((x(p+q))y)p = x(pyp) + ((xq)y)p = -((xq)\bar{p})\bar{y} = ((\bar{q}\bar{x})\bar{p})\bar{y} = (\overline{p(xq)})\bar{y} = 0,$$

using (2.13) and (4.11). Similarly, we obtain $((x(p+q)y)q = ((y(p+q))x)p = ((y(p+q)x)q = 0$. Therefore, $[x,y]^* \in J$, and so $R^0(p,q)^{*2} \subseteq J$ and by Proposition 2.9(ii) $R^{*3}(p,q) = 0$. Since $\dim R^0(p,q) = 6$, there is an $x \in R^0(p,q)$ with $n(x) \neq 0$. But, $x * x = x(p-q)x = t(x(p-q))x - n(x)(\overline{p-q}) = -n(x)(\overline{p-q}) \neq 0$, hence $R^0(p,q)^{*2} \neq 0$. □

For case (iii) of Lemma 4.8, it turns out that $C(p,q)/R^0(p,q)$ is a Malcev–admissible algebra such that $(C(p,q)/R^0(p,q))^-$ is isomorphic to the unique non–Lie, Malcev algebra of dimension 4. For completeness, we here include a proof that determines this unique algebra.

Recall that an anticommutative algebra M over F with multiplication $[x,y]$ is called a Malcev algebra if it satisfies the Malcev identity (I.4.12) which is equivalent to (I.4.13):

$$J(x,y,[x,z]) = [J(x,y,z),x] \tag{4.17}$$

where J is the Jacobian $J(x,y,z) = [[x,y],z] + [[y,z],x] + [[z,x],y]$ (Lemma I.4.6). We note from Theorem I.4.11 that the J–nucleus N of M is an ideal of M (see (I.4.21)). It is well known that the linear span $J(M,M,M)$ of all $J(x,y,z)$ is an ideal of M and any two elements of M generate a Lie subalgebra of M [**S1**, p.429]. We first show [**E1**]:

Lemma 4.11. *If B is a one–dimensional ideal of a Malcev algebra M over F, then $[B, M^2] = J(M,M,B) = 0$, and in particular, $B \subseteq N(M)$, where $M^2 = [M,M]$.*

Proof. If we let $B = Fb$, then for any $x \in M$, $[b,x] = \lambda_x b$ $(\lambda_x \in F)$, so the map $\lambda : M \to F : x \mapsto \lambda_x$ is a linear functional. Using the identity $m(x,y,z,t) = 0$ given by (I.4.14), for $x,y,z \in M$ we compute $m(b,y,x,z) = 0$ as

$$\lambda_x\lambda_{[y,z]} - \lambda_{[[x,y],z]} + \lambda_{[x,z]}\lambda_y = 0. \tag{4.18}$$

Interchanging x, y in (4.18) we obtain $\lambda_{[[x,y],z]} = 0$ and hence (4.18) reduces to

$$\lambda_y\lambda_{[x,z]} + \lambda_x\lambda_{[y,z]} = 0. \tag{4.19}$$

If $\lambda = 0$ then clearly $MB = J(M, M, B) = 0$. Thus, we assume $\lambda \neq 0$ or $\lambda_x \neq 0$ for some $x \in M$. Putting $y = x$ in (4.18) we have $\lambda_{[x,z]} = 0$ for any $z \in M$ and by (4.19) $\lambda_{M^2} = 0$ or $[B, M^2] = 0$. It follows from this that $J(b, x, y) = (\lambda_x\lambda_y - \lambda_{[x,y]} - \lambda_x\lambda_y)b = 0$, so $J(B, M, M) = J(M, M, B) = 0$. □

If M is a Malcev algebra of dimension ≤ 3, then $\dim J(M, M, M) \leq 1$ since the Jacobian is a skew–symmetric function on its arguments. But, if $\dim J(M, M, M) = 1$, then $J(M, M, M) \subseteq N$ by Lemma 4.11, a contradiction, since $\dim M \leq 3$. Therefore, we have:

Corollary 4.12. *Any Malcev algebra over F (of characteristic $\neq 2$) of dimension ≤ 3 is a Lie algebra.* □

Proposition 4.13. (i) *There is, up to isomorphism, a unique non–Lie, Malcev algebra of dimension 4 over a field F of characteristic $\neq 2, 3$ whose multiplication is given by*

(4.20)

	u	v	w	t
u	0	$-v$	$-w$	t
v	v	0	t	0
w	w	$-t$	0	0
t	$-t$	0	0	0

(ii) *There do not exist non–Lie, Malcev algebras of dimension 4 over a field of characteristic 3.*

Proof. (i) Let F be a field of characteristic $\neq 2$ and let M be a non–Lie, Malcev algebra over F of dimension 4 with basis $\{a, b, c, d\}$. Suppose $\dim J(M, M, M) \geq 2$. Since any two elements of M generate a Lie subalgebra, we may assume that $J(a, b, c)$ and $J(a, b, d)$ are linearly independent, and $J(a, b, [a, b]) = 0$. From this, we find that $[a, b]$ is a linear combination of a and b; in fact, if $[a, b] = \alpha a + \beta b + \gamma c + \delta d$, then $0 = J(a, b, [a, b]) = \gamma J(a, b, c) + \delta J(a, b, d)$ and so $[a, b] = \alpha a + \beta b$. It follows from this and (4.17) that

$$J(a, b, [a, c]) = [J(a, b, c), a] = J(a, [a, b], c) = \beta J(a, b, c).$$

This implies $[a, c] \in Fa + Fb + Fc \equiv L$, and similarly, $[b, c] \in L$. Therefore, L is a subalgebra of M which is by Corollary 4.12 a Lie algebra, a contradiction, since $J(a, b, c) \neq 0$. Thus, it must be that $\dim J(M, M, M) = 1$. In view of Lemma 4.11, $J(M, M, M) \subseteq N = N(M)$ and $[J(M, M, M), M^2] = 0$. Since $\dim M = 4$, it is easily seen that if $\dim N \geq 2$, then M is a Lie algebra. Hence, $N = J(M,M,M)$ is a one–dimensional ideal of M.

Let $\{u,v,w,t\}$ be a basis of M with $N = Ft$, so $N = F(J(u,v,w))$. If $[N,M] = 0$, then by (4.17)

$$J(u,v,[u,v]) = 0 = [J(u,w,v),u] = J(u,w,[u,v]),$$
$$J(v,w,[u,v]) = [v,J(v,w,u)] = 0.$$

Hence, $[u,v] \in N$ and similarly, $[u,w]$ and $[v,w]$ are in N. This implies $M^2 \subseteq N$ and $[M^2,M] = 0$, forcing M to be a Lie algebra. Therefore, if we let $B = \{x \in M : [N,x] = 0\}$, then $M^2 \subseteq B$ and $\dim B = 3$. It is now possible to choose a basis $\{u,v,w,t\}$ of M with $N = Ft$ and $B = Fv + Fw + Ft$, so that

$$[v,t] = [w,t] = 0, \ \ [u,t] = t. \tag{4.21}$$

If $[v,w] = 0$, then $B^2 = 0$ by (4.21) and $J(u,v,w) = [[u,v],w] + [[w,u],v] \in B^2 = 0$ since $M^2 \subseteq B$. Thus, $[v,w] \neq 0$, and we can choose v and w so that $[v,w] = t$, since

$$J(u,v,[v,w]) = [J(u,v,w),v] = 0 = [w,J(u,v,w)] = J(u,w,[v,w]).$$

Since $J(u,[u,v],w) = [J(u,v,w),u] = -J(u,v,w) \neq 0$ and $J(u,[u,v],v) = 0$, we obtain $[u,v] = -v + \gamma t$ for some $\gamma \in F$ and so by (4.21) $[u, v - \frac{\gamma}{2}t] = -(v - \frac{\gamma}{2}t)$. Similarly, from $J(u,[u,w],v) = -[J(u,v,w),u] = J(u,v,w)$ and $J(w,u,[u,w]) = 0$ we get $[u,w] = -w + \delta t$ for $\delta \in F$. Replacing $v - \frac{\gamma}{2}t$ and $w - \frac{\delta}{2}t$ by v and w, respectively, we arrive at the multiplication given by (4.20). It is straightforward to see that Table (4.20) gives a Malcev algebra such that $J(M,M,M) = F(J(u,v,w))$ and $J(u,v,w) = -3t$, which together with the argument above proves part (ii). □

Returning to case (iii) of Lemma 4.8, we now prove:

Theorem 4.14. *Let C be a Cayley–Dickson algebra over F and let p,q be fixed elements of C such that p and q are linearly independent with $\bar{p}q \neq 0$ and $Fp + Fq$ is totally isotropic. Then,*

(i) *$C(p,q)/R^0(p,q)$ is isomorphic to a simple algebra of dimension 4 over F with multiplication given by*

(4.22)

	a_1	a_2	a_3	a_4
a_1	a_1	$a_1 + a_2$	0	0
a_2	0	a_2	a_3	a_4
a_3	a_3	0	0	$-(a_1 + a_2)$
a_4	a_4	0	$a_1 + a_2$	0

Moreover, if the characteristic of F is not 3, then $(C(p,q)/R^0(p,q))^-$ is isomorphic to the unique Malcev algebra given by (4.20).

(ii) $R^0(p,q)^{*2} = 0$.

Proof. (i) Since $(\bar{p}q)\bar{p} = t(\bar{p}q)\bar{p} = 0 = (\bar{p}q)\bar{q} = n(q)\bar{p}$, by (4.12) $\bar{p}, \bar{q} \in R^0(p,q) = (\bar{p}q)C$. Hence, there exist $z_1, z_2 \in C$ such that

$$(\bar{p}q)z_1 = \bar{p}, \ (\bar{p}q)z_2 = -\bar{q}. \tag{4.23}$$

Using this and (2.13) we obtain $0 = ((\bar{p}q)z_1)p = t(z_1p)\bar{p}q$, so $t(z_1p) = 0$. Similarly, from $((\bar{p}q)z_1)q = t(z_1q)\bar{p}q = \bar{p}q$, $((\bar{p}q)z_2)p = -\bar{q}p = t(z_2p)\bar{p}q$ and $((\bar{p}q)z_2)q = t(z_2q)\bar{p}q = -\bar{q}q = 0$, one has

$$t(z_1q) = t(z_2p) = 1, \ t(z_1p) = t(z_2q) = 0. \tag{4.24}$$

If we let $V = \{x \in C : t(xp) = t(xq) = (x,\bar{p}) = (x,\bar{q}) = 0\}$, then $(\bar{p}q)C = (\bar{p}q)V$ and hence there is a basis $\{u_i = (\bar{p}q)z_i : i = 1,2,3,4\}$ of $R^0(p,q)$ with $t(z_ip) = t(z_iq) = 0$, $i = 3,4$. From this, (4.23) and (4.24), we obtain

$$\begin{aligned} u_ip &= t(z_ip)\bar{p}q = \delta_{i2}\bar{p}q, \\ u_iq &= t(z_iq)\bar{p}q = \delta_{i1}\bar{p}q. \end{aligned} \tag{4.25}$$

Since $R^0(p,q) = \ker L_{\bar{p}q}$ is totally isotropic and the Witt index of (,) is 4, there is a totally isotropic subspace $W = Fx_1 + Fx_2 + Fx_3 + Fx_4$ such that $C = W \oplus R^0(p,q)$ and $t(x_i\bar{u}_j) = 2(x_i,u_j) = \delta_{ij}$ (see [**J4**, p.368]). Since $R^0(p,q)$ is totally isotropic, denoting $\hat{x} = x + R^0(p,q)$ we get from this

$$\hat{x}_i * \hat{x}_j = \sum_{k=1}^{4} t\left((x_i * x_j)\bar{u}_k\right)\hat{x}_k. \tag{4.26}$$

For further reduction of (4.26), we use (2.12) and (4.25) to compute

$$\begin{aligned} t(((x_ip)x_j)\bar{u}_k) &= t((\bar{u}_k(x_ip)x_j) = -t((\bar{x}_i(u_kp))x_j) + t(x_i\bar{u}_k)t(px_j) \\ &= -\delta_{k2}t(\bar{x}_i(\bar{p}q)x_j) + \delta_{ik}\delta_{ij}, \\ t((x_jq)x_i)\bar{u}_k) &= t(\bar{u}_k(x_jq)x_i) = -t((\bar{x}_j(u_kq))x_i) + t(x_j\bar{u}_k)t(qx_i) \\ &= -\delta_{k1}t(\bar{x}_j(\bar{p}q)x_i) - \delta_{jk}\delta_{2i}, \end{aligned}$$

since $p = \bar{u}_1$ and $q = -\bar{u}_2$. Let $\alpha_{ij} = t((\bar{x}_i(\bar{p}q))x_j)$. Since $t((\bar{x}_i(\bar{p}q))x_j) = t(\bar{x}_j((\overline{\bar{p}q})x_i)) = -t((\bar{x}_j(\bar{p}q))x_i)$, we get $\alpha_{ij} = -\alpha_{ji}$ and so $\alpha_{ii} = 0$ for all i. Therefore, (4.26) becomes

$$\hat{x}_i * \hat{x}_j = \sum_{k=1}^{4} \left[-(\delta_{1k} + \delta_{2k})\alpha_{ij} + \delta_{ik}\delta_{1j} + \delta_{jk}\delta_{2i}\right]\hat{x}_k. \tag{4.27}$$

Since $(\bar{p}q)x_j \in R^0(p,q)$, $(\bar{p}q)x_j = \sum_{h=1}^4 \lambda_{jh}u_h$, but

$$\alpha_{ij} = t((\bar{x}_i(\bar{p}q))x_j) = t(\bar{x}_i((\bar{p}q)x_j)) = t(\bar{x}_i \sum_{h=1}^{4} \lambda_{jh}u_h) = \lambda_{ji}.$$

From $((\bar{p}q)x_j)p = t(x_jp)\bar{p}q = \delta_{ij}\bar{p}q$ and $((\bar{p}q)x_j)p = \sum_{h=1}^4 \lambda_{jh}u_hp = \lambda_{j2}\bar{p}q$ by (4.2), we get $\delta_{1j} = \lambda_{j2}$, and similarly, $\lambda_{j1} = -\delta_{2j}$ by computing $((\bar{p}q)x_j)q$. Therefore,

$$\alpha_{ij} = \lambda_{ji},\ \lambda_{j2} = \delta_{1j},\ \lambda_{j1} = -\delta_{2j} \tag{4.28}$$

for $i,j = 1,2,3,4$. Letting $\gamma = \alpha_{34} = -\alpha_{43}$, from (4.28) we find the matrix (α_{ij}) to be

$$(\alpha_{ij}) = \begin{pmatrix} 0 & -1 & 0 & 0 \\ 1 & 0 & 0 & 0 \\ 0 & 0 & 0 & \gamma \\ 0 & 0 & -\gamma & 0 \end{pmatrix}.$$

From this and (4.27), it is straightforward to check that the multiplication in $C(p,q)/R^0(p,q)$ is given by

	$\hat{x}_1$	$\hat{x}_2$	$\hat{x}_3$	$\hat{x}_4$
$\hat{x}_1$	$\hat{x}_1$	$\hat{x}_1 + \hat{x}_2$	0	0
$\hat{x}_2$	0	$\hat{x}_2$	$\hat{x}_3$	$\hat{x}_4$
$\hat{x}_3$	$\hat{x}_3$	0	0	$-\gamma(\hat{x}_1 + \hat{x}_2)$
$\hat{x}_4$	$\hat{x}_4$	0	$\gamma(\hat{x}_1 + \hat{x}_2)$	0

If $\gamma = 0$, then $F\hat{x}_3$ is an ideal of $C(p,q)/R^0(p,q)$, but this is impossible since $R^0(p,q)$ is the unique maximal ideal of $C(p,q)$ by Theorem 4.6. Letting $a_i = \hat{x}_i$ $(i \neq 3)$ and $a_3 = \gamma^{-1}\hat{x}_3$, we obtain the multiplication given by (4.22). It follows from (4.22) that

$$[[a_1,a_3],a_4] + [[a_3,a_4],a_1] + [[a_4,a_1],a_3] = 6(a_1 + a_2).$$

Hence, if the characteristic of F is not 2, nor 3, then $(C(p,q)/R^0(p,q))^-$ is a non-Lie, Malcev algebra of dimension 4, which must be, by Proposition 4.13, isomorphic to the algebra given by (4.20).

(ii) For any $x \in C$, we have

$$((\bar{p}q)x)p = -((\bar{q}p)x)p = -\bar{q}(pxp) = -\bar{q}t(xp)p = t(xp)\bar{p}q \quad ,$$
$$((\bar{p}q)x)q = \bar{p}(qxq) = t(xq)\bar{p}q.$$

Hence, for any $x, y \in R^0(p,q) = (\bar{p}q)C$, we get $x = (\bar{p}q)a$, $y = (\bar{p}q)b$ for some $a, b \in C$, and $(xp)y = (((\bar{p}q)a)p)y = t(ap)(\bar{p}q)y = 0$ since $y \in \ker L_{\bar{p}q}$. Similarly, $(yq)x = 0$, and hence $x * y = 0$, i.e., $R^0(p,q)^{*2} = 0$. □

Unlike cases (i) and (ii) of Lemma 4.8, we do not know whether there exists a subalgebra of $C(p,q)$ that complements $R^0(p,q)$. It remains to treat the case $p = q$. The following is useful for this case.

Lemma 4.15. *Let C be a Cayley–Dickson algebra over F and let $0 \neq p$ be a fixed element of C with $n(p) = 0$. Then, $l(p) = \{x \in C : xp = 0\}$ is a 4–dimensional ideal of the p–homotope $C(p)$ and $l(p) = C\bar{p}$.*

Proof. Let $x \cdot y$ denote the product in $C(p)$. For any $x \in C$ and $y \in l(p)$, $(x \cdot y)p = ((xp)y)p = x(pyp) = 0$ and $(y \cdot x)p = ((yp)x)p = 0$, so $l(p)$ is an ideal of $C(p)$. Since $(x\bar{p})p = x(\bar{p}p) = n(p)x = 0$ for $x \in C$, $C\bar{p} \subseteq l(p)$. Also, $l(p)$ is totally isotropic since every element of $l(p)$ is a zero divisor, and hence $\dim C(p) \leq \dim l(p) \leq 4$. Assume $t(p) \neq 0$. If $x \in l(p)$, then $t(p)x = x(p + \bar{p}) = x\bar{p}$ and $x \in C\bar{p}$, hence $l(p) = C\bar{p}$. But,

$$8 \geq \dim l(p) + \dim l(\bar{p}) = \dim \operatorname{Im} R_{\bar{p}} + \dim \ker R_{\bar{p}} = 8,$$

showing $\dim l(p) = 4$. If $t(p) = 0$, then $\bar{p} = -p$ and so $C\bar{p} = \operatorname{Im} R_p$. Hence, in this case, $8 = \dim \operatorname{Im} R_p + \dim \ker R_p = \dim C\bar{p} + \dim l(p) \leq 4 + 4$, implying $l(p) = C\bar{p}$ and $\dim l(p) = 4$. □

The following result determines the structure of $C(p,p)$.

Theorem 4.16. *Let C be a Cayley–Dickson algebra over F and let p be a fixed element of C.*

(i) *If $p \neq 0$ and $n(p) \neq 0$, then $C(p,p)$ is a Malcev algebra isomorphic to C^-.*

(ii) *If $p \neq 0$ and $n(p) = 0$, then $C(p,p)$ is a non–nilpotent solvable Malcev algebra such that $C(p,p)^{*(3)} = 0$.*

(iii) *If $p = 0$, then $C(p,p)$ is a zero algebra.*

Proof. (i) Since $n(p) \neq 0$, p is invertible in C and by Proposition 2.2 $C(p)$ is isomorphic to C. Hence, $C(p,p) = C(p)^- = Z(C(p)) \oplus [C(p), C(p)]$ is isomorphic to C^-, where $[C(p), C(p)]$ is a simple Malcev algebra.

(ii) Note that $l(p)$ is an ideal of $C(p,p)$ contained in $R^0(p,p) = R^0(p,0)$ (Lemma 4.15). Since $\dim R^0(p,0) = 7$ and $\dim l(p) = 4$ by Lemma 4.15, we choose an $u \in R^0(p,p)$ with $u \notin l(p)$. Let $x \in C$ be such that $t(xp) = 1$. Then,

$$(x * u)p = [(xp)u - (up)x]\,p = x(pup) - u(pxp) = -u(t(xp)p) = -up$$

by (2.10). Thus, $x * u + u \in l(p)$ but $x * u \notin l(p)$. Continuing this process, we have $x_1 * (\cdots (x_k * u) \cdots) \notin l(p)$ for all $k \geq 1$ where $x_k = x$ $(k = 1, 2, \ldots)$. This shows that $C(p,p)$ is not nilpotent. Since $\dim C(p,0)/R^0(p,0) = 1$ and $C(p,p) = C(p,0)^-$,

$$C(p,p) * C(p,p) = [C(p,0), C(p,0)]^{\circledast} \subseteq R^0(p,0) = R^0(p,p) \tag{4.29}$$

and $R^0(p,p) * R^0(p,p) = [R^0(p,0), R^0(p,0)]^{\circledast} \subseteq l(p)$ since for $x, y \in R^0(p,p)$, $(x * y)p = [(xp)y - (yp)x]p = x(pyp) - y(pxp) = 0$, where $[\ ,\ ]^{\circledast}$ denotes the commutator in $C(p,0)$. Since $l(p) * l(p) = [l(p), l(p)]^{\circledast} = 0$, we conclude from (4.29) that $C(p,p)$ is solvable with $C(p,p)^{*(3)} = 0$.

(iii) is clear. □

From Theorems 4.5, 4.9, 4.10, 4.14 and 4.16, we summarize the structure of $C(p,q)$ and $R^0(p,q)$ in:

Theorem 4.17. *Let C be a Cayley–Dickson algebra over F and let p, q be fixed elements of C.*

(i) *If $p \neq q$ and $Fp + Fq$ is not totally isotropic, then $C(p,q)$ is simple.*

(ii) *If $p \neq q$, p and q are linearly dependent and $Fp+Fq$ is totally isotropic, then $R^0(p,q)$ is the unique maximal ideal of $C(p,q)$ of dimension 7, with $R^0(p,q)^{*4} = 0$ and there is a subalgebra S of $C(p,p)$ with $S \cong F$ such that $C(p,q) = R^0(p,q) \oplus S$.*

(iii) *If $p \neq q$, p and q are linearly independent with $\bar{p}q = 0$ and $Fp + Fq$ is totally isotropic, then $R^0(p,q)$ is the unique maximal ideal of $C(p,q)$ of dimension 6 with $R^0(p,q)^{*3} = 0$ and there is a simple subalgebra T of $C(p,q)$ of dimension 2 which is isomorphic to the twisted mutation algebra given by (4.14) and which complements $R^0(p,q)$.*

(iv) *If $p \neq q$, p and q are linearly independent with $\bar{p}q \neq 0$ and $Fp + Fq$ is totally isotropic, then $R^0(p,q)$ is the unique maximal ideal of $C(p,q)$ of dimension 4 with $R^0(p,q)^{*2} = 0$ and $C(p,q)/R^0(p,q)$ is a simple algebra of dimension 4 with multiplication given by (4.22). In this case, if the characteristic of F is not 3, then $(C(p,q)/R^0(p,q))^-$ is isomorphic to the unique non–Lie, Malcev algebra of dimension 4 given by (4.20).*

(v) *If $p = q$ and $n(p) \neq 0$, then $C(p,q)$ is isomorphic to the Malcev algebra C^-.*

(vi) *If $p = q \neq 0$ and $n(p) = 0$, then $C(p,q)$ is a solvable Malcev algebra with $C(p,q)^{*(3)} = 0$.*

(vii) *If $p = q = 0$, then $C(p,q)$ is a zero algebra.* □

5. Simple and prime mutations of alternative algebras

The purpose of this section is to extend the two main results (Theorems II.3.7 and II.3.13) in Section II.3 for the associative case to mutations of simple or prime alternative algebras. Recall that an (nonassociative) algebra B is called *prime* if, for any two ideals I, J of B, $IJ = 0$ implies $I = 0$ or $J = 0$.

For an algebra B over a field F, let $Z = Z(B)$ and $\Gamma = \Gamma(B)$ denote the center and the centroid of B, respectively. Then, Z is a commutative associative subring of B and Γ is an associative algebra over F with a unit element 1 (identify $\alpha = \alpha 1$, $\alpha \in F$). We note that $\gamma \in \Gamma$ if and only if

$$\gamma(xy) = (\gamma x)y = x(\gamma y) \tag{5.1}$$

for all $x, y \in B$. Let S be a (nonempty) multiplicative subset ($ab \in S$ for $a, b \in S$) of Z which does not contain zero divisors in B, that is, if $\alpha x = 0$ for $\alpha \in S$, $x \in B$, then $\alpha = 0$ or $x = 0$. One can construct the ring $S^{-1}B$ of quotients of B modulo S [**ZSSS1**, p. 185] which is also isomorphic to $S^{-1}Z \otimes_Z B$ [**ZSSS1**, p.192]. In particular, if $Z^* = Z - \{0\}$ contains no zero divisors in B, then $K \equiv (Z^*)^{-1}Z$ is the field of quotients of Z and $(Z^*)^{-1}B \cong K \otimes_Z B$. It is shown that the map φ given by

$$\varphi : B \longrightarrow K \otimes_Z B \ : \ x \mapsto 1 \otimes x \tag{5.2}$$

is an injection [**ZSSS1**].

For an F-subalgebra $S \neq 0$ of $\Gamma = \Gamma(B)$, B is said to be S-torsion free if, for any $\alpha \in S$ and $x \in B$, $\alpha x = 0$ implies $\alpha = 0$ or $x = 0$. If $S \neq 0$ is an F-subalgebra of Γ which is an integral domain (not necessarily with 1), and such that B is S-torsion free, then as above, $K \otimes_S B$ is made into a K-algebra where $K = (S^*)^{-1}S$ is the field of quotients of S. The following is useful for our discussion.

Lemma 5.1. *Let B be any algebra over a field F with center Z and centroid Γ.*

(i) *If $B^2 \neq 0$ and $S \neq 0$ is an F-subalgebra of Γ such that B is S-torsion free, then S is an integral domain. In this case, the map $\varphi : B \to K \otimes_S B$ given by (5.2) is injective where $K = (S^*)^{-1}S$, and any element $a \in K \otimes_S B$ has the form*

$$a = \nu^{-1}(1 \otimes x) = \nu^{-1} \otimes x \tag{5.3}$$

for some $\nu \in S$ and $x \in B$.

(ii) *If B is prime, then Γ is an integral domain with 1 and B is Γ-torsion free. Moreover, the map $Z \to \Gamma \ : \ z \mapsto L_z = R_z$ is a ring isomorphism into Γ and hence Z is regarded as an F-subalgebra of Γ. If B has a unit element 1, then $Z = \Gamma$.*

(iii) *Let S satisfy the same conditions as in* (i). *Then, B is a prime F–algebra if and only if $K \otimes_S B$ is a prime K–algebra.*

Proof. (i) Let $\lambda, \mu \in S$. Since $B^2 \neq 0$, there are elements $x, y \in B$ with $xy \neq 0$, and by (5.1) $(\lambda\mu)(xy) = \lambda[\mu(xy)] = (\mu x)(\lambda y) = \mu[x(\lambda y)] = (\mu\lambda)(xy)$. Since B is S–torsion free, $\lambda\mu = \mu\lambda$ and so S is commutative. If $\lambda\mu = 0$, then $(\lambda\mu)x = (\mu\lambda)x = \lambda(\mu x) = 0$ for all $x \in B$ and hence $\lambda = 0$ or $\mu = 0$, since B is S–torsion free. Thus, S is an integral domain. The injectivity of the map φ follows from [**AM1**, Chapt. 3, Ex. 12]. For $a \in K \otimes_S B$, let $a = \alpha_1 \otimes x_1 + \cdots + \alpha_n \otimes x_n$ with $\alpha_i \in K$ and $x_i \in B$, $i = 1, \ldots, n$. If we let $\alpha_i = \nu_i^{-1}\mu_i$ for $\mu_i, \nu_i \in S$, then

$$\nu a = 1 \otimes ((\nu_2 \cdots \nu_n)\mu_1 x_1 + \cdots + (\nu_1 \cdots \nu_{n-1})\mu_n x_n)$$

with $\nu = \nu_1 \cdots \nu_n \neq 0 \in S$. This gives (5.3).

(ii) Assume $\lambda x = 0$ for $\lambda \in \Gamma$ and $x \in B$. If we let $I = \lambda B$ and $J = \{y \in B : \lambda y = 0\}$, then by (5.1) I and J are ideals of B with $IJ = 0$. Since B is prime, $I = 0$ or $J = 0$, i.e., $\lambda = 0$ or $x = 0$. Therefore, B is Γ–torsion free, and by part (i) Γ is an integral domain with 1. If $u, v \in Z$ then clearly $L_{uv} = L_u L_v$ and $L_u L_x = L_x L_u$, $L_u R_x = R_x L_u$ for any $x \in B$. If $L_u = 0$ for $u \in Z$ then $\mathbb{Z}u$ is an ideal of B with $(\mathbb{Z}u)^2 = 0$, so $u = 0$ since B is prime. Hence, the map is a ring isomorphism of Z into Γ. If B has a unit element 1, then, for any $\lambda \in \Gamma$, $L_u = \lambda$ with $u = \lambda(1)$. Using (5.1) it is easy to see that $u \in Z$, hence $Z = \Gamma$.

(iii) Assume $K \otimes_S B$ is prime. If I and J are ideals of B with $IJ = 0$, then $(K \otimes_S I)(K \otimes_S J) = 0$ and so $K \otimes_S I = 0$ or $K \otimes_S J = 0$, since $K \otimes_S I$ and $K \otimes_S J$ are ideals of the K–algebra $K \otimes_S B$. This implies $I = 0$ or $J = 0$, hence B is prime. Suppose now that B is prime, and let U, V be nonzero ideals of $K \otimes_S B$ with $UV = 0$. We regard B as an F–subalgebra of $K \otimes_S B$ via the map φ. From (5.3) it follows that $U \cap B \neq 0$ and $V \cap B \neq 0$. But, since $U \cap B$ and $V \cap B$ are ideals of B with $(U \cap B)(V \cap B) = 0$, $U \cap B = 0$ or $V \cap B = 0$. This shows that if B is prime then so is $K \otimes_S B$. □

We apply the results above to an alternative algebra A over a field F. Assume that the center $Z = Z(A)$ is not zero and contains no zero divisors in A. Let $K = (Z^*)^{-1}Z$ be the field of quotients of Z. If the K–algebra $K \otimes_Z A$ is a Cayley–Dickson algebra over K, then A is called a *Cayley–Dickson ring*. If A is prime, then $Z \neq 0$ [**ZSSS1**, p.190] and by Lemma 5.1(ii) Z is an F–subalgebra of Γ such that A is Z–torsion free. Therefore, from Lemma 5.1(iii) and the result in [**ZSSS1**, p.194], we conclude:

Corollary 5.2. *Let A be a prime alternative algebra over F with center Z. Then, $K \otimes_Z A$ is a prime alternative algebra over K. If the characteristic of F is* *not 3 and A is not associative, then $K \otimes_Z A$ is a Cayley–Dickson algebra over K.* □

Therefore, we first treat the primeness of $C(p,q)$ for a Cayley–Dickson algebra C.

Lemma 5.3. *Let C be a Cayley–Dickson algebra over F and let p,q be fixed elements of C with $p \neq q$. Then, $C(p,q)$ is prime if and only if $R^0(p,q)=0$.*

Proof. If $R^0(p,q) = 0$, then by Theorem 4.5 $C(p,q)$ is simple and hence is prime. Assume now that $R^0(p,q) \neq 0$, or equivalently, $Fp+Fq$ is totally isotropic (see Theorem 4.5). We now consider the three cases in Lemma 4.8. If p and q are linearly independent with $\bar{p}q \neq 0$, then $R^0(p,q) * R^0(p,q) = 0$ by Theorem 4.17 and hence $C(p,q)$ is not prime.

Suppose p and q are linearly independent with $\bar{p}q = 0$. For $x \in C$,

$$\begin{aligned} x * \bar{p} &= (xp)\bar{p} - (\bar{p}q)x = n(p)x - (\bar{p}q)x = 0, \\ \bar{p} * x &= (\bar{p}p)x - (xq)\bar{p} = -(xq)\bar{p} = -t(xq)\bar{p} + (\bar{q}\bar{x})\bar{p} \\ &= -t(xq)\bar{p} - (\bar{q}p)x - t(xp)\bar{q} = -t(xq)\bar{p} - t(xp)\bar{q}, \end{aligned}$$

using (2.13). Hence, we have $x*\bar{p}, \bar{p}*x \in J = F\bar{p}+F\bar{q}$, and similarly, $x*\bar{q}, \bar{q}*x \in J$. This shows that J and $R^0(p,q)$ are nonzero ideals of $C(p,q)$ with $J * R^0(p,q) = 0$ (Proposition 2.9), and hence $C(p,q)$ is not prime. Finally, assume that p and q are linearly dependent, say $p \neq 0$ and $q = \alpha p$ ($\alpha \in F$). Then, by Lemma 4.15 $l(p) = \{x \in C : xp = 0\}$ has dimension 4. For any $x \in C$ and $y \in l(p)$, we have

$$\begin{aligned} (x*y)p &= ((xp)y)p - ((yq)x)p = x(pyp) - \alpha((yp)x)p = 0, \\ (y*x)p &= ((yp)x)p - ((xq)y)p = -\alpha x(pyp) = 0. \end{aligned}$$

Therefore, $l(p)$ is an ideal of $C(p,q)$, and clearly $l(p) * l(p) = 0$, hence $C(p,q)$ is not prime. This completes the proof. □

We now extend Theorem II.3.7 to mutations of an alternative algebra.

Theorem 5.4. *Let A be an alternative algebra over a field F of characteristic $\neq 2,3$ and let p,q be fixed elements of A with $p \neq q$.*

(i) *$A(p,q)$ is simple if and only if A is simple and $R^0(p,q)=0$.*

(ii) *$A(p,q)$ is prime if and only if A is prime and $R^0(p,q)=0$.*

Proof. (i) If $A(p,q)$ is simple, then A is simple and by Theorem I.3.17 A is either associative or a Cayley–Dickson algebra over its center Z, an extension field of F. In the former case, by Theorem II.3.7, $R^0(p,q) = 0$. For the latter, $A(p,q)$ is regarded as an algebra over Z and $R^0(p,q)$ is an ideal of the Z–algebra A. Hence, by Theorem 4.5, $R^0(p,q) = 0$ also. Assume now that A is simple and $R^0(p,q) = 0$. If A is associative, then the result follows from Theorem II.3.7. If A is a Cayley–Dickson algebra over Z, then we regard $A(p,q)$ as an algebra over Z which is simple by Theorem 4.5, since $R^0(p,q)$ is a Z–ideal of $A(p,q)$. But, $A(p,q)$ is the F–descent of the Z–algebra $A(p,q)$, and hence is simple over F (Lemma I.2.1(iii)).

(ii) Recall first that if A is a prime algebra over F, then either A is associative, or $K \otimes_Z A$ is a Cayley–Dickson algebra over K (Corollary 5.2), where $K = (Z^*)^{-1}Z$ is the field of quotients of Z. For the former, the result follows from Theorem II.3.7 again. For the latter, identifying $x = 1 \otimes x$ for $x \in A$ via the map φ of (5.2), we note from Lemma 5.1(iii) that $A(p,q)$ is prime if and only if

$$K \otimes_Z A(p,q) = (K \otimes_Z A)(1 \otimes p, 1 \otimes q) \tag{5.4}$$

is prime, since Z is contained in the centroid of $A(p,q)$. Let $\Lambda(p,q)$ denote the set given in Definition 1.1 (Section III.1). Then, $\Lambda(1 \otimes p, 1 \otimes q) = 1 \otimes \Lambda(p,q)$, and if $a \in R^0(1 \otimes p, 1 \otimes q)$, then by (5.3) $\nu a = 1 \otimes x$ for some $\nu \in Z^*$, $x \in A$, and

$$1 \otimes \Lambda(p,q)x = \Lambda(a \otimes p, 1 \otimes q)(\nu a) = \nu\Lambda(1 \otimes p, 1 \otimes q)a = 0,$$

hence $\Lambda(p,q)x = 0$ or $x \in R^0(p,q)$. This shows that $a = \nu^{-1} \otimes x \in K \otimes_Z R^0(p,q)$ and hence

$$K \otimes_Z R^0(p,q) = R^0(1 \otimes p, 1 \otimes q). \tag{5.5}$$

If $A(p,q)$ is prime, then as in Lemma II.3.1 A is prime. But, by Lemma 5.3 and (5.4) $R^0(1 \otimes p, 1 \otimes q) = 0$, and by (5.5) $K \otimes_Z R^0(p,q) = 0$, so $R^0(p,q) = 0$. If A is prime and is not associative, and if $R^0(p,q) = 0$, then it follows from (5.4), (5.5) and Lemma 5.3 that $K \otimes_Z A(p,q)$ is prime. Hence, by Lemma 5.1 (iii) $A(p,q)$ is prime also. □

As a consequence of Theorem 5.4, we have (cf. Corollary II.3.8):

Corollary 5.5. *Let A and p,q be the same as in Theorem 5.4. Assume that there are elements $\alpha, \beta \in \Gamma$, the centroid of A, such that $\alpha p + \beta q$ is not a zero divisor in A. Then, $A(p,q)$ is prime (or simple) if and only if A is prime (or simple). If A has, in particular, a unit element, then $A(p,q)$ is prime (or simple) if and only if A is prime (or simple).*

Proof. As in the associative case, we notice that

$$R^0(p,q) \subseteq \{x \in A : (\alpha p + \beta q)x(\alpha p + \beta q) = 0\} = 0,$$

and if $A(p,q)$ has a unit element, then $p - q$ is invertible in A (Theorem 3.5). □

Let $\Gamma_{p,q}$ denote the centroid of $A(p,q)$. Since the multiplication algebra of $A(p,q)$ is contained in the multiplication algebra of A, we have $\Gamma \subseteq \Gamma_{p,q}$, and hence $A(p,q)$ is regarded as an algebra over Γ. In view of Theorem 5.4(i), the proof for the associative case (Proposition 3.11) applies to show:

Proposition 5.6. *Let A be an alternative algebra over F and let p, q be fixed elements of A with $p \neq q$. If $A(p,q)$ is simple, then $\Gamma = \Gamma_{p,q}$.* □

We now extend Theorem II.3.13 to the alternative case.

Theorem 5.7. *Let A be an alternative algebra over F of characteristic $\neq 2,3$ and let p, q be fixed elements of A with $p \neq q$. If $A(p,q)$ is prime, then $\Gamma = \Gamma_{p,q}$.*

Proof. Since $A(p,q)$ is prime, so is A. Hence, by Corollary 5.2 A is either associative or a Cayley–Dickson ring. In the former case, the result follows from Theorem II.3.13. In the latter case, by Lemma 5.1, A is embedded into the Cayley–Dickson algebra $K \otimes_Z A$ over K, and since $K \otimes_Z A$ is central simple over K, its centroid is $K \otimes id$ (Theorem I.2.5). For $\gamma \in \Gamma_{p,q}$, $1 \otimes \gamma$ belongs to the centroid of $K \otimes_Z A(p,q) = (K \otimes_Z A)(1 \otimes p, 1 \otimes q)$, which by Proposition 5.6 equals $K \otimes id$. Hence, there is a $\delta \in K$ such that $1 \otimes \gamma = \delta \otimes id$. For any $x, y \in A$,

$$1 \otimes \gamma(xy) = \delta \otimes xy = (\delta \otimes x)(1 \otimes y) = (1 \otimes \gamma(x))\,(1 \otimes y) = 1 \otimes (\gamma(x))\,y$$

and by Lemma 5.1(i) we have $\gamma(xy) = (\gamma(x))y$. Similarly, $\gamma(xy) = x(\gamma(y))$, and from (5.1) we deduce that $\gamma \in \Gamma$ and $\Gamma = \Gamma_{p,q}$. □

We close this section with a result on division mutations of artinian alternative algebras.

Theorem 5.8. *Let A be an artinian alternative algebra over F and let p, q be fixed elements of A.*

(i) *If $A(p,q)$ is a division algebra, then so is A.*

(ii) *If A is a Cayley–Dickson algebra C over F, then $C(p,q)$ is a division algebra if and only if C is a division algebra and $n(p) \neq n(q)$.*

Proof. (i) If $A(p,q)$ is a division algebra then it is simple and hence A is simple also. Thus, A is either associative or a Cayley–Dickson algebra over its center. If A is associative, then the result follows from Theorem II.4.10. In the latter case, if A is not a division algebra, then A is the split Cayley–Dickson algebra $C(K)$ over an extension field K of F. Thus, there is a $0 \neq x \in A$ such that $n(x) = t(x(p-q)) = 0$ and hence $x*x = x(p-q)x - t(x(p-q))x - n(x)\overline{(p-q)} = 0$ by (2.10), a contradiction. Thus, if $A(p,q)$ is a division algebra then so is A.

(ii) Let A be a Cayley–Dickson algebra C over F. If C is a division algebra over F and $n(p) \neq n(q)$, then the same proof as in Theorem II.4.11 applies to show that $C(p,q)$ is a division algebra. Thus, assume $C(p,q)$ is a division algebra over F. By part (i), C is a division algebra over F. If $p = -q$, then $C(p,q) = C(p)^+$ and so $C(p,q)$ is not a division algebra since C is isomorphic to $C(p)$ (Proposition 2.2). Suppose $n(p) = n(q)$ with $p \neq -q$, and let x be a nonzero element of C such that $t(xp) = t(xq) = 0$. Then,

$$\begin{aligned}(\bar{p}+\bar{q}) * x &= ((\bar{p}+\bar{q})p)x - (xq)(\bar{p}+\bar{q}) \\ &= n(p)x - n(q)x + (\bar{q}p)x - (xq)\bar{p} = (\bar{q}p)x - (xq)\bar{p} \\ &= -(\bar{q}\bar{x})\bar{p} + t(xp)\bar{q} - (xq)\bar{p}, \quad \text{by (2.13)} \\ &= -(xq + \overline{xq})\bar{p} = -t(xq)\bar{p} = 0,\end{aligned}$$

a contradiction. This shows that if $C(p,q)$ is a division algebra then so is C and $n(p) \neq n(q)$. □

6. Mutations of artinian alternative algebras

Let A be an alternative algebra over F and let p, q be fixed elements of A. As in Definition 1.1, consider the subspaces

$$R(p,q) = \{x \in A : \Lambda(p,q)x \subseteq R(A)\}, \tag{6.1}$$

$$J(p,q) = \{x \in A : x * A \subseteq R(p,q) \text{ and } A * x \subseteq R(p,q)\}, \tag{6.2}$$

where $R(A)$ denotes the radical of A (Definition I.1.4). Note that if $R(A) = 0$ then $R(p,q) = R^0(p,q)$, and that if A is artinian then

$$J(A) = R(A) \subseteq R(A(p,q)) \tag{6.3}$$

(Proposition 1.1(ii)). Hence, if A is semisimple artinian, then $R(p,q) = R^0(p,q)$.

This section is devoted to extensions of the results discussed in Section II.5 for the associative case to artinian alternative algebras. Specifically, we investigate relationships among $R(A)$, $R(p,q)$, $J(p,q)$ and $R(A(p,q))$, and the structure of $A(p,q)/J(p,q)$ for an artinian alternative algebra A. We begin our discussion by showing that $R(p,q)$ is an ideal of $A(p,q)$ and $J(p,q)$ is the solvable radical of $A(p,q)$. In what follows, we assume that A denotes an artinian alternative algebra over a field F. Thus, by Wedderburn–Artin–Zhevlakov Theorem, $A/R(A)$ is the direct sum

$$A/R(A) = A_1 \oplus \cdots \oplus A_n \tag{6.4}$$

of simple artinian alternative algebras A_i over F, where each A_i is isomorphic either to $M_m(D)$ for a division algebra D over F or to a Cayley–Dickson algebra over its center (an extension field of F). For fixed elements $p, q \in A$, let

$$\tilde{p} = p + R(A) = p_1 + \cdots + p_n, \qquad \tilde{q} = q + R(A) = q_1 + \cdots + q_n \tag{6.5}$$

with $p_i, q_i \in A_i$, $i = 1, \ldots n$. As in Lemma II.5.2, we have the decomposition

$$A(p,q)/R(A) = (A/R(A))\,(\tilde{p},\tilde{q}) = A_1(p_1,q_1) \oplus \cdots \oplus A_n(p_n,q_n). \tag{6.6}$$

Lemma 6.1. *Let C be a Cayley–Dickson algebra over F and let p, q be fixed elements of C.*

(i) *If $p \neq q$, then $R^0(p,q) = R(p,q) = J(p,q) = R(C(p,q))$.*

(ii) *If $p = q \neq 0$ and $n(p) = 0$, then $R(p,q) \subset J(p,q) = R(C(p,q)) = C(p,q)$.*

(iii) *If $p = q \neq 0$ and $n(p) \neq 0$, then*

$$0 = R(p,q) \subset J(p,q) = R(C(p,q)) = Z(C(p,q)) = F\bar{p}.$$

Proof. For any algebra B, arguing the same as in the proof of Lemma II.1.3(i), we find that all nilpotent ideals of B are contained in the radical $R(B)$.

(i) By Theorem 4.6, $R^0(p,q) = R(p,q)$ is the only maximal ideal of $C(p,q)$. Since $C(p,q)/R(p,q)$ is simple by Theorem 4.17, $(C(p,q)/R(p,q))^2 \neq 0$ and $C(p,q) * C(p,q) \not\subseteq R(p,q)$. This implies $R(p,q) = J(p,q) = R(C(p,q))$.

(ii) $C(p,q)$ is solvable by Theorem 4.16(ii) and by (4.28) $C(p,q) * C(p,q) \subseteq R(p,q)$. Since $R(p,q)$ is nilpotent in $C(p,q)$, from the remark above we find that there are no maximal ideals M of $C(p,q)$ such that $C(p,q) * C(p,q) \not\subseteq M$. Hence, $C(p,q) = J(p,q) = R(C(p,q)) \supset R(p,q)$.

(iii) Since p is invertible, $R(p,q) = 0$ and

$$C(p,q) = C(p)^- = Z(C(p)) \oplus [C(p), C(p)] = Z(C(p)) \oplus C(p,q) * C(p,q),$$

where $C(p,q)*C(p,q)$ is simple. Moreover, $Z(C(p))$ and $C(p,q)*C(p,q)$ are the only proper ideals of $C(p,q)$ (Theorem I.4.17). Hence, $R(C(p,q)) = Z(C(p)) = J(p,q)$. In this case, $C(p)$ is a Cayley–Dickson algebra with unit element $p^{-1} = n(p)^{-1}\bar{p}$ (Proposition 2.1) and so $Z(C(p)) = F\bar{p}$. □

Lemma 6.2. *Let A be a simple artinian alternative algebra over F and let p, q be fixed elements of A.*

(i) *$R^0(p,q) = R(p,q)$ is an ideal of $A(p,q)$ and $R(A(p,q))$ coincides with the radical of $A(p,q)$ as an algebra over the center $Z = Z(A)$.*

(ii) *The relation*

$$R(A) \subseteq R(p,q) \subseteq J(p,q) \subseteq R(A(p,q)) \tag{6.7}$$

holds, and in particular, $R(p,q) = J(p,q)$ *if and only if* $p \neq q$.

Proof. (i) If A is associative, then $R^0(p,q)$ is clearly an ideal of $A(p,q)$. If A is not associative, then A is a Cayley–Dickson algebra over $Z = Z(A)$ and by Theorem 2.7 $R(p,q)$ is an ideal of $A(p,q)$ over Z, hence $R(p,q)$ is an ideal of $A(p,q)$ over F. For the second part, note that M is a maximal ideal of $A(p,q)$ with $A(p,q)*A(p,q) \not\subseteq M$ if and only if it is a maximal Z–ideal of $A(p,q)$ with $A(p,q)*A(p,q) \not\subseteq M$. For, if M is a maximal ideal of $A(p,q)$ with $A(p,q)*A(p,q) \not\subseteq M$ but is not a maximal Z–ideal of $A(p,q)$, then $M \subseteq ZM = A(p,q)$ and $A(p,q)*A(p,q) = (ZM)*A(p,q) = M*A(p,q) \subseteq M$, a contradiction.

(ii) Note first that $R(p,q)$ and $J(p,q)$ are independent of the base field of the algebra. If A is associative, then the result follows from Proposition II.1.5 and Lemma II.5.1. If A is not associative, then it is a consequence of Lemma 6.1 and part (i). □

The second part of Lemma 6.2(ii) is an extension of Lemma II.5.1 to the alternative case. We now extend Lemma 6.2 to artinian alternative algebras.

Proposition 6.3. *Let A be an artinian alternative algebra over F and p,q be fixed elements of A.*

(i) *$R(p,q)$ and $J(p,q)$ are ideals of $A(p,q)$, and*

$$\begin{aligned} R(p,q)/R(A) &= R_1(p_1,q_1) \oplus \cdots \oplus R_n(p_n,q_n), \\ J(p,q)/R(A) &= J_1(p_1,q_1) \oplus \cdots \oplus J_n(p_n,q_n), \end{aligned} \tag{6.8}$$

where $R_i(p_i,q_i) = R_i^0(p_i,q_i)$ and $J_i(p_i,q_i)$ is defined by (6.2) in A_i, $i = 1,\dots,n$.

(ii) *The relation (6.7) holds for $A(p,q)$.*

Proof. It suffices to show that $R(p,q)$ is an ideal of $A(p,q)$, since the remaining parts are the same as for the associative case (Lemma II.5.2). We use decompositions (6.4)–(6.6). Let $\pi : A \longrightarrow A/R(A)$ be the natural projection. Then, π induces the natural projection : $A(p,q) \longrightarrow A(p,q)/R(A) = \sum_{i=1}^n A_i(p_i,q_i)$. By Lemma 6.2, $\sum_{i=1}^n R_i(p_i,q_i)$ is an ideal of $A(p,q)/R(A)$, since each A_i is simple artinian and $R_i(p_i,q_i)$ is an ideal of $A_i(p_i,q_i)$. It is easy to show that $R(p,q) = \pi^{-1}\left(\sum_{i=1}^n R_i(p_i,q_i)\right)$ (see the proof of Lemma II.5.2), hence $R(p,q)$ is an ideal of $A(p,q)$.

(ii) Since $R(A) \subseteq R(A(p,q)) \subseteq J(p,q)$ by the definition, it suffices to verify $J(p,q) \subseteq R(A(p,q))$. Using the decomposition (6.6) it follows from Proposition

I.1.5 and Lemma II.5.9 that

$$(6.9)\qquad \begin{aligned} R(A(p,q)/R(A)) &= R(A(p,q))/R(A) \\ &= R(A_1(p_1,q_1)) \oplus \cdots \oplus R(A_n(p_n,q_n)). \end{aligned}$$

Since A_i is simple artinian, by (6.7) $J_i(p_i,q_i) \subseteq R(A_i(p_i,q_i))$, and by (6.8) and (6.9), $J(p,q)/R(A) \subseteq R(A(p,q))/R(A)$, hence $J(p,q) \subseteq R(A(p,q))$. □

We use these results to show:

Theorem 6.4. *Let A be an artinian alternative algebra over F and let p,q be fixed elements of A. Then, $J(p,q)$ is the solvable radical of $A(p,q)$ which contains all solvable ideals of $A(p,q)$, and $J(p,q)^{*(3)} \subseteq R(A)$.*

Proof. Assume first that A is simple. By Theorem II.4.2 and Lemma 4.1 $R(p,q)^{*4} = 0$ and so $R(p,q)^{*(2)} = 0$. Since $J(p,q)^{*(1)} \subseteq R(p,q)$ by (6.2), $J(p,q)$ is solvable with $J(p,q)^{*(3)} = 0$. It follows from Theorems II.4.2, II.4.7, Lemma II.5.1 and Lemmas 6.1, 6.2 that $J(p,q)$ is the solvable radical of $A(p,q)$. Hence, for the general case, from (6.6) and (6.8) we find $J(p,q)/R(A)$ to be the solvable radical of $A(p,q)/R(A)$ with $(J(p,q)/R(A))^{*(3)} = 0$. Therefore, since $R(A)$ is a nilpotent ideal of $A(p,q)$, $J(p,q)$ is the solvable radical of $A(p,q)$ with $J(p,q)^{*(3)} \subseteq R(A)$. □

Theorem 6.4 generalizes Theorem II.5.3(i) to the alternative case. We now determine the structure of the (solvable) semisimple algebra $A(p,q)/J(p,q)$. The following result is an extension of the structure theorem, Theorem II.5.3, to an artinian alternative algebra.

Theorem 6.5. *Let A be an artinian alternative algebra over F and let p,q be fixed elements of A. Then, $A(p,q)/J(p,q)$. is a finite direct sum of ideals, each of which is isomorphic to one of the algebras :*

(i) *A simple twisted mutation algebra over F described in Definition II.1.2.*

(ii) *A semisimple Lie algebra over F of the form $M_m(D)^-/Z(M_m(D))$ for some division algebra D over F.*

(iii) *A simple mutation of a Cayley–Dickson algebra over its center.*

(iv) *A simple algebra of dimension 4 over an extension field of F with multiplication given by (4.22).*

(v) *A simple non–Lie, Malcev algebra of dimension 7 over an extension field of F (for* $\operatorname{char} F \neq 3$*).*

(vi) *A simple Lie algebra of type A_2 over an extension field of F (for* $\operatorname{char} F = 3$*).*

Proof. We use decompositions (6.4)–(6.6). As for the associative case (Theorem II.5.3), from (6.8) we obtain

$$A(p,q)/J(p,q) = A_1(p_1,q_1)/J_1(p_1,q_1) \oplus \cdots \oplus A_n(p_n,q_n)/J_n(p_n,q_n). \tag{6.10}$$

If a simple summand A_i is associative with $p_i \neq q_i$, then by Theorem II.5.3 $A_i(p_i,q_i)/J_i(p_i,q_i)$ is isomorphic to an algebra given by (i), whereas for $p_i = q_i$, it is isomorphic to an algebra described by (ii). Thus, we may assume that A is simple artinian and is not associative, so that A is a Cayley–Dickson algebra over its center $K = Z(A)$. Consider first the case $p \neq q$. By Lemma 6.2(ii) $R(p,q) = R^0(p,q) = J(p,q)$, and hence $A(p,q)/J(p,q)$ is isomorphic respectively to the algebra over K described by (iii), (i) and (iv), according as the conditions in (i), (ii) and (iii), and (iv) of Theorem 4.17. For the case $p = q$, we let $A(p,p) = C(p,p)$ over K. If $n(p) = 0$, then by Lemma 6.1(ii) $A(p,p)/J(p,p) = 0$. Assume $n(p) \neq 0$. By Theorem 4.16(i) and Lemma 6.1(iii), $C(p,p) = Z(C(p)^-) \oplus [C(p), C(p)]$ with $J(p,p) = Z(C(p)^-) = K\bar{p}$. Hence, if the characteristic of F is not 3, then $A(p,p)/J(p,p)$ is isomorphic to the simple non–Lie, Malcev algebra $[C(p), C(p)]$, as described in (v). If the characteristic of F is 3, then it is well known [**E2**] that it is isomorphic to a Lie algebra of type A_2. □

We remark that if the characteristic of F is 3, then the Lie algebra of type A_2 in (vi) do not appear as algebras described in (ii).

Theorem 6.6. *Let A be an artinian alternative algebra over F and let p, q be fixed elements of A.*

(i) *$J(p,q) = R(A)$ if and only if $R(p,q) = R(A)$ and there are no maximal ideals M of A such that $A^2 \not\subseteq M$ and $p - q \in M$.*

(ii) *If $A(p,q)$ has a unit element, then $R(A) = R(p,q) = J(p,q) = R(A(p,q))$.*

(iii) *$R(p,q) = J(p,q)$ if and only if $p - q \notin M$ for any maximal ideal M of A with $A^2 \not\subseteq M$. In this case, $R(p,q) = J(p,q) = R(A(p,q))$.*

(iv) *If $p - q$ is invertible in A, then $R(p,q) = J(p,q) = R(A(p,q))$.*

Proof. If $J(p,q) = R(A)$, then by (6.8) $J_i(p_i,q_i) = 0$ for $i = 1,\ldots,n$, and hence by Lemma 6.1(i) and Theorem II.4.7, $p_i \neq q_i$ for all $i = 1,\ldots,n$. Thus, part (i) can be proven by using Lemma 6.2(ii) and the same arguments as in Proposition II.5.5. From (6.8) and Lemma 6.2(ii), it follows that $R(p,q) = J(p,q)$ if and only if $p_i \neq q_i$ for $i = 1,\ldots,n$. Hence, if $J(p,q) = R(p,q)$, then by Theorem 4.17 each $A_i(p_i,q_i)/R_i(p_i,q_i)$ is simple and so $A(p,q)/R(p,q)$ is semisimple, showing $R(A(p,q)) \subseteq R(p,q)$ (Corollary I.1.8). Parts (iii) and (iv) follow from this and the same arguments as for Theorem II.5.7. For part (ii), we observe that if $A(p,q)$ has a unit element, then by (6.6) each $A_i(p_i,q_i)$ has a unit element. Since A_i is simple for $i = 1, \ldots, n$, by Corollary 5.5 $A_i(p_i, q_i)$ is simple also. Hence, $A(p,q)/J(p,q)$ is semisimple and as before, $R(A(p,q)) \subseteq R(A)$. Part (ii) is now a consequence of Proposition 6.3(ii). □

Theorem 6.6 extends Proposition II.5.5, Corollary II.5.6, Theorem II.5.7 and Corollary II.5.8 to artinian alternative algebras. We end this section with a structure theorem for $R(A(p,q))$ relative to $R(A)$.

Theorem 6.7. *Let A be a finite-dimensional alternative algebra over F and let p, q be fixed elements of A with decompositions (6.4) and (6.5) such that $A_i(p_i, q_i)/J_i(p_i, q_i)$ is simple for $1 \le i \le s$ and is isomorphic to $M_m(D)^-/Z(M_m(D))$, for $s < i \le n$, with* $\operatorname{char} F$ *dividing the degree of $M_m(D)$. Then, the radical $R(A(p,q))$ of $A(p,q)$ is determined by*

$$R(A(p,q))/R(A) = \\ J_1(p_1,q_1) \oplus \cdots \oplus J_s(p_s,q_s) \oplus A_{s+1}(p_{s+1},q_{s+1}) \oplus \cdots \oplus A_n(p_n,q_n).$$

In particular, if the characteristic of F is zero, then $R(A(p,q)) = J(p,q)$, the solvable radical of $A(p,q)$.

Proof. We use decomposition (6.9). If $A_i(p_i,q_i)/J_i(p_i,q_i)$ is simple, then by Lemma II.5.1, Theorem II.5.7 and Lemma 6.1 we have $R(A_i(p_i,q_i)) = J_i(p_i,q_i)$. If $s < j \le n$, then under the assumption we must have $p_j = q_j$, and by Proposition II.5.11(ii) $R(A_j(p_j,q_j)) = A_j(p_j,q_j)$. Hence, the desired relation follows from (6.9). If the characteristic of F is zero, then it must be that $s = n$, hence by (6.8) and (6.9) $R(A(p,q)) = J(p,q)$. □

If C is a Cayley–Dickson algebra over F and p is a fixed element of C with $n(p) = 0$, then by Lemma 6.1(ii) $R(C(p,p)) = C(p,p)$. But, this case is not included in Theorem 6.7. On the other hand, by Lemma 6.1(iii), the case $p = q$ with $n(p) \neq 0$ arises in the decomposition of Theorem 6.7 with summand $J(p,p)$ for $1 \le i \le s$. Theorem 6.7 is also an extension of Theorem II.5.12 to the alternative case, with exclusion of some summands $R(A(p,p)) = J(p,p)$ where $A(p,p)/J(p,p)$ is isomorphic to $M_m(D)^-/Z(M_m(D))$ with $\operatorname{char} F$ not dividing the degree of $M_m(D)$ (Proposition II.5.11(i)).

CHAPTER IV

AUTOMORPHISMS AND DERIVATIONS

1. Introduction

A central topic of interest in the theory of nonassociative algebras is the study of automorphisms and derivations of algebras. As is well known, many linear algebraic and Lie groups and their Lie algebras are obtained from the automorphism groups and derivation algebras of certain nonassociative algebras. If B is a finite–dimensional algebra over a field F, then there is a natural relationship between the automorphism group $\operatorname{Aut} B$ and the derivation algebra $\operatorname{Der} B$ of B. The Lie algebra of $\operatorname{Aut} B$, viewed as a linear algebraic group over F, is a subalgebra of $\operatorname{Der} B$. If F is the field of real numbers or complex numbers, then $\operatorname{Aut} B$ is a Lie subgroup of the linear group $GL(B)$ and $\operatorname{Der} B$ is the Lie algebra of $\operatorname{Aut} B$ [**SW1**].

In the theory of mutations $A(p,q)$ of alternative algebras A over a field F, it is natural to study $\operatorname{Aut} A(p,q)$ and $\operatorname{Der} A(p,q)$ in relation with $\operatorname{Aut} A$ and $\operatorname{Der} A$. Our primary concern in this chapter is to determine automorphism groups and derivation algebras of simple mutations $A(p,q)$ for a central simple, artinian alternative algebra A over F. Thus, if A is not associative, then A is a Cayley–Dickson algebra over F. In this case, by Theorem III.5.4 the simplicity of $A(p,q)$ is equivalent to the condition that $p \neq q$ and $R^0(p,q) = 0$. If A is a composition algebra over F, then by Theorem III.4.5 this is equivalent to the condition that $p \neq q$ and $\alpha p + \beta q$ is invertible in A for some $\alpha, \beta \in F$.

In the associative case, we first investigate isomorphisms between prime mutation algebras under some restrictions on the fixed elements. Our main results then follows as applications of the results on prime mutation algebras. For this we define:

Definition 1.1. Let A be an associative algebra over F. A pair (a,b) of elements $a, b \in A$ is called *normal* in A if

$$A = aA + bA = Aa + Ab.$$

□

If A has a unit element, then it follows easily that for any invertible element in A of the form $\alpha a + \beta b$ for $\alpha, \beta \in Z(A)$, the pair (a,b) is normal in A. For semisimple artinian A, one easily proves:

Lemma 1.1. *Let A be a semisimple artinian associative algebra over F and let $p, q \in A$. The pair (p,q) is normal if and only if $R^0(p,q) = 0$.*

Proof. From Lemma II.4.1, $R^0(p,q) = l(p,q) + r(p,q) = l(f) + r(e)$ for some idempotents $e, f \in A$ such that $Ap + Aq = Ae$ and $pA + qA = fA$. As in the proof of Theorem II.4.2, we have $l(f) = A(1-f)$ and $r(e) = (1-e)A$. Hence, $R^0(p,q) = 0$ if and only if $A(1-f) = (1-e)A = 0$ if and only if $e = f = 1$. □

Let B be a prime noncommutative associative algebra over F with unit element, and let (a,b) be a normal pair in B with $a \neq b$. In Section IV.3 we determine explicitly all isomorphisms of $A(p,q)$ to $B(a,b)$ for any associative algebra A with $p \neq q$. For an associative algebra A over F with unit element 1, let $A^\times$ denote the group of all invertible elements in A. For fixed elements $p,q \in A$, the sets defined by

$$(1.1) \qquad H = \{(a,b) \in A^\times \times A^\times : pa = bp,\ qa = bq\},$$
$$(1.2) \qquad K = \{(a,a) \in A^\times \times A^\times : a \in Z(A)\}$$

are subgroups of $A^\times \times A^\times$ with $K \subseteq H$, and $K \cong Z(A)^\times$. For $(a,b) \in H$, let $\Phi_{a,b}$ be the linear map given by

$$(1.3) \qquad \Phi_{a,b} : A(p,q) \to A(p,q) \ : \ x \mapsto axb^{-1}.$$

Lemma 1.2. *Let H, K and $\Phi_{a,b}$ be the same as in (1.1)- (1.3). Then,*

$$(1.4) \qquad \Phi : H \to \operatorname{Aut} A(p,q) \ : \ (a,b) \mapsto \Phi_{a,b}$$

is a group homomorphism with $\ker \Phi = K$. *Thus, $\Phi(H)$ is a subgroup of* $\operatorname{Aut} A(p,q)$ *isomorphic to H/K.*

Proof. We first show $\Phi_{a,b} \in \operatorname{Aut} A(p,q)$ for any $(a,b) \in H$. Since $pa = bp$ and $qa = bq$, for $x, y \in A$

$$\begin{aligned}
\Phi_{a,b}(x * y) &= axpyb^{-1} - ayqxb^{-1} \\
&= (axb^{-1})p(ayb^{-1}) - (ayb^{-1})q(axb^{-1}) \\
&= \Phi_{a,b}(x) * \Phi_{a,b}(y)
\end{aligned}$$

and hence $\Phi_{a,b} \in \operatorname{Aut} A(p,q)$. If $(a,b),(c,d) \in H$, then

$$\Phi_{a,b} \circ \Phi_{c,d}(x) = acxd^{-1}b^{-1} = acx(bd)^{-1} = \Phi_{ac,bd}(x)$$

for $x \in A$. Thus, Φ is a homomorphism. If $\Phi_{a,b} = id$ for $(a,b) \in H$, then $\Phi_{a,b}(1) = a1b^{-1} = 1$, so $a = b$. Since $\Phi_{a,a}(x) = axa^{-1} = x$ for all $x \in A$, $a \in Z(A)$ and hence $K = \ker \Phi$. □

In Section IV.3, we show that if a central simple artinian associative algebra A with $A(p,q)$ simple does not possess an anti–automorphism, then any automorphism of $A(p,q)$ is of the form $\Phi_{a,b}$, that is, $\operatorname{Aut} A(p,q) = \Phi(H)$. Otherwise, it turns out that either $\operatorname{Aut} A(p,q) = \Phi(H)$ or $[\operatorname{Aut} A(p,q) : \Phi(H)] = 2$. For the latter case, let A be an associative algebra over F with unit element 1. Assume that A has an anti–automorphism τ, and let (a,b) be an element of $A^\times \times A^\times$ such that

$$(1.5) \qquad b\tau(p) = -qa, \quad b\tau(q) = -pa.$$

Define the linear map $\tau(a,b)$ by

$$\tau(a,b) : A(p,q) \to A(p,q) \ : \ x \mapsto a\tau(x)b^{-1}. \tag{1.6}$$

For $x, y \in A$, we have

$$\begin{aligned}
\tau(a,b)(x * y) &= a(\tau(xpy) - \tau(yqx))b^{-1} \\
&= a\tau(y)\tau(p)\tau(x)b^{-1} - a\tau(x)\tau(q)\tau(y)b^{-1} \\
&= -a\tau(y)b^{-1}qa\tau(x)b^{-1} + a\tau(x)b^{-1}pa\tau(y)b^{-1} \\
&= \tau(a,b)(x) * \tau(a,b)(y)
\end{aligned}$$

using (1.5), and hence $\tau(a,b) \in \operatorname{Aut} A(p,q)$. If τ is an anti–automorphism of A, then denote by M_τ the set of elements $\tau(a,b) \in \operatorname{Aut} A(p,q)$ for all $(a,b) \in A^\times \times A^\times$ satisfying (1.5).

Proposition 1.3. *Let A be an associative algebra with 1 and let $T(A)$ be the union of M_τ's for all anti–automorphisms τ of A. Assume $T(A) \neq \emptyset$.*

(i) *If $\tau(a,b) \in T(A)$, then $\tau(a,b)^{-1} = \tau^{-1}(\tau^{-1}(b), \tau^{-1}(a)) \in T(A)$.*

(ii) *If A is not commutative, then $T(A) \cap \Phi(H) = \emptyset$.*

(iii) *Each element of $T(A)^2$ is an automorphism of $A(p,q)$ of the form $\sigma(x)c$ ($x \in A$) for some $\sigma \in \operatorname{Aut} A$ and some $c \in A^\times$ such that $\sigma(p) = cp$ and $\sigma(q) = cq$.*

Proof. (i) If $\tau(a,b) \in T(A)$, then

$$\begin{aligned}
\tau(a,b) \circ \tau^{-1}(\tau^{-1}(b), \tau^{-1}(a))(x) &= a\tau(\tau^{-1}(b)\tau^{-1}(x)\tau^{-1}(a^{-1}))b^{-1} \\
&= a(a^{-1}xb)b^{-1} = x
\end{aligned}$$

for all $x \in A$. Since $\tau^{-1}(a)\tau^{-1}(p) = \tau^{-1}(pa) = -q\tau^{-1}(b)$ by (1.5) and similarly, $\tau^{-1}(a)\tau^{-1}(q) = -p\tau^{-1}(b)$, one has $\tau^{-1}(\tau^{-1}(b), \tau^{-1}(a)) \in T(A)$.

(ii) Assume A is not commutative. If $T(A) \cap \Phi(H) \neq \emptyset$, then $\tau(a,b) = \Phi_{c,d}$ for some $\tau(a,b) \in T(A)$ and $(c,d) \in H$. Hence, $a\tau(x)b^{-1} = cxd^{-1}$ for all $x \in A$, and in particular, $a^{-1}cd^{-1}b = d^{-1}ba^{-1}c = 1$ since $\tau(1) = 1$. Thus,

$$\tau(xy) = a^{-1}cxyd^{-1}b = (a^{-1}cxd^{-1}b)(a^{-1}cyd^{-1}b) = \tau(x)\tau(y).$$

This implies $\tau \in \operatorname{Aut} A$, which forces A to be commutative, a contradiction.

(iii) Let $\tau_1(a,b)$, $\tau_2(d,e) \in T(M)$, so that

$$\begin{aligned}
&\tau_1(p) = -b^{-1}qa, \quad \tau_1(q) = -b^{-1}pa, \\
&\tau_2(p) = -e^{-1}qd, \quad \tau_2(q) = -e^{-1}pd.
\end{aligned} \tag{1.7}$$

For $x \in A$, we find

$$\begin{aligned} \tau_2(d,e) \circ \tau_1(a,b)(x) &= d\tau_2(a\tau_1(x)b^{-1})e^{-1} \\ &= d\tau_2(b^{-1})\tau_2 \circ \tau_1(x)\tau_2(a)e^{-1} \\ &= d\tau_2(b^{-1})\tau_2 \circ \tau_1(x)\tau_2(b)d^{-1}(d\tau_2(ab^{-1})e^{-1}). \end{aligned}$$

Letting $\sigma(x) = d\tau_2(b^{-1})\tau_2 \circ \tau_1(x)\tau_2(b)d^{-1}$ and $c = d\tau_2(ab^{-1})e^{-1}$, it follows that $\sigma \in \operatorname{Aut} A$ and $\tau_2(d,e) \circ \tau_1(a,b)(x) = \sigma(x)c$. Moreover,

$$\begin{aligned} \sigma(p) &= -d\tau_2(b^{-1})\tau_2(b^{-1}qa)\tau_2(b)d^{-1} = -d\tau_2(b^{-1})\tau_2(a)\tau_2(q)d^{-1} \\ &= d\tau_2(b^{-1})\tau_2(a)e^{-1}p = cp \end{aligned}$$

using (1.7), and similarly, $\sigma(q) = cq$. □

If A is not commutative and is central simple artinian with $A(p,q)$ simple, then it is proven in Section IV.3 that either $\operatorname{Aut} A(p,q) = \Phi(H)$ or $\operatorname{Aut} A(p,q) = \Phi(H) \cup T(A)$ (disjoint) with $[\operatorname{Aut} A(p,q) : \Phi(H)] = 2$. Consequently, if $\operatorname{Aut} A(p,q) \neq \Phi(H)$ then $\operatorname{Aut} A(p,q)$ is isomorphic to a group having H/K as a subgroup of index 2.

Section IV.4 is devoted to the structure of $\operatorname{Der} A(p,q)$. The main task is to describe any derivation d of $A(p,q)$ as $d = \hat{d} + R_b$ for some $\hat{d} \in \operatorname{Der} A$ and $b \in A$, and then to identify $\operatorname{Der} A(p,q)$ with a Lie subalgebra of $A^- \times A^-$ modulo $Z(A)$. The former case is possible when A is a noncommutative associative algebra with 1 and $p \neq q$ are fixed elements of A such that $Ap + Aq = A$ and $A(p,q)$ is prime. For the latter, we need a stronger restriction on A. Let A be associative algebra with 1. For fixed elements $p, q \in A$, consider the subset L of $A \times A$ given by

$$L = \{(a,c) \in A \times A : pa = cp,\ qa = cq\}. \tag{1.8}$$

We identify the center $Z(A)$ of A with the subset $\{(a,a) \in A \times A : a \in Z(A)\}$ via $a \mapsto (a,a)$.

Lemma 1.4. *Let A be an associative algebra with 1 and let p, q be fixed elements in A.*

(i) *The set L of (1.8) is a Lie subalgebra of $A^- \times A^-$ and $Z(A)$ is an ideal of L contained in $Z(L)$.*

(ii) *For each $(a,c) \in L$, the map $d_{a,c}$ defined by*

$$d_{a,c} : A(p,q) \to A(p,q) \ : \ x \mapsto d_{a,c}(x) = ax - xc \tag{1.9}$$

is a derivation of $A(p,q)$.

(iii) *The map ρ given by*

$$\rho : L \to \operatorname{Der} A(p,q) \ : \ (a,c) \mapsto d_{a,c} \tag{1.10}$$

is a Lie algebra homomorphism such that $\ker \rho = Z(A)$. *Thus,* $\rho(L) \cong L/Z(A)$.

Proof. (i) This is immediate from (1.8).

(ii) For $(a,c) \in L$ and $x, y \in A$, we obtain

$$\begin{aligned} d_{a,c}(x) * y + x * d_{a,c}(y) &= (ax - xc)py - yq(ax - xc) \\ &\quad + xp(ay - yc) - (ay - yc)qx \\ &= a(xpy - yqx) - (xpy - yqx)c = d_{a,c}(x * y), \end{aligned}$$

using $xcpy = xpay$ and $yqax = ycqx$. Thus, $d_{a,c} \in \operatorname{Der} A(p,q)$.

(iii) If $(a,c),\ (d,e) \in L$, then $\rho([(a,c),(d,e)]) = \rho([a,d],[c,e])$, and

$$\begin{aligned} [d_{a,c}, d_{d,e}](x) &= d_{a,c}(dx - xe) - d_{d,e}(ax - xc) \\ &= a(dx - xe) - (dx - xe)c - d(ax - xc) + (ax - xc)e \\ &= (ad - da)x - x(ce - ec) = d_{[a,d],[c,e]}(x). \end{aligned}$$

Hence, ρ is a Lie algebra homomorphism. If $\rho(a,c) = 0$, then $ax = xc$ for all $x \in A$, so $a = c$ for $x = 1$. But, $ax = xa$ for all $x \in A$ implies $a \in Z(A)$. □

For a finite–dimensional central simple associative algebra A with $A(p,q)$ simple, we show in Section IV.4 that the map ρ of (1.10) is surjective and hence $\operatorname{Der} A(p,q) \cong L/Z(A)$.

For any algebra B over F and a subset S of B, define

$$\begin{aligned} \operatorname{Aut}_S B &= \{\varphi \in \operatorname{Aut} B : \varphi(x) = x \text{ for all } x \in S\}, \\ \operatorname{Der}_S B &= \{d \in \operatorname{Der} B : d(x) = 0 \text{ for all } x \in S\}. \end{aligned}$$

Clearly, $\operatorname{Aut}_S B$ is a subgroup of $\operatorname{Aut} B$ and $\operatorname{Der}_S B$ is a subalgebra of $\operatorname{Der} B$. If $S = \{u\}$, then we write $\operatorname{Aut}_S B = \operatorname{Aut}_u B$ and $\operatorname{Der}_S B = \operatorname{Der}_u B$. If T is the subalgebra of B generated by u, then it is immediate that $\operatorname{Aut}_T B = \operatorname{Aut}_u B$ and $\operatorname{Der}_T B = \operatorname{Der}_u B$.

Sections IV.5 and IV.6 are concerned with $\operatorname{Aut} A(p,q)$ and $\operatorname{Der} A(p,q)$ for composition algebras A over F. Composition properties of A result in a more concrete description of $\operatorname{Aut} A(p,q)$ and $\operatorname{Der} A(p,q)$. For a generalized quaternion algebra Q over F with $Q(p,q)$ simple, we show in Section IV.5 that $\operatorname{Aut} Q(p,q)$ is isomorphic either to the isotropic subgroup $\operatorname{Aut}_{\bar{p}q} Q$ of $\operatorname{Aut} Q$ at $\bar{p}q$ or to an extension group of $\operatorname{Aut}_{\bar{p}q} Q$ by index 2, and $\operatorname{Der} Q(p,q) \cong \operatorname{Der}_{\bar{p}q} Q$. However, we obtain similar results for Cayley–Dickson algebras C under the stronger condition that $p+q$ is invertible in C with $p \neq q$ (so $C(p,q)$ is simple).

Proofs of the main results in Sections IV.3 and IV.4 make use of some results on Jordan homomorphisms and derivations of associative algebras. We present these in Section IV.2 with proofs. Throughout this chapter, the base field F is assumed to be of characteristic $\neq 2$.

2. Jordan homomorphisms and derivations

For an associative algebra A with multiplication xy over a field F, denote by $x \circ y = \frac{1}{2}(xy + yx)$ the multiplication in the Jordan algebra A^+. A map φ of A into an associative algebra B over F is called a *Jordan homomorphism*, or a *J–homomorphism*, if it is an algebra homomorphism of A^+ into B^+. A *Jordan derivation* or *J–derivation* of A is defined to be a derivation of A^+. Proofs of the main results in Sections IV.3 and IV.4 invoke two classical theorems on J–homomorphisms and J–derivations: Any surjective J–homomorphism of A to B with B prime is either a homomorphism or an anti–homomorphism of A onto B, and that any J–derivation of a prime associative algebra A is a derivation of A. In this section we give complete proofs of these theorems. Our proofs are essentially drawn from [**H1**] with a modified organization.

Let A, B be associative algebras over F and let $\varphi : A \to B$ be a J–homomorphism. Since $x^2 = x \circ x$ and $xyx = 2(y \circ x) \circ x - y \circ x^2$, applying φ to these and expanding it, we obtain

$$\varphi(x^2) = \varphi(x)^2, \quad \varphi(xyx) = \varphi(x)\varphi(y)\varphi(x) \tag{2.1}$$

for $x, y \in A$. The last identity in (2.1) is linearized to

$$\varphi(xyz) + \varphi(zyx) = \varphi(x)\varphi(y)\varphi(z) + \varphi(z)\varphi(y)\varphi(x). \tag{2.2}$$

We put

$$\begin{aligned} g(x,y) &= \varphi(xy) - \varphi(x)\varphi(y), \\ h(x,y) &= \varphi(xy) - \varphi(y)\varphi(x), \end{aligned} \tag{2.3}$$

and denote by $g(A,A)$ and $h(A,A)$ the linear spans of all $g(x,y)$ and all $h(x,y)$, respectively. Since $g(x,x) = h(x,x) = 0$ for $x \in A$, g and h are skew–symmetric bilinear maps : $A \times A \to B$.

Let B^{op} denote the opposite algebra of B with multiplication $x \cdot y = yx$. The linear map $\varphi : A \to B^{op}$ is also a J–homomorphism, since $(B^{op})^+ = B^+$. If g^{op} and

h^{op} denote the bilinear maps given by (2.3) associated with $\varphi : A \to B^{op}$, then by (2.3)

$$g^{op} = h\,, \quad h^{op} = g\,. \tag{2.4}$$

Lemma 2.1. *Let A and B be associative algebras over F and let $\varphi : A \to B$ be a J-homomorphism. For $x, y, z, t, u \in A$, we have*

$$g(x,y)h(x,y) = h(x,y)g(x,y) = 0, \tag{2.5}$$

$$g(x,y)h(x,z) + g(x,z)h(x,y) = h(x,y)g(x,z) + h(x,z)g(x,y) = 0, \tag{2.6}$$

$$\begin{aligned} &g(x,y)h(t,z) + g(t,y)h(x,z) + g(x,z)h(t,y) + g(t,z)h(x,y) = 0,\\ &h(x,y)g(t,z) + h(t,y)g(x,z) + h(x,z)g(t,y) + h(t,z)g(x,y) = 0, \end{aligned} \tag{2.7}$$

$$\begin{aligned} &(\varphi(x)\circ\varphi(y))\varphi(u)\varphi(y)\varphi(x) + \varphi(x)\varphi(y)\varphi(u)(\varphi(x)\circ\varphi(y))\\ &\qquad = \varphi((x\circ y)uyx + xyu(x\circ y)), \end{aligned} \tag{2.8}$$

$$g(x,y)\varphi(u)h(x,y) + h(x,y)\varphi(u)g(x,y) = 0, \tag{2.9}$$

$$\begin{aligned} &g(x,y)\varphi(u)h(x,z) + g(x,z)\varphi(u)h(x,y) + h(x,y)\varphi(u)g(x,z)\\ &\qquad + h(x,z)\varphi(u)g(x,y) = 0. \end{aligned} \tag{2.10}$$

Proof. In light of (2.4) with a J–homomorphism $\varphi : A \to B^{op}$, for (2.5)–(2.7) it suffices to verify the first identities. For (2.5), we compute

$$\begin{aligned} g(x,y)h(x,y) &= \varphi(xy)^2 - \varphi(x)\varphi(y)\varphi(xy)\\ &\qquad -\varphi(xy)\varphi(y)\varphi(x) + \varphi(x)\varphi(y^2)\varphi(x)\\ &= \varphi(xyxy) - \varphi(xy(xy) + (xy)yx) + \varphi(x)\varphi(y^2)\varphi(x)\\ &= \varphi(xyxy) - \varphi(xyxy) - \varphi(xy^2x) + \varphi(xy^2x) = 0, \end{aligned}$$

using (2.1)–(2.3). Linearization of (2.5) relative to y gives (2.6) which is in turn linearized to (2.7). For (2.8), define the triple product

$$\{abc\} = abc + cba.$$

From (2.2), we have $\varphi(\{abc\}) = \{\varphi(a)\varphi(b)\varphi(c)\}$. Also,

$$\begin{aligned} (x\circ y)uyx + xyu(x\circ y) &= \frac{1}{2}(xyuyx + yxuyx + xyuxy + xyuyx)\\ &= \frac{1}{4}\{x\{yuy\}x\} + \frac{1}{2}\{x\{xuy\}y\} - \frac{1}{2}\{x^2uy^2\}. \end{aligned}$$

Applying φ to this and expanding it, one has (2.8).

For (2.9), by skew–symmetry of g and h,

$$g(x,y)\varphi(u)h(x,y)+h(x,y)\varphi(u)g(x,y)=g(x,y)\varphi(u)h(x,y)+h(y,x)\varphi(u)g(y,x)$$

and by (2.3) the right side is expanded to

$$\begin{aligned}
&\varphi(xy)\varphi(u)\varphi(xy)-\varphi(xy)\varphi(u)\varphi(y)\varphi(x)-\varphi(x)\varphi(y)\varphi(u)\varphi(xy)\\
&\qquad+\varphi(x)\varphi(y)\varphi(u)\varphi(y)\varphi(x)+\varphi(yx)\varphi(u)\varphi(yx)-\varphi(yx)\varphi(u)\varphi(y)\varphi(x)\\
&\qquad\quad-\varphi(x)\varphi(y)\varphi(u)\varphi(yx)+\varphi(x)\varphi(y)\varphi(u)\varphi(y)\varphi(x)\\
&=\varphi(xyuxy+yxuyx+2xyuyx)\\
&\quad-2\left[\varphi(x\circ y)\varphi(u)\varphi(y)\varphi(x)+\varphi(x)\varphi(y)\varphi(u)\varphi(x\circ y)\right]\\
&=\varphi\left[xyuxy+yxuyx+2xyuyx-2((x\circ y)uyx+xyu(x\circ y))\right]\\
&=\varphi(0)=0,
\end{aligned}$$

using (2.1) and (2.8). Identity (2.10) is a linearization of (2.9). □

Lemma 2.2. *Let A be an associative algebra over F.*

(i) *If A is a nilalgebra of index 2, i.e., $x^2=0$ for all $x\in A$, then $A^3=0$.*

(ii) *If A is semiprime and u is an element of A such that $xux=0$ for all $x\in A$, then $u=0$.*

Proof. (i) Since $xy=-yx$ for all $x,y\in A$, $xyz=-yxz=yzx=-xyz$ for all $x,y,z\in A$ and hence $xyz=0$, or $A^3=0$.

(ii) By the assumption, the u–homotope $A(u)$ is a nilalgebra of index 2. Hence, by (i) $A(u)^3=AuAuA=(AuA)^2=0$. Since A is semiprime, $u=0$. □

Using Lemmas 2.1 and 2.2, we now prove our principal result on J–homomorphisms.

Theorem 2.3. *Let A and B be associative algebras over F where B is prime. Then, any surjective J–homomorphism φ of A to B is either a homomorphism or an anti–homomorphism of A to B.*

Proof. It suffices to show $g(A,A)=0$ or $h(A,A)=0$. Suppose $g(A,A)\neq 0\neq h(A,A)$. Since φ is surjective, by (2.5) and (2.9) one gets $g(x,y)^2Bh(x,y)=0$, which implies $g(x,y)^2=0$ or $h(x,y)=0$ by primeness of B. Assume $g(x,y)^2\neq 0$ for some $x,y\in A$, so $h(x,y)=0$. It follows from (2.6) that

$$g(x,y)h(x,A)=h(x,A)g(x,y)=0.$$

Multiplying $g(x,y)$ to (2.10) from the left, this implies $g(x,y)^2Bh(x,A) = 0$. Since B is prime and $g(x,y)^2 \neq 0$, $h(x,A) = 0$. Interchanging x and y in the above arguments, we get $h(y,A) = 0$ as well. Therefore, from (2.7)

$$g(x,y)h(A,A) = 0. \tag{2.11}$$

On the other hand, modulo $h(A,A)$, by (2.3) we have for $a,b,c \in A$

$$\begin{aligned}\varphi(c)h(a,b) &= \varphi(c)\varphi(ab) - \varphi(c)\varphi(b)\varphi(a)\\ &\equiv \varphi(abc) - \varphi(c)\varphi(b)\varphi(a)\\ &\equiv \varphi(bc)\varphi(a) - \varphi(c)\varphi(b)\varphi(a) = h(b,c)\varphi(a).\end{aligned}$$

Thus, $Bh(A,A) \subseteq h(A,A) + h(A,A)B$, and this implies that $h(A,A) + h(A,A)B$ is an ideal of B and $(Bg(x,y)B)(h(A,A) + h(A,A)B) = 0$ using (2.11). Since B is prime and $h(A,A) \neq 0$, it must be that $g(x,y) = 0$, contrary to the assumption $g(x,y) \neq 0$.

Therefore, $g(x,y)^2 = 0$ for all $x,y \in A$, and applying this to $g^{op} = h$ with J–homomorphism $\varphi : A \to B^{op}$, we get $h(x,y)^2 = 0$ for all $x,y \in A$. It follows from this and (2.5) that

$$[\varphi(x),\varphi(y)]^2 = (h(x,y) - g(x,y))^2 = 0,$$

so $[a,b]^2 = 0$ for all $a,b \in B$. For $a = g(x,y) \neq 0$ for some $x,y \in A$ and for any $b \in B$, since $[a,b]^2 = a^2 = 0$,

$$ababa = a([a,b]^2 - abab + ab^2a + ba^2b) = 0. \tag{2.12}$$

If we let $S = aB$ and $T = \{x \in S : xS = 0\}$, then by (2.12) T is an ideal of S such that $x^2 \in T$ for all $x \in S$. Hence, S/T is a nilalgebra of index 2 and by Lemma 2.2(i) $S^3 \subseteq T$, so $S^4 = 0$. This gives $(BaB)^4 = 0$ and $a = 0$ since B is prime, a contradiction. Therefore, $g(A,A) = 0$ or $h(A,A) = 0$. □

We now turn our attention to J–derivations d of an associative algebra A over F. Applying d to the identities $x^2 = x \circ x$ and $xyx = 2(y \circ x) \circ x - y \circ x^2$, we have

$$d(x^2) = d(x)x + xd(x),\ d(xyx) = d(x)yx + xd(y)x + xyd(x) \tag{2.13}$$

for all $x,y \in A$. Linearizing the second expression of (2.13) and using the triple product $\{xyz\} = xyz + zyx$, one finds

$$d(\{xyz\}) = \{d(x)yz\} + \{xd(y)z\} + \{xyd(z)\} \tag{2.14}$$

for $x,y,z \in A$. We let

$$f(x,y) = d(xy) - d(x)y - xd(y) \tag{2.15}$$

and denote by $f(A,A)$ the linear span of all $f(x,y)$. Since $f(x,x) = 0$, f is skew–symmetric.

Lemma 2.4. *Let d be a J-derivation of an associative algebra A over F. Then, for any $x, y, z \in A$,*

$$f(x,y)[x,y] = 0, \tag{2.16}$$

$$f(x,y)[x,z] + f(x,z)[x,y] = 0. \tag{2.17}$$

Proof. Applying d to $\{x,y,xy\} = (xy)^2 + xy^2x$ and using (2.13), (2.14), we obtain

$$\begin{aligned}\{d(x),y,xy\} + \{x,d(y),xy\} + \{x,y,d(xy)\}\\ = d(xy)xy + xyd(xy) + d(x)y^2x\\ +x[d(y)y + yd(y)]x + xy^2d(x).\end{aligned}$$

Expanding the left side of this gives (2.16), and (2.17) follows from a linearization of (2.16). □

Lemma 2.5. *Let A be an associative algebra over F. Then, for any $x \in A$, $[x, A]A$ is an ideal of A.*

Proof. If $a, b \in A$, then $a[x,b] = [x,ab] - [x,a]b$ and so $A[x,A] \subseteq [x,A] + [x,A]A$. Hence, $A[x,A]A \subseteq [x,A]A$ and $[x,A]A$ is an ideal of A. □

We let $V = \{x \in A : f(x,A) = 0\}$, and note that d is a derivation of A if and only if $V = A$. We show that this is the case if A is prime.

Lemma 2.6. *Let A be a prime associative algebra over F and let V be as above.*

(i) *If x and y are elements of A such that $[x,y] = 0$, then $f(x,y) = 0$.*

(ii) *$Z(A) \subseteq V$ and $f(A,A)[V,V] = 0$.*

(iii) *If $xy = 0$ for $x, y \in A$, then $yx \in V$.*

Proof. (i) By (2.17) $f(x,y)[x,A] = 0$ and so $f(x,y)[x,A]A = 0$. Since $[x,A]A$ is an ideal of A by Lemma 2.5 and A is prime, we have either $f(x,y) = 0$ or $x \in Z(A)$, and similarly, either $f(x,y) = 0$ or $y \in Z(A)$. Suppose $f(x,y) \neq 0$. Then, $x, y \in Z(A)$, and so $xy = x \circ y$, which implies $d(xy) = d(x \circ y) = d(x) \circ y + x \circ d(y) = d(x)y + xd(y)$. Thus, $f(x,y) = 0$, a contradiction.

(ii) It is immediate from part (i) that $Z(A) \subseteq V$. If $z \in V$, then by (2.17) $f(x,y)[x,z] = 0$ for any $x, y \in A$. Linearizing this gives

$$f(x,y)[u,z] + f(u,y)[x,z] = 0$$

for any $x, y, u \in A$, which reduces to $f(x,y)[u,z] = 0$ for $x, y \in A$ and $u, z \in V$.

(iii) We first show $f(x,y)=0$. If $yx=0$, then this follows from part (i). If $yx\neq 0$, then since $xy=0$, by (2.13)

$$0=d(xyzyx)=d(x)yzyx+xd(y)zyx=-f(x,y)zyx$$

for $z\in A$. This implies $f(x,y)Ayx=0$, and hence $f(x,y)=0$ since A is prime. But, for $z\in A$, one finds by (2.14)

$$\begin{aligned} d(yxz)=d(\{yxz\}) &= \{d(y)xz\}+\{yd(x)z\}+\{yxd(z)\} \\ &= d(y)xz+zxd(y)+yd(x)z+zd(x)y+yxd(z) \\ &= [d(y)x+yd(x)]z+z[d(x)y+xd(y)]+yxd(z) \\ &= [d(yx)+f(y,x)]z-zf(x,y)+yxd(z) \\ &= d(yx)z+yxd(z). \end{aligned}$$

Hence, $f(yx,A)=0$, i.e., $yx\in V$. □

Theorem 2.7. *Any J–derivation d of a prime associative algebra A over F is a derivation of A.*

Proof. By (2.16), $f(x,y)[x,y]A=0$ for $x,y\in A$, and hence by Lemma 2.6(iii) $[x,y]Af(x,y)\subseteq V$. Assume $f(A,A)\neq 0$, and let $f(x,y)\neq 0$ for some $x,y\in A$. Then, by Lemma 2.6(ii)

$$\begin{aligned} 0 &= f(x,y)[[x,z]Af(x,z),[x,y]Af(x,y)] \\ &= f(x,y)[x,z]Af(x,z)[x,y]Af(x,y) \end{aligned}$$

for any $z\in A$, using (2.16). This implies that either $f(x,y)[x,z]=0$ or $f(x,z)[x,y]=0$. Since by (2.17) these two expressions add to zero, we have $f(x,y)[x,A]=0$, so $f(x,y)[x,A]A=0$. Since $[x,A]A$ is an ideal of A by Lemma 2.5 and A is prime, $[x,A]A=0$, so $[x,A]=0$, i.e., $x\in Z(A)$, a contradiction by Lemma 2.6(ii). Therefore, $f(A,A)=0$ and d is a derivation of A. □

3. Automorphisms of mutations of associative algebras

Let A and B be associative algebras over F with multiplications denoted by juxtaposition. For fixed elements $p,q\in A$ and $a,b\in B$, denote by $A(p,q)$ and $B(a,b)$ the (p,q)–mutation of A and (a,b)–mutation of B with multiplications given by "$*$". In this section, we investigate isomorphisms of $A(p,q)$ to $B(a,b)$ and, in particular, $\operatorname{Aut} A(p,q)$. Isomorphisms of $A(p,q)$ to $B(a,b)$ are determined in terms of isomorphisms or anti–isomorphisms of A to B, when B is a prime, noncommutative associative algebra with 1 and (a, b) is a normal pair in B with $a \neq b$. This enables us to determine $\operatorname{Aut} A(p,q)$ for any central simple artinian associative algebra A with $A(p,q)$ simple. It turns out in this case that either $\operatorname{Aut} A(p,q) = \Phi(H)$ as in Lemma 1.2, or $\operatorname{Aut} A(p,q)$ is an extension group of $\Phi(H)$ by index 2.

Let

$$\phi : A(p,q) \to B(a,b)$$

an (algebra) homomorphism. We define the bilinear maps $g, h : A \times A \to B$ by

$$(3.1) \qquad \begin{aligned} g(x,y) &= \phi(xpy) - \phi(x)a\phi(y) = \phi(yqx) - \phi(y)b\phi(x), \\ h(x,y) &= \phi(xpy) + \phi(y)b\phi(x) = \phi(yqx) + \phi(x)a\phi(y) \end{aligned}$$

for $x, y \in A$, and denote by $g(A,A)$ and $h(A,A)$ the linear spans of all $g(x,y)$ and $h(x,y)$, respectively. The skew–symmetric bilinear maps $g^-, h^- : A \times A \to B$ are defined by

$$(3.2) \qquad \begin{aligned} g^-(x,y) &= g(x,y) - g(y,x) = \phi(x(p-q)y) - \phi(x)(a-b)\phi(y), \\ h^-(x,y) &= h(x,y) - h(y,x) = \phi(x(p-q)y) - \phi(y)(a-b)\phi(x). \end{aligned}$$

The right sides of (3.2) follow from (3.1). Note that $g^-(A,A) = 0$ or $h^-(A,A) = 0$ if and only if ϕ is a homomorphism or an anti–homomorphism of $A(p-q)$ into $B(a-b)$, where $A(p-q)$ is the $(p-q)$–homotope of A.

Since $B(a,b) = B^{op}(-b,-a)$, if we let g^{op} and h^{op} denote the bilinear maps defined by (3.1) associated with the homomorphism $\phi : A(p,q) \to B^{op}(-b,-a)$, then it is clear from (3.1) that

$$(3.3) \qquad g^{op} = h\,, \quad h^{op} = g.$$

A crucial step for our main results is to show that if B is not commutative and $B(a,b)$ is prime with $a \neq b$, then either $g(A,A) = 0$ or $h(A,A) = 0$. We verify this result by steps.

Lemma 3.1. *Let $\phi : A(p,q) \to B(a,b)$ be a surjective homomorphism, and let $M = g(A,A) + g(A,A)aB$ and $N = h(A,A) + h(A,A)aB$. Then, M and N are ideals of $B(a,b)$, and moreover,*

$$\begin{aligned} M &= g(A,A) + Bag(A,A) = g(A,A) + g(A,A)bB = g(A,A) + Bbg(A,A), \\ N &= h(A,A) + Bah(A,A) = h(A,A) + h(A,A)bB = h(A,A) + Bbh(A,A). \end{aligned}$$

Proof. In view of (3.3) with $\phi : A(p,q) \to B^{op}(-b,-a)$, it suffices to verify the first relation. We note from (3.1) that $\phi(xpy) \equiv \phi(x)a\phi(y)$ and $\phi(xqy) \equiv \phi(x)b\phi(y)$

modulo $g(A,A)$. Therefore, for $x, y, z \in A$, we compute

$$\begin{aligned}
\phi(z)ag(x,y) &= \phi(z)a\phi(xpy) - \phi(z)a\phi(x)a\phi(y)\\
&\equiv \phi(zpxpy) - \phi(z)a\phi(x)a\phi(y)\\
&\equiv \phi(zpx)a\phi(y) - \phi(z)a\phi(x)a\phi(y) = g(z,x)a\phi(y),\\
\phi(z)bg(x,y) &= \phi(z)b\phi(xpy) - \phi(z)b\phi(x)a\phi(y)\\
&\equiv \phi(zqxpy) - \phi(z)b\phi(x)a\phi(y)\\
&\equiv \phi(zqx)a\phi(y) - \phi(z)b\phi(x)a\phi(y) = g(x,z)a\phi(y),\\
\phi(z)ag(x,y) &= \phi(z)a\phi(yqx) - \phi(z)a\phi(y)b\phi(x)\\
&\equiv \phi(zpy)b\phi(x) - \phi(z)a\phi(y)b\phi(x) = g(z,y)b\phi(x),\\
\phi(z)bg(x,y) &= \phi(z)b\phi(yqx) - \phi(z)b\phi(y)b\phi(x)\\
&\equiv \phi(zqy)b\phi(x) - \phi(z)b\phi(y)b\phi(x) = g(y,z)b\phi(x).
\end{aligned}$$

It follows from these relations that $g(A,A)aB \equiv Bag(A,A) \equiv g(A,A)bB \equiv Bbg(A,A)$ modulo $g(A,A)$. This gives the first relation, which also implies that M is an ideal of $B(a,b)$. □

Lemma 3.2. *Let $\phi : A(p,q) \to B(a,b)$ be a homomorphism. Then, for $x, y, z \in A$*

$$g(x,y)(a-b)\phi(z) - \phi(z)(a-b)g(x,y) = g^-(z, xpy + yqx), \tag{3.4}$$

$$\phi(z)(a-b)h(x,y) - h(x,y)(a-b)\phi(z) = h^-(z, xpy + yqx). \tag{3.5}$$

Proof. We only need to verify (3.4), since (3.5) follows from this with $g^{op} = h$ and $\phi : A(p,q) \to B^{op}(-b,-a)$. We first note in $A(p,q)$ that

$$(x * z) * y - x * (z * y) + (yqx) * z + z * (xpy) = z(p-q)(xpy + yqx) \tag{3.6}$$

for $x, y, z \in A$. Applying ϕ to this gives

$$\begin{aligned}
&(\phi(x) * \phi(z)) * \phi(y) - \phi(x) * (\phi(z) * \phi(y))\\
&+\phi(yqx) * \phi(z) + \phi(z) * \phi(xpy) = \phi(z(p-q)(xpy + yqx)).
\end{aligned} \tag{3.7}$$

In $B(a,b)$, (3.6) reads as

$$\begin{aligned}
&(\phi(x) * \phi(z)) * \phi(y) - \phi(x) * (\phi(z) * \phi(y)) + (\phi(y)b\phi(x)) * \phi(z)\\
&+\phi(z) * (\phi(x)a\phi(y)) = \phi(z)(a-b)(\phi(x)a\phi(y) + \phi(y)b\phi(x)).
\end{aligned}$$

Subtracting this from (3.7), we obtain by (3.1) and (3.2)

$$\begin{aligned}
&g(x,y) * \phi(z) + \phi(z) * g(x,y)\\
&\quad = \phi(z(p-q)(xpy + yqx)) - \phi(z)(a-b)(\phi(x)a\phi(y) + \phi(y)b\phi(x))\\
&\quad = g^-(z, xpy + yqx)\\
&\qquad -\phi(z)(a-b)(-\phi(xpy) + \phi(x)a\phi(y) - \phi(yqx) + \phi(y)b\phi(x))\\
&\quad = g^-(z, xpy + yqx) + 2\phi(z)(a-b)g(x,y),
\end{aligned}$$

which implies (3.4). □

If $u \in A$, then denote

$$R^0(u) = \{x \in A : uxu = 0\},$$

$l(u) = \{x \in A : xu = 0\}$ and $r(u) = \{x \in A : ux = 0\}$.

Lemma 3.3. *Let A and B be prime associative algebras over F. If $\phi : A(p,q) \to B(a,b)$ is an isomorphism, then*

(i) $\phi(R^0(p-q)) = R^0(a-b)$.

(ii) *If $z \in l(p-q)$, then $g^-(z,A) \subseteq r(a-b)$ and $h^-(z,A) \subseteq l(a-b)$.*

(iii) *If $z \in l(p-q)$, then $\phi(z)(a-b)\phi(xpy+yqx)(a-b)\phi(z) = 0$ for all $x, y \in A$.*

Proof. (i) We note from Lemma 2.2(ii) that $x \in R^0(p-q)$ if and only if $y(p-q)x(p-q)y = 0$ for any $y \in A$. If $\{x,y\}^*$ denotes the product in $A(p,q)^+ = A(p-q)^+$, then

$$y(p-q)x(p-q)y = 2\{\{x,y\}^*,y\}^* - \{x,\{y,y\}^*\}^*.$$

Since ϕ is an isomorphism, this shows that $x \in R^0(p-q)$ if and only if

$$2\{\{\phi(x),\phi(y)\}^*,\phi(y)\}^* - \{\phi(x),\{\phi(y),\phi(y)\}^*\}^* = 0$$

for all $y \in A$ if and only if $\phi(z) \in R^0(a-b)$.

(ii) If $z \in l(p-q)$, then by (3.2) $g^-(z,x) = \phi(z(p-q)x) - \phi(z)(a-b)\phi(x) = -\phi(z)(a-b)\phi(x)$ for any $x \in A$. But, since $z \in R^0(p-q)$ and $\phi(z) \in R^0(a-b)$ by part (i), $(a-b)g^-(z,x) = -(a-b)\phi(z)(a-b)\phi(x) = 0$, as desired. The second relation similarly follows.

(iii) If $z \in l(p-q)$, then $\phi(z) \in R^0(a-b)$ and by (3.4)

$$\begin{aligned}\phi(z)(a-b)g(x,y)(a-b)\phi(z) &= \phi(z)(a-b)\phi(z)(a-b)g(x,y)\\ &\quad +\phi(z)(a-b)g^-(z,xpy+yqx) = 0.\end{aligned}$$

Similarly, $\phi(z)(a-b)h(x,y)(a-b)\phi(z) = 0$ using (3.5), and hence

$$\begin{aligned}0 &= \phi(z)(a-b)(g(x,y)+h(x,y))(a-b)\phi(z)\\ &= \phi(z)(a-b)\phi(xpy+yqx)(a-b)\phi(z).\end{aligned}$$ □

For a prime associative algebra A, the following result gives a condition for $A(p,q)$ to be commutative.

Lemma 3.4. *Let A be a prime associative algebra over F. Then, for $p, q, u \in A$,*

(i) *$A(u)$ is commutative if and only if either $u = 0$ or A is commutative.*

(ii) *$A(p, q)$ is commutative if and only if either $p + q = 0$ or A is commutative.*

(iii) *If $Z(A)$ contains a nonzero ideal of A, then A is commutative.*

Proof. (i) If $xuy = yux$ for any $x, y \in A$, then $xu^2 = u^2x$, so $u^2 \in Z(A)$. If $u^2 = 0$, then $xu(yu) = (yu)ux = 0$ and $(AuA)^2 = 0$, so $u = 0$ since A is prime. Otherwise, $uxu^2 = u(u^2x) = u^2ux = xuu^2$ for any $x \in A$, hence $[u, x]u^2 = 0$. Since A is prime, $Z(A)$ does not contain zero divisors. Thus, $[u, x] = 0$ or $u \in Z(A)$ and so $[x, y]u = xuy - yux = 0$ for any $x, y \in A$. This shows that A is commutative. The converse is clear.

(ii) If $A(p, q)$ is commutative, then $xpy - yqx = ypx - xqy$ and so $x(p+q)y = y(p + q)x$ for all $x, y \in A$. Thus, $A(p + q)$ is commutative, and the result follows from part (i).

(iii) If $Z(A)$ contains a nonzero ideal I of A, then for $0 \neq a \in I$ we have $a[x, y] = [ax, y] \in [I, A] = 0$ and so $[x, y] = 0$ for all $x, y \in A$. □

Corollary 3.5. *Let A be an associative algebra over F and let p, q be distinct elements of A.*

(i) *If A is semiprime, then A is commutative if and only if A^+ is associative.*

(ii) *If $A(p, q)$ is prime, then A is commutative if and only if $A(p, q)$ is commutative and associative.*

Proof. (i) If $(x, y, z)^+$ denotes the associator in A^+, then

$$\begin{aligned}(x, y, z)^+ &= \frac{1}{4}[(xy + yx)z + z(xy + yx) - x(yz + zy) - (yz + zy)x] \\ &= \frac{1}{4}[y, [x, z]].\end{aligned}$$

Hence, A^+ is associative if and only if $[A, [A, A]] = 0$ if and only if A is commutative by Lemma I.4.15.

(ii) If A is commutative, then $A(p, q) = A(p - q)$ is clearly commutative and associative. Conversely, if $A(p, q)$ is commutative and associative, then by Lemma 3.4(ii), either A is commutative or $p+q = 0$. If $p+q = 0$, then since $A(p, q) = A(2p)^+$ is a prime associative algebra, $A(2p)$ is commutative by part (i). Since A is also prime, by Lemma 3.4(i) A is commutative. □

For isomorphisms between prime mutation algebras, we have:

Proposition 3.6. *Let A and B be associative algebras over F where B is not commutative. Let $p, q \in A$ and $a, b \in B$ with $p \neq q$ and $a \neq b$. If $B(a,b)$ is prime and $\phi : A(p,q) \to B(a,b)$ is an isomorphism, then either $g(A,A) = 0$ or $h(A,A) = 0$.*

Proof. Assume that $g(A,A) \neq 0 \neq h(A,A)$. We first show that this assumption implies $R^0(p-q) = 0 = R^0(a-b)$. If $p+q=0$, then $R^0(p-q) = R^0(p,q) = 0$ by Corollary II.3.7(i), since $A(p,q)$ is prime. Assume then that $p+q \neq 0$. By Lemma II.3.2(i), there is a nonzero ideal D of A contained in the linear span of $\{xpy + yqx : x, y \in A\}$, and by Lemma 3.3(iii) $\phi(D)$ is a nonzero ideal of $B(a,b)$ such that

$$\phi(z)(a-b)\phi(D)(a-b)\phi(z) = 0 \tag{3.8}$$

for any $z \in l(p-q)$. Since $B(a,b)$ is prime, $R^0(a,b) = 0$ and hence by Proposition II.3.5 $\phi(D)$ must contain a nonzero ideal of B. Since B is prime, this together with (3.8) implies that $\phi(z)(a-b) = 0$ or $(a-b)\phi(z) = 0$. In the former case, by (3.2)

$$g^-(z,x) = \phi(z(p-q)x) - \phi(z)(a-b)\phi(x) = 0 \tag{3.9}$$

for any $x \in A$ and hence by (3.4) $g(A,A)(a-b)\phi(z) = 0$. Therefore,

$$(g(A,A) + Bag(A,A))(a-b)\phi(z) = 0. \tag{3.10}$$

Since $g(A,A) + Bag(A,A)$ is a nonzero ideal of $B(a,b)$ by Lemma 3.1 and since $R^0(a,b) = 0$, it contains a nonzero ideal of B (Proposition II.3.5). Thus, from (3.10) we have $(a-b)\phi(z) = 0$. Applying the same arguments to $g^{op} = h$ with $\phi : A(p,q) \to B^{op}(-b,-a)$, we find that if $(a-b)\phi(z) = 0$, then $h^-(z,A) = 0$ and so $\phi(z)(a-b) = 0$. Therefore,

$$\phi(l(p-q)) \subseteq l(a-b) \cap r(a-b). \tag{3.11}$$

Let $l^{op}(p-q)$ denote the left annihilator of $p-q$ in A^{op}. Then, $r(p-q) = l^{op}(p-q)$, $l(p-q) = r^{op}(p-q)$, and $\phi : A^{op}(q,p) \to B^{op}(b,a)$ is an isomorphism. Applying the argument above to these gives

$$\phi(r(p-q)) \subseteq l(a-b) \cap r(a-b). \tag{3.12}$$

Now, for $z \in l(p-q)$, $g^-(x,z) = \phi(x(p-q)z) - \phi(x)(a-b)\phi(z) = \phi(x(p-q)z)$ by (3.11), but $g^-(x,z) = -g^-(z,x) = 0$ by (3.9). Hence, $\phi(x(p-q)z) = 0$, i.e., $A(p-q)z = 0$ and so $z \in r(p-q)$ since A is prime. This shows that $l(p-q) \subseteq r(p-q)$, and similarly, $r(p-q) \subseteq l(p-q)$ using (3.12). Therefore, $l(p-q) = r(p-q)$ is an ideal of A, and $r(p-q)A(p-q) \subseteq r(p-q)(p-q) = l(p-q)(p-q) = 0$, so $r(p-q) = 0 = l(p-q)$. This forces $R^0(p-q) = 0$, which implies $R^0(a-b) = \phi(R^0(p-q)) = 0$ by Lemma 3.3(i).

Therefore, from Corollary II.3.7(i) we conclude that $A(p,q)^+ = A(p-q)^+$ and $B(a,b)^+ = B(a-b)^+$ are prime (so are $A(p-q)$ and $B(a-b)$). Since $\phi : A(p-q) \to B(a-b)$ is a J–homomorphism, by Theorem 2.3 ϕ is either an isomorphism or an anti–isomorphism of $A(p-q)$ to $B(a-b)$. Assume that $\phi : A(p-q) \to B(a-b)$ is an isomorphism, or equivalently, $g^-(A,A) = 0$ (see (3.2)). Thus, by (3.4) $g(A,A) \subseteq Z(B(a-b))$. Using the first two relations in the proof of Lemma 3.1, we obtain

$$\phi(z)(a-b)g(x,y) \equiv g^-(z,x)a\phi(y) = 0$$

modulo $g(A,A)$, and so $B(a-b)g(A,A) \subseteq g(A,A)$. Similarly, the last two relations in that proof imply $g(A,A)(a-b)B \subseteq g(A,A)$. Hence, $g(A,A)$ is a nonzero ideal of $B(a-b)$ contained in $Z(B(a-b))$. This together with Lemma 3.4(i),(iii) forces B to be commutative, a contradiction. If ϕ is an anti–isomorphism of $A(p-q)$ to $B(a-b)$, then we deduce the same conclusion using the isomorphism $\phi : A(p-q) \to B^{op}(a-b)$ with $g^{op} = h$. This completes the proof of the theorem. □

The following example shows that Proposition 3.6 is not true if B is commutative.

Example 3.7. Let $B = F[X]$ be the polynomial algebra in X over F. Note that B is an integral domain and so is a prime algebra. We choose $p = X$, $q = 0$, $a = 1$, $b = 1 - X^2$, and let $A = BX$, the ideal generated by X. The map

$$\phi : A(p,q) \to B(a,b) \ : \ uX \mapsto u$$

is an isomorphism since

$$\phi(uX * vX) = \phi(uvX^3) = uvX^2 = uv - v(1-X^2)u = \phi(uX) * \phi(vX).$$

But, $g(X,X) = \phi(XpX) - \phi(X)a\phi(X) = X^2 - 1 \neq 0$ and $h(X,X) = \phi(XpX) + \phi(X)b\phi(X) = X^2 + (1-X^2) = 1 \neq 0$. Since $R^0(a,b) = 0$, $B(a,b)$ is prime. Moreover, $B(a,b) = B(a-b,0) = B(a-b)$ and $g^-(uX,vX) = \phi(uXpvX) - \phi(uX)(a-b)\phi(vX) = uvX^2 - uX^2v = 0$. □

From Example 3.7, we observe that if A and B are commutative, then there are infinitely many choices of pairs (p,q) or (a,b) which give rise to the same mutation algebra. Note also that if B is commutative then we can regard the mutation $B(a,b)$ as $B(a-b,0)$ or $B(a-b)$.

Proposition 3.8. *Let B be a prime commutative associative algebra over F and let A be any associative algebra over F. Let $p,q \in A$ and $a,b \in B$ with $p \neq q$ and $a \neq b$. If $\phi : A(p,q) \to B(a,b)$ is an isomorphism, then A is commutative and ϕ is an isomorphism of $A(p-q)$ to $B(a-b)$.*

Proof. Since B is commutative, $B(a,b) = B(a-b)$ is commutative and associative. Since B is prime, $a-b \neq 0$ is not a zero divisor of B and so $R^0(a-b) = 0$. Hence, $B(a,b) = B(a-b,0)$ is prime (Corollary II.3.7). Since ϕ is an isomorphism, we find that $A(p,q)$ is a prime commutative associative algebra, and hence A is commutative by Corollary 3.5(ii). □

The following result determines isomorphisms between mutations of prime associative algebras under some restrictions.

Theorem 3.9. *Let B be a prime noncommutative associative algebra over F with unit element 1 and let (a,b) be a normal pair in B with $a \neq b$. Let A be any associative algebra over F and let $p,q \in A$ with $p \neq q$. If ϕ is an isomorphism of $A(p,q)$ to $B(a,b)$, then there exist an invertible element $c \in B$, and a map $\sigma : A \to B$ such that*

$$\phi(x) = \sigma(x)c \tag{3.13}$$

for $x \in A$, where σ is either an isomorphism of A to B satisfying

$$\sigma(p) = ca, \quad \sigma(q) = cb, \tag{3.14}$$

or is an anti–isomorphism of A to B satisfying

$$\sigma(p) = -cb, \quad \sigma(q) = -ca. \tag{3.15}$$

Conversely, for any isomorphism or anti–isomorphism σ of A to B satisfying (3.14) or (3.15) and any invertible element $c \in B$, the map ϕ defined by (3.13) is an isomorphism of $A(p,q)$ to $B(a,b)$.

Proof. Since (a,b) is normal in B and B has a unit element,

$$R^0(a,b) \subseteq \{x \in B : BxB = 0\} = 0$$

and so $B(a,b)$ is prime. Hence, by Proposition 3.6, either $g(A,A) = 0$ or $h(A,A) = 0$. Assume $g(A,A) = 0$, so that $\phi(xpy) = \phi(x)a\phi(y)$ and $\phi(xqy) = \phi(x)b\phi(y)$ for any $x,y \in A$. Let $e \in A$ be such that $\phi(e) = 1$. Then,

$$\phi(e(px+qy)) = \phi(epx) + \phi(eqy) = a\phi(x) + b\phi(y)$$

for $x,y \in A$, and hence $B = aB + bB = \phi(e(pA+qA))$, so $e(pA+qA) = A$ and similarly, $A = (Ap+Aq)e$. Therefore, $A = Ae = eA$, which implies that A has a unit element and e is invertible (an element u with $e = ue$ is a unit element). Consequently, (p,q) is normal in A. If we let $c = \phi(1) \in B$, then

$$B = \phi(A) = \phi(1pA + 1qA) = c(aB + bB) = cB,$$

and similarly, $Bc = B$. Hence, c is invertible in B.

We define the linear map σ by

$$\sigma : A \to B \ : \ x \mapsto \phi(x)c^{-1}. \tag{3.16}$$

For any $x, y \in A$, we obtain

$$\sigma(xpy) = \phi(xpy)c^{-1} = \phi(x)a\phi(y)c^{-1} = (\phi(x)c^{-1})ca(\phi(y)c^{-1}) = \sigma(x)ca\sigma(y),$$

and also $\sigma(xqy) = \sigma(x)cb\sigma(y)$. If $x \in A$ and $y = py_1 + qy_2$ for $y_1, y_2 \in A$, then

$$\begin{aligned} \sigma(xy) = \sigma(xpy_1) + \sigma(xqy_2) &= \sigma(x)ca\sigma(y_1) + \sigma(x)cb\sigma(y_2), \\ \sigma(y) = \sigma(1py_1) + \sigma(1qy_2) &= \sigma(1)ca\sigma(y_1) + \sigma(1)cb\sigma(y_2) \\ &= ca\sigma(y_1) + cb\sigma(y_2), \end{aligned}$$

using the relations above and (3.16). Hence, $\sigma(xy) = \sigma(x)\sigma(y)$ for $x \in A$, $y \in pA + qA$, and σ is an isomorphism of A to B. Moreover, $\sigma(p) = \sigma(1p1) = \sigma(1)ca\sigma(1) = ca$ and similarly, $\sigma(q) = cb$, giving (3.14).

Conversely, if σ is an isomorphism of A to B satisfying (3.14) and c is any invertible element of B, then let $\phi : A(p,q) \to B(a,b)$ be the map defined by (3.13). For $x, y \in A$, we get

$$\begin{aligned} \phi(xpy - yqx) &= \sigma(xpy)c - \sigma(yqx)c = \sigma(x)\sigma(p)\sigma(x)c - \sigma(y)\sigma(q)\sigma(x)c \\ &= \sigma(x)ca\sigma(y)c - \sigma(y)cb\sigma(x)c = \phi(x)a\phi(y) - \phi(y)b\phi(x), \end{aligned}$$

hence ϕ is an isomorphism.

If $h(A, A) = 0$, then by (3.1) $\phi(xpy) = -\phi(y)b\phi(x)$ and $\phi(yqx) = -\phi(x)a\phi(y)$ for $x, y \in A$. Using the isomorphism $\phi : A(p,q) \to B^{op}(-b, -a)$ with $g^{op} = h$, the arguments above imply that the map σ given by (3.16) is an isomorphism of A to B^{op}, and so is an anti–isomorphism of A to B. But, $\sigma(xpy) = -\phi(y)b\phi(x)c^{-1} = -\sigma(y)cb\sigma(x)$ and $\sigma(xqy) = -\phi(y)a\phi(x)c^{-1} = -\sigma(y)ca\sigma(x)$. Hence, σ satisfies (3.15). □

The following result is immediate from the proof of Theorem 3.9.

Corollary 3.10. *Let $A, B, (p,q)$ and (a,b) be the same as in Theorem 3.9. If ϕ is an isomorphism of $A(p,q)$ to $B(a,b)$, then A has a unit element 1, (p,q) is normal in A, and the invertible element c of B satisfying (3.13)–(3.15) can be taken to be $\phi(1)$.* □

Note in Example 3.7 that $B = F[X]$ and $A = BX$ satisfy the conditions in Theorem 3.9, but $A \not\cong B = B^{op}$. Thus, Theorem 3.9 is not true when B is commutative. However, when an isomorphism $\phi : A(p,q) \to B(a,b)$ is constrained to the condition $\phi(1) = 1$, we have:

Corollary 3.11. *Let A and B be prime associative algebras over F with unit element. Let $p, q \in A$ with $p \neq q$ and (a, b) be a normal pair in B with $a \neq b$. If $\phi : A(p,q) \to B(a,b)$ is an isomorphism with $\phi(1) = 1$, then ϕ is either an isomorphism or an anti-isomorphism of A to B.*

Proof. If B is not commutative, then since $c = \phi(1) = 1$ and so $\phi = \sigma$, the result follows from Theorem 3.9. Assume B is commutative. By Proposition 3.8 ϕ becomes an isomorphism of $A(p-q)$ to $B(a-b)$, both being associative and commutative. Hence, for $x, y \in A$,

$$\phi(x(p-q)y) = \phi(x)(a-b)\phi(y) = \phi(x)\phi(y)(a-b),$$
$$\phi(x(p-q)y) = \phi(xy(p-q)1) = \phi(xy)(a-b)\phi(1) = \phi(xy)(a-b),$$

since A is commutative (Proposition 3.8). Since B is prime and commutative, $a-b$ is not a zero divisor in B. Hence, $\phi(xy) = \phi(x)\phi(y)$ for $x, y \in A$ and ϕ is an isomorphism of A to B. □

Theorem 3.9 can be applied to determine surjective homomorphisms between prime mutation algebras in a form given by (3.13)–(3.15).

Corollary 3.12. *Let A and B be associative algebras over F where A has a unit element, and let (p, q) be normal in A with $p \neq q$. Assume B is not commutative and $B(a, b)$ is prime for $a \neq b$ in B. If $\phi : A(p,q) \to B(a,b)$ is a surjective homomorphism, then there exist an invertible element $c \in B$, and a map $\sigma : A \to B$ satisfying (3.13) and the conditions that σ is either a homomorphism satisfying (3.14) or an anti-homomorphism of A to B satisfying (3.15). Conversely, if $\sigma : A \to B$ is any map satisfying these conditions and c is invertible in B, then the map $\phi : A(p,q) \to B(a,b)$ defined by (3.13) is a surjective homomorphism.*

Proof. Let D be the largest ideal of A contained in $\ker\phi$ and let $\bar{A} = A/D$. Then, ϕ induces a surjective homomorphism $\bar{\phi} : \bar{A}(\bar{p}, \bar{q}) \to B(a,b)$, where $\bar{p} = p + D$ and $\bar{q} = q + D$. If $\bar{p} = \bar{q}$, then $\bar{A}(\bar{p}, \bar{q})$ is anticommutative and so is $\bar{\phi}(\bar{A}(\bar{p}, \bar{q})) = B(a,b)$. Hence, $0 = x * x = x(a-b)x$ for all $x \in B$ and by Lemma 2.2(ii) $a - b = 0$, a contradiction. Therefore, $\bar{p} \neq \bar{q}$, and it suffices to prove the theorem for $\bar{\phi} : \bar{A}(\bar{p}, \bar{q}) \to B(a,b)$. For, if $\bar{\sigma} : \bar{A} \to B$ is a map satisfying (3.13) with condition (3.14) or (3.15), then $\sigma = \bar{\sigma} \circ \pi$ would give a desired map : $A \to B$ for ϕ, where $\pi : A \to \bar{A}$ is the natural projection. Thus, we may assume that $\ker\phi$ does not contain proper ideals of A.

Let I and J be ideals of A with $IJ = 0$. Then, either $JI = 0$, so $I * J = 0$, or $(JI)^2 = 0$, so $(JI)*(JI) = 0$. In any case, we find ideals I, J of A with $I * J = 0$, and hence $\phi(I) * \phi(J) = 0$. Since $B(a,b)$ is prime, I or J is contained in $\ker\phi$, so $I = 0$

or $J = 0$. This shows that A is prime. Hence, by Proposition II.3.5 $\ker\phi \subseteq R^0(p,q)$, but

$$AR^0(p,q)A = (Ap + Aq)R^0(p,q)(pA + qA) = 0,$$

so $R^0(p,q) = 0$ since A is prime. Thus, $\ker\phi = 0$ and ϕ is an isomorphism. The result now follows from Theorem 3.9. □

The next example illustrates that the condition of normal pair (p,q) in A or (a,b) in B is essential for Theorem 3.9 and its corollaries.

Example 3.13. Let A be the free associative algebra over F with unit element generated by $\{X,Y\}$, that is, A is the algebra of noncommutative polynomials in X, Y over F. Thus, a basis of A consists of the monomials

$$X^{r_1}Y^{s_1}X^{r_2}Y^{s_2}\cdots X^{r_n}Y^{s_n},$$

for n, r_i, s_i $(i = 1,\dots,n)$ in $\mathbb{N}$, the set of nonnegative integers, with $r_{i+1} \neq 0 \neq s_i$ for $i = 1,\dots n-1$.

Let $k : \mathbb{N} \to \mathbb{N}$ be given by $k(0) = 0$, $k(1) = 2$, $k(2) = 1$ and $k(n) = n$ for $n > 2$, and define a bijective linear map $\phi : A \to A$ by

$$\phi(X^{r_1}Y^{s_1}X^{r_2}Y^{s_2}\cdots X^{r_n}Y^{s_n}) = Y^{r_1}X^{k(s_1)}Y^{r_2}X^{k(s_2)}\cdots Y^{r_n}X^{k(s_n)},$$

that is, ϕ interchanges X and Y, and modifies the exponents by means of k. Clearly, $\phi(uXv) = \phi(u)Y\phi(v)$ for any $u, v \in A$, hence

$$\phi : A(X) = A(X,0) \to A(Y) = A(Y,0)$$

is an isomorphism with $\phi(1) = 1$. Note that $(Y,0)$ is not normal in A, i.e., $AY \neq A \neq YA$. Thus, if the conclusion of Theorem 3.9 were valid in this case, then ϕ would be either an automorphism or an anti–automorphism of A, but $\phi(Y) = X^2$ and $\phi(Y^2) = X \neq \phi(Y)^2$. □

However, for simple mutations of artinian associative algebras, Theorem 3.9 and its corollaries hold without the normality condition.

Corollary 3.14. *Let B be an artinian associative algebra over F such that $B(a,b)$ is simple for $a,b \in B$. Let A be an associative algebra over F and let $p,q \in A$. If $\phi : A(p,q) \to B(a,b)$ is an isomorphism, then there exist an invertible element $c \in B$ and a map $\sigma : A \to B$ which satisfy the relations in Theorem 3.9. Conversely, for any such map $\sigma : A \to B$ and an invertible element $c \in B$, the map $\phi : A(p,q) \to B(a,b)$ defined by (3.13) is an isomorphism.*

Proof. Since $B(a,b)$ is simple, so is B. Since B is artinian, $a \neq b$ by Theorems II.3.14(iv) and II.4.7 and so $R^0(a,b) = 0$ (Corollary II.3.7). Thus, it follows from Lemma 1.1 that (a,b) is normal in B. If $p = q$, then $B(a,b)$ is anticommutative and by Lemma 2.2(ii) we would have $a - b = 0$. Therefore, $p \neq q$, and if B is not commutative, then the result follows from Theorem 3.9.

Assume B is commutative. Then, $B(a,b) = B(a-b) = B(a-b,0)$ is simple, so $R^0(a-b) = 0$ and $(a-b,0)$ is a normal pair in B (Lemma 1.1). By Proposition 3.8, A is commutative, and ϕ is an isomorphism of $A(p-q)$ to $B(a-b)$. Hence, the proof of Theorem 3.9 applies to this case, regarding $A(p,q) = A(p-q,0)$ and $B(a,b) = B(a-b,0)$, i.e., replacing p by $p-q$, q by 0, a by $a-b$ and b by 0 (note $(g(A,A) = 0)$. □

When $A = B$ is central simple artinian with $A(p,q)$ simple, Theorem 3.9 or Corollary 3.14 enables us to determine $\operatorname{Aut} A(p,q)$ in terms of the group $\Phi(H)$ or its extension by index 2, as discussed in Section IV.1. Let $A^\times$ denote the group of invertible elements in A, and let H, K be the subgroups of $A^\times \times A^\times$ given by (1.1), (1.2). As in (1.3) and (1.4), let

$$\Phi_{a,b} : A(p,q) \longrightarrow A(p,q) \ : \ x \mapsto axb^{-1},$$
$$\Phi : H \longrightarrow \operatorname{Aut} A(p,q) \ : \ (a,b) \mapsto \Phi_{a,b}$$

for $(a,b) \in H$. Recall from Lemma 1.2 that $\ker \Phi = K$ and $\Phi(H) \cong H/K$.

Theorem 3.15. *Let A be a central simple noncommutative artinian associative algebra over F such that $A(p,q)$ is simple. Let H, K and Φ be as above.*

(i) *If there do not exist anti-automorphisms of A, then* $\operatorname{Aut} A(p,q) = \Phi(H) \cong H/K$.

(ii) *If there exist anti-automorphisms of A, then either* $\operatorname{Aut} A(p,q) = \Phi(H)$ *or* $[\operatorname{Aut} A(p,q) : \Phi(H)] = 2$.

Proof. (i) In view of Lemma 1.2, it suffices to show that Φ is surjective. Let $\phi \in \operatorname{Aut} A(p,q)$ be any element. Since A does not have anti-automorphisms, by Corollary 3.14 ϕ is given by $\phi(x) = \sigma(x)c$ ($x \in A$) for some $c \in A^\times$ and $\sigma \in \operatorname{Aut} A$, satisfying the relations $\sigma(p) = cp$ and $\sigma(q) = cq$. But, since A is central simple artinian, by the theorem of Noether–Skolem [**P1**, p.230], there is an $a \in A^\times$ such that $\sigma(x) = axa^{-1}$ for any $x \in A$. Letting $b = c^{-1}a$, we have $\sigma(p) = apa^{-1} = cp$, so $bp = pa$, and similarly, $bq = qa$. Hence $(a,b) \in H$ (see (1.1)), and $\phi(x) = \sigma(x)c = axa^{-1}c = axb^{-1}$, i.e., $\phi = \Phi_{a,b}$. Therefore, Φ is surjective.

(ii) Assume A has an anti-automorphism τ. If any $\phi \in \operatorname{Aut} A(p,q)$ is given by (3.13) with $\sigma \in \operatorname{Aut} A$ satisfying (3.14), then as above we have $\operatorname{Aut} A(p,q) = \Phi(H)$. Otherwise, by Corollary 3.14, there is a $\phi \in \operatorname{Aut} A(p,q)$ such that $\phi(x) = \sigma(x)c$

($x \in A$) for some $c \in A^\times$ and an anti-automorphism σ of A satisfying $\sigma(p) = -cq$ and $\sigma(q) = -cp$. But, since $\sigma \circ \tau^{-1} \in \operatorname{Aut} A$, $\sigma \circ \tau^{-1}(x) = axa^{-1}$ for some $a \in A^\times$ and so $\sigma(x) = \sigma \circ \tau^{-1} \circ \tau(x) = a\tau(x)a^{-1}$ for any $x \in A$. Hence,

$$\sigma(p) = a\tau(p)a^{-1} = -cq, \qquad c^{-1}a\tau(p) = -qa,$$
$$\sigma(q) = a\tau(q)a^{-1} = -cp, \qquad c^{-1}a\tau(q) = -pa.$$

Letting $b = c^{-1}a$, from this we obtain $b\tau(p) = -qa$, $b\tau(q) = -pa$, and

$$\phi(x) = \sigma(x)c = a\tau(x)a^{-1}c = a\tau(x)b^{-1}$$

for any $x \in A$. Therefore, $\phi = \tau(a,b) \in T(A) \subseteq \operatorname{Aut} A(p,q)$, given by (1.5) and (1.6) (also see Proposition 1.3). Since A is not commutative, it follows from this and Proposition 1.3 that $\operatorname{Aut} A(p,q) = \Phi(H) \cup T(A)$ (disjoint) with $T(A) \circ T(A)^{-1} \subseteq \Phi(H)$. Therefore, $[\operatorname{Aut} A(p,q) : \Phi(H)] = 2$. □

If the algebra A in Theorem 3.15 is commutative, then the result trivially holds, since $A \cong F \cong A(p,q)$, so $\operatorname{Aut} A(p,q) = \{id\}$ and $H = K$. Note that the case $[\operatorname{Aut} A(p,q) : \Phi(H)] = 2$ occurs if and only if $T(A) \neq \emptyset$. If $T(A) \neq \emptyset$, then for any $\tau(a,b) \in T(A)$, $\operatorname{Aut} A(p,q) = \Phi(H) \cup \tau(a,b) \circ \Phi(H)$.

4. Derivations of mutations of associative algebras

Let A be an associative algebra over F. For any derivation

$$d : A(p,q) \longrightarrow A(p,q),$$

define the bilinear map $f : A \times A \to A$ by

$$f(x,y) = d(xpy) - d(x)py - xpd(y) = d(yqx) - d(y)qx - yqd(x)$$

for $x, y \in A$ and denote by $f(A,A)$ the linear span of $f(x,y)$ for all $x, y \in A$. Let

$$f^-(x,y) = f(x,y) - f(y,x) = d(x(p-q)y) - d(x)(p-q)y - x(p-q)d(y).$$

Note that $f^-(A,A) = 0$ if and only if d is a derivation of the $(p-q)$–homotope $A(p-q)$ of A.

In this section we determine derivations of prime mutations under similar restrictions as in Theorem 3.9. A crucial step is to show that $f(A,A) = 0$ for any derivation d of a prime mutation algebra $A(p,q)$. The proof of this invokes Theorem 2.7. Finally, $\operatorname{Der} A(p,q)$ is described in terms of the Lie algebra L given by (1.8) for any finite–dimensional central simple associative algebra with $A(p,q)$ simple.

Lemma 4.1. *Let A and f be the same as above, and let $M = f(A,A) + f(A,A)pA$. Then,*

$$M = f(A,A) + f(A,A)qA = f(A,A) + Apf(A,A) = f(A,A) + Aqf(A,A)$$

and M is an ideal of $A(p,q)$.

Proof. We have $d(xpy) \equiv d(x)py + xpd(y)$ and $d(xqy) \equiv d(x)qy + xqd(y)$ modulo $f(A,A)$. Therefore, for $x, y, z \in A$,

$$\begin{aligned} zqf(x,y) &= zqd(xpy) - zqd(x)py - zqxpd(y) \\ &\equiv d(zqxpy) - d(z)qxpy - zqd(x)py - d(zqxpy) + d(zqx)py \\ &= f(x,z)py, \end{aligned}$$

$$zqf(x,y) = zqd(yqx) - zqd(y)qx - zqyqd(x) \equiv f(y,z)qx. \tag{4.1}$$

Similarly, we obtain

$$zpf(x,y) \equiv f(y,z)qx, \quad zpf(x,y) \equiv f(z,x)py. \tag{4.2}$$

Hence, $Apf(A,A) \equiv Aqf(A,A) \equiv f(A,A)pA \equiv f(A,A)qA$ modulo $f(A,A)$ and the result follows easily. □

Lemma 4.2. *For any $x, y, z \in A$, the following identity holds:*

$$f(x,y)(p-q)z - z(p-q)f(x,y) = f^{-}(z, xpy + yqx). \tag{4.3}$$

Proof. Applying d to (3.6) in the proof of Lemma 3.2, one has

$$\begin{aligned} & d((x*z)*y - x*(z*y)) + d(yqx)*z \\ & +(yqx)*d(z) + d(z)*(xpy) + z*d(xpy) \\ & = d(z(p-q)(xpy + yqx)). \end{aligned} \tag{4.4}$$

Substituting in (3.6) $d(x)$ for x, $d(y)$ for y and $d(z)$ for z, and adding the resulting identities, we obtain

$$\begin{aligned} & d((x*z)*y - x*(z*y)) + (d(y)qx)*z + (yqd(x))*z \\ & \quad +(yqx)*d(z) + z*(d(x)py) + z*(xpd(y)) + d(z)*(xpy) \\ & = d(z)(p-q)(xpy + yqx) + z(p-q)(d(x)py + yqd(x)) \\ & \quad +z(p-q)(xpd(y) + d(y)qx). \end{aligned}$$

Subtracting this from (4.4) gives

$$f(x,y)*z + z*f(x,y) = f^{-}(z, xpy + yqx) + 2z(p-q)f(x,y)$$

which is equivalent to (4.3). □

Proposition 4.3. *Let A be a noncommutative associative algebra over F such that $A(p,q)$ is prime for $p,q \in A$ with $p \neq q$. Then, for any derivation d of $A(p,q)$, $f(A,A)=0$.*

Proof. If $p+q=0$, then $A(p,q)=A(p)^+$ is prime and so is $A(p)$ (Corollary II.3.7(iv)). Hence, by Theorem 2.7 d is a derivation of $A(p)$, so $f(A,A)=0$ in this case. Assume then that $p+q \neq 0$. By Lemma II.3.2(i), there is a nonzero ideal D of A contained in the linear span of $\{xpy+yqx : x,y \in A\}$. Let z be any element of $l(p-q)$. For $x,y \in A$, by (4.3) we obtain

$$f^-(z, xpy+yqx) = f(x,y)(p-q)z \in l(p-q). \tag{4.5}$$

Thus, $f^-(z,D) \in l(p-q)$, but since $f^-(z,D) = -d(z)(p-q)D$,

$$d(z)(p-q)D(p-q) = 0.$$

Since A is prime, this implies $d(z)(p-q)=0$, and so $d(l(p-q)) \subseteq l(p-q)$. By the same argument, one gets $d(r(p-q)) \subseteq r(p-q)$. Hence, for $z \in l(p-q)$, $f^-(z,A) = -d(z)(p-q)A = 0$ which by (4.5) implies $f(x,y)(p-q)z=0$ for all $x,y \in A$. From this we deduce

$$(f(A,A)+Apf(A,A))(p-q)l(p-q) = 0. \tag{4.6}$$

Suppose $f(A,A) \neq 0$. Then, by Lemma 4.1, $f(A,A)+Apf(A,A)$ is a nonzero ideal of $A(p,q)$ and contains a nonzero ideal I of A by Proposition II.3.5. By (4.6), $I(p-q)Al(p-q)=0$ and so $l(p-q)=0$ by primeness of A. From $d(r(p-q)) \subseteq r(p-q)$, we similarly argue that $r(p-q)=0$. But, this is equivalent to $R^0(p-q)=0$. Thus, by Corollary II.3.7(iii), $A(p-q)$ is prime. Since $A(p,q)^+ = A(p-q)^+$ and d is also a derivation of $A(p-q)^+$, it follows from Theorem 2.7 that d is a derivation of $A(p-q)$, or equivalently, $f^-(A,A)=0$. Thus, by (4.3), $f(A,A) \subseteq Z(A(p-q))$. But, from (4.1) and (4.2), $z(p-q)f(x,y) \equiv 0$ modulo $f(A,A)$, which implies that $f(A,A)$ is a nonzero ideal of $A(p-q)$ contained in $Z(A(p-q))$. Since $A(p-q)$ and A are prime, it now follows from Lemma 3.4(i),(iii) that A is commutative, a contradiction. This proves $f(A,A)=0$. □

Let $A=F[X]$ be the polynomial algebra over F, and let $p=X$ and $q=X-1$. Then, $A(p,q)=A(p-q)=A$, being prime, and for the usual derivation d of A with $d(X)=1$, we have $f(X,1)=d(X^2)-d(X)X=X \neq 0$. Thus, Proposition 4.3 is not valid for commutative algebras. The following result determines derivations of prime mutation algebras.

Theorem 4.4. *Let A be a noncommutative associative algebra with unit element 1 over F, and let $p,q \in A$ be fixed elements such that $A(p,q)$ is prime with*

$p \neq q$ and $Ap + Aq = A$. A linear map $d : A(p,q) \to A(p,q)$ is a derivation of $A(p,q)$ if and only if d is expressed as

$$d = \hat{d} + R_b \tag{4.7}$$

for some $\hat{d} \in \operatorname{Der} A$ and some $b \in A$ satisfying the relations $\hat{d}(p) = bp$ and $\hat{d}(q) = bq$, where R_b is the right multiplication in A by b. If A is commutative, then the result holds with the conditions $\hat{d}(p-q) = b(p-q)$ and $A = (p-q)A$.

Proof. We note from Proposition 4.3 that $f(A, A) = 0$ for a linear map $d : A(p,q) \to A(p,q)$ if and only if

$$d(xpy) = d(x)py + xpd(y), \quad d(xqy) = d(x)qy + xqd(y) \tag{4.8}$$

for all $x, y \in A$, if and only if $d \in \operatorname{Der} A(p,q)$. If A is commutative then $A(p,q) = A(p-q) = A(p-q, 0)$. Thus, if $d \in \operatorname{Der} A(p,q)$, then d is a derivation of $A(p-q)$ and $d(x(p-q)y) = d(x)(p-q)y + x(p-q)d(y)$. Hence, (4.8) holds in this case by replacing p by $p-q$ and q by 0. Accordingly, if A is commutative, then we may assume $q = 0$.

Let $d \in \operatorname{Der} A(p,q)$. Since $A = Ap + Aq$, any $x \in A$ is expressed as $x = x_1p + x_2q$. Hence, by (4.8)

$$\begin{aligned} d(x) &= d(x_1p1) + d(x_2q1) = d(x_1)p + d(x_2)q + (x_1p + x_2q)d(1) \\ &= d(x_1)p + d(x_2)q + xd(1). \end{aligned}$$

If we let $\hat{d} = d - R_b$ with $b = d(1)$, then $\hat{d}(x) = d(x_1)p + d(x_2)q$ and for $y \in A$

$$\begin{aligned} \hat{d}(xy) &= d(xy) - xyb = d(x_1py + x_2qy) - xyb \\ &= (d(x_1)p + d(x_2)q)y + (x_1p + x_2q)d(y) - xyb \\ &= \hat{d}(x)y + x(d(y) - yb) = \hat{d}(x)y + x\hat{d}(y). \end{aligned}$$

Hence, $\hat{d} \in \operatorname{Der} A$, and $\hat{d}(p) = d(p) - pb = d(1p1) - pd(1) = d(1)p + pd(1) - pd(1) = bp$ and similarly, $\hat{d}(q) = bq$.

Conversely, if d is given by (4.7) for some $\hat{d} \in \operatorname{Der} A$ and $b \in A$ with $\hat{d}(p) = bp$ and $\hat{d}(q) = bq$, then

$$\begin{aligned} d(xpy) &= \hat{d}(xpy) + xpyb = \hat{d}(x)py + x\hat{d}(p)y + xp\hat{d}(y) + xpyb \\ &= \hat{d}(x)py + xbpy + xp\hat{d}(y) + xpyb = d(x)py + xpd(y) \end{aligned}$$

and similarly, $d(xqy) = d(x)qy + xqd(y)$. Thus, $d \in \operatorname{Der} A(p,q)$. □

If A is artinian and $A(p,q)$ is simple, then as in the proof of Corollary 3.14 A is simple with $p \neq q$ and (p,q) is normal in A (Lemma 1.1). Hence, as a consequence of Theorem 4.4, we have:

Corollary 4.5. *If A is artinian and $A(p,q)$ is simple, then $d : A(p,q) \to A(p,q)$ is a derivation of $A(p,q)$ if and only if d is given by (4.7) for some $\hat{d} \in \operatorname{Der} A$ and $b \in A$ with $\hat{d}(p) = bp$ and $\hat{d}(q) = bq$.* □

Let K be the Lie subalgebra of $A^- \times A^-$ given by (1.8) where we identify $Z(A) = \{(a,a) \in A \times A : a \in Z(A)\}$. As in (1.9) and (1.10), for $(a,c) \in L$, let

$$d_{a,c} : A(p,q) \to A(p,q) \ : \ x \mapsto ax - xc, \tag{4.9}$$

$$\rho : L \to \operatorname{Der} A(p,q) \ : \ (a,c) \mapsto d_{a,c}. \tag{4.10}$$

Recall from Lemma 1.4 that, for any associative algebra A with 1, ρ is a Lie algebra homomorphism such that $\ker \rho = Z(A)$, so $\rho(L) \cong L/Z(A)$. We now prove that if A is finite–dimensional, central simple with $A(p,q)$ simple, then ρ is surjective, so $\operatorname{Der} A(p,q) \cong L/Z(A)$. Our proof is based on Corollary 4.5 and the well known result that $\operatorname{Der} A$ is inner, i.e., for any $d \in \operatorname{Der} A$, $d = ad_a : A \to A \ : \ x \mapsto ad_a(x) = [a,x]$ for some $a \in A$.

Theorem 4.6. *Let A be a finite–dimensional, central simple associative algebra over F, and let $p, q \in A$ be such that $A(p,q)$ is simple. Then, $\operatorname{Der} A(p,q) = \rho(L) \cong L/Z(A)$.*

Proof. In view of Lemma 1.4, it suffices to show that the map $\rho : L \to \operatorname{Der} A(p,q)$ is surjective. Thus, let d be any derivation of $A(p,q)$. By Corollary 4.5, $d = \hat{d} + R_b$ for some $\hat{d} \in \operatorname{Der} A$ and $b \in A$ such that $\hat{d}(p) = bp$ and $\hat{d}(q) = bq$. But, there is an element $a \in A$ such that $\hat{d}(x) = ax - xa$ for any $x \in A$ [**P1**, p.230]. Hence, $bp = \hat{d}(p) = ap - pa$, so $(a-b)p = pa$, and also $(a-b)q = qa$. If we let $c = a - b$, then $(a,c) \in L$ and $d(x) = ax - xa + xb = ax - xc = d_{a,c}(x)$ for any $x \in A$, so $\rho((a,c)) = d$. Thus, ρ is surjective. □

5. Composition algebras: The associative case

This section and the next are devoted to the investigation of $\operatorname{Aut} A(p,q)$ and $\operatorname{Der} A(p,q)$ for composition algebras A over F. The rich properties of the composition law, $n(xy) = n(x)n(y)$ in A, enable us to determine $\operatorname{Aut} A(p,q)$ and $\operatorname{Der} A(p,q)$ in a more concrete form when $A(p,q)$ is simple. A key fact is that if $A(p,q)$ is simple, then $Fp + Fq$ is not totally isotropic (Lemma III.2.6 and Proposition III.2.9), and hence there is an invertible element s in A of the form

$$s = \alpha p + \beta q, \tag{5.1}$$

or equivalently $n(s) \neq 0$, for some $\alpha, \beta \in F$. If $\dim A = 1, 2,$ or 4 and $A(p,q)$ is simple, then we show that $\operatorname{Der} A(p,q) = \operatorname{Der}_w A$ for some $w \in A$ and that $\operatorname{Aut} A(p,q)$

is isomorphic either to $\mathrm{Aut}_w A$ or to an extension group of $\mathrm{Aut}_w A$ by index 2. However, for the case $\dim A = 8$, i.e., the only nonassociative case, we obtain in Section IV.6 a similar result under a stronger restriction that $p + q$ is invertible in A.

If $\dim A = 1$ or 2, then A is commutative, associative and $A(p,q) = A(p-q)$. In this case, it suffices to work with the u–homotope $A(u)$ of A. If $\dim A = 2$ then either $A \cong F \oplus F$ or $A \cong F(v)$, a separable quadratic extension field of F. The following result treats all cases of $\dim A \leq 2$.

Theorem 5.1. *Let A be a composition algebra over F of dimension ≤ 2 and let $u \in A$ be a fixed element.*

(i) *If $A \cong F$ and $u \neq 0$, so $A(u) \cong F$, then* $\mathrm{Aut}\, A(u) = \{id\}$ *and* $\mathrm{Der}\, A(u) = 0$.

(ii) *If $A \cong F$ and $u = 0$, so $A(u)$ is a trivial algebra, then* $\mathrm{Aut}\, A(u) \cong F^\times$ *and* $\mathrm{Der}\, A(u) \cong F$.

(iii) *If A is a quadratic extension field of F and $u \neq 0$, so $A(u) \cong A$, then* $\mathrm{Aut}\, A(u) \cong \mathbb{Z}_2$ *and* $\mathrm{Der}\, A(u) = 0$.

(iv) *If A is a quadratic extension field of F and $u = 0$, so $A(u)$ is trivial, then* $\mathrm{Aut}\, A(u) \cong GL(2,F)$ *and* $\mathrm{Der}\, A(u) \cong M_2(F)$.

(v) *If $A \cong F \oplus F$ and u is invertible, then $A(u) \cong A$,* $\mathrm{Aut}\, A(u) \cong \mathbb{Z}_2$ *and* $\mathrm{Der}\, A(u) = 0$.

(vi) *If $A \cong F \oplus F$ and $u \neq 0$ is not invertible, then $A(u) = Fa + Fb$ with $aua = a$ and $A(u)ub = 0$. In this case,* $\mathrm{Aut}\, A(u) \cong F^\times$ *and* $\mathrm{Der}\, A(u) \cong F$.

(vii) *If $A \cong F \oplus F$ and $u = 0$, so $A(u)$ is trivial, then* $\mathrm{Aut}\, A(u) \cong GL(2,F)$ *and* $\mathrm{Der}\, A(u) \cong M_2(F)$.

Proof. Parts (i), (ii), (iv) and (vii) are clear. For parts (iii) and (v), we have $A(u) \cong F1 + Fv$ where v satisfies a separable quadratic equation $X^2 + \alpha X + \beta = 0$. Hence, $d(1) = d(v) = 0$ for any $d \in \mathrm{Der}\, A(u)$ and $\sigma(v) = v$ or $\bar{v}$, the conjugate of v, for any $\sigma \in \mathrm{Aut}\, A(u)$. Therefore, the results follow. For part (vi), let $A = F \oplus F$ and let $u = (\alpha, \beta)$ with $n(u) = \alpha\beta = 0$. We may assume that $\alpha = 0$ but $\beta \neq 0$. Letting $a = (0, \beta^{-1})$ and $b = (1,0)$, $\{a,b\}$ gives a desired basis of $A(u)$. Relative to this basis, any $\sigma \in \mathrm{Aut}\, A(u)$ has the matrix $\sigma = e_{11} + \mu e_{22}$ for $\mu \in F^\times$, and the matrix of any $d \in \mathrm{Der}\, A(u)$ is given by $d = \mu e_{22}$ for $\mu \in F$. □

In the remainder of this section, we investigate automorphisms and derivations of $Q(p,q)$ for a generalized quaternion algebra Q over F. Let n, t and $x \mapsto \bar{x} = t(x)1 - x$ denote the norm, trace and standard involution in Q. For $x \in Q$, R_x denotes the right multiplication in Q by x. We begin with:

Lemma 5.2. *Let Q be a generalized quaternion algebra over F and let $p, q \in Q$ be fixed elements with $p \neq q$ such that $Q(p,q)$ is simple. Then, there exist scalars $\alpha, \beta \in F$ and elements $a, b \in Q$ with $\alpha a + \beta b = 1$ such that $Q(p,q)$ is isomorphic to $Q(a,b)$. Moreover, (α, β) can be chosen to be $(1,0), (0,1)$ or $(1,1)$.*

Proof. If $Q(p,q)$ is simple, then as noted above $Fp + Fq$ is not totally isotropic with respect to the norm n in Q. Thus, there is an invertible element s of the form (5.1) for some $\alpha, \beta \in F$. If we let $a = s^{-1}p$ and $b = s^{-1}q$, then $\alpha a + \beta b = 1$. As in (II.2.9), the right multiplication R_s in Q gives an algebra isomorphism,

$$R_s : Q(p,q) \to Q(a,b) \ : \ x \mapsto xs. \tag{5.2}$$

Since $Fp + Fq$ is not totally isotropic if and only if one of $n(p)$, $n(q)$ and $n(p+q)$ is not zero, we can take $(\alpha, \beta) = (1,0), (0,1)$ or $(1,1)$. □

Let A be an associative algebra over F with unit element 1. For a subset S of A, the centralizer $Z_A(S)$ of S in A defined by

$$Z_A(S) = \{x \in A : xy = yx \text{ for all } y \in S\}$$

is a subalgebra of A, since it contains $Z(A) \supseteq F1 = F$. Also, $Z_A(A)^\times$ is a subgroup of $A^\times$ containing $Z(A)^\times \supseteq F^\times$. For $a \in A^\times$, let i_a denote the inner automorphism of A given by $i_a(x) = axa^{-1}$ for $x \in A$. Then, the inner automorphism group $\mathrm{IAut}\, A$ and inner derivation algebra $\mathrm{IDer}\, A$ of A given by

$$\begin{aligned} \mathrm{IAut}\, A &= \{i_a : a \in A^\times\} = i_{A^\times}, \\ \mathrm{IDer}\, A &= \{ad_a : a \in A\} = ad_A \end{aligned}$$

are respectively a normal subgroup of $\mathrm{Aut}\, A$ and an ideal of $\mathrm{Der}\, A$, since $\sigma \circ i_a \circ \sigma^{-1} = i_{\sigma(a)}$ for $\sigma \in \mathrm{Aut}\, A$ and $[\partial, ad_a] = ad_{\partial(a)}$ for $\partial \in \mathrm{Der}\, A$ (also see Definition I.4.3).

Define the maps

$$\begin{aligned} i : Z_A(S)^\times &\to \mathrm{Aut}_S\, A \ : \ a \mapsto i_a, \\ ad : Z_A(S)^- &\to \mathrm{Der}_S\, A \ : \ a \mapsto ad_a. \end{aligned}$$

Then, i and ad are group and Lie algebra homomorphisms with $\ker i = \{a \in Z_A(S)^\times : i_a = id\} = Z(A)^\times$ and $\ker ad = \{a \in Z_A(S)^- : ad_a = 0\} = Z(A)$. Hence,

$$i(Z_A(S)^\times) \cong Z_A(S)^\times / Z(A)^\times, \ ad(Z_A(S)^-) \cong Z_A(S)^- / Z(A).$$

If A is finite-dimensional central simple over F, then, as is well known, $\mathrm{Aut}\, A = \mathrm{IAut}\, A$ and $\mathrm{Der}\, A = \mathrm{IDer}\, A$ and so the maps i and ad are surjective. Since $Z(A) = F$ in this case, the isomorphism relations above become

$$\mathrm{Aut}_S\, A \cong Z_A(S)^\times / F^\times, \ \mathrm{Der}_S\, A \cong Z_A(S)^- / F. \tag{5.3}$$

It is clear that $\mathrm{Aut}_S\, A = \mathrm{Aut}\, A$ or $\mathrm{Der}_S\, A = \mathrm{Der}\, A$ if and only if $S \subseteq F\,(= Z(A))$. Therefore, we have:

Proposition 5.3. *Let A be a finite-dimensional central simple associative algebra over F, and let S be a subset of A. Then, the isomorphism relations (5.3) hold for A, and $\operatorname{Aut}_S A = \operatorname{Aut} A$ or $\operatorname{Der}_S A = \operatorname{Der} A$ if and only if $S \subseteq F$.* □

Since Q is central simple over F, Proposition 5.3 applies to Q with $S = \{w\}$.

Corollary 5.4. *Let Q be a generalized quaternion algebra over F and let w be a fixed element of Q. Then, $\operatorname{Aut}_w Q \cong Z_Q(w)^\times / F^\times$ and $\operatorname{Der}_w Q \cong Z_Q(w)^- / F$. Moreover, $\operatorname{Aut}_w Q = \operatorname{Aut} Q$ if and only if $w \in F$, and $\operatorname{Der}_w Q = \operatorname{Der} Q$ if and only if $w \in F$.* □

Therefore, $\operatorname{Aut}_w Q \neq \operatorname{Aut} Q$ or $\operatorname{Der}_w Q \neq \operatorname{Der} Q$ if and only if $w \notin F$ if and only if $S = F1 + Fw$ is a subalgebra of Q of dimension 2. In this case, $Z_Q(w) = Z_Q(S)$ and $S \subseteq Z_Q(S)$. If there is an element $x \in Z_Q(S)$ with $x \notin S$, then $S \oplus Fx$ would become an abelian subalgebra of Q^-, which is impossible. Thus, $S = Z_Q(S)$. There are three possibilities for S: (1) S is a quadratic extension field of F; (2) $S \cong F \oplus F$; (3) n is degenerate on S. In case (1), $n(x) \neq 0$ for all $0 \neq x \in S$ while in case (2) n is nondegenerate on S but $n(x) = 0$ for some $0 \neq x \in S$. In case (2), if we let $S = F \oplus F$, then $S^\times = \{(\alpha, \beta) \in S : \alpha\beta \neq 0\}$ and the map $Z_Q(w)^\times = S^\times \to F^\times : (\alpha, \beta) \mapsto \alpha\beta^{-1}$ is a surjective homomorphism with kernel $F^\times$. Therefore, $\operatorname{Aut}_w Q \cong S^\times / F^\times \cong F^\times$. For case (3), we have a nonzero element $x \in S$ such that $t(x) = n(x) = 0$. Hence, $S = F1 + Fx$, and $S^\times = \{\alpha + \beta x : \alpha, \beta \in F \text{ with } \alpha \neq 0\}$ is the direct product of $F^\times$ and the subgroup $\{1 + \beta x : \beta \in F\} \cong (F, +)$. Thus, $\operatorname{Aut}_w Q \cong S^\times / F^\times \cong (F, +)$. In all cases, we find that $\operatorname{Der}_w Q$ is a one–dimensional Lie algebra over F.

The next result is instrumental for our discussion.

Lemma 5.5. *Let Q be a generalized quaternion algebra over F, and let $a, b \in Q$ be such that $a \neq b$ and $\alpha a + \beta b = 1$ for some $\alpha, \beta \in F$. Let*

$$U = \{(u, v) \in Q^\times \times Q^\times : v\bar{a} = -bu, v\bar{b} = -au\}. \tag{5.4}$$

Then, $U \neq \emptyset$ if and only if either

$$n(a) = n(b), \text{ or} \tag{5.5}$$

$$n(a) = -n(b) \neq 0 = t(a\bar{b}). \tag{5.6}$$

Proof. Assume that $U \neq \emptyset$ and $n(a) \neq n(b)$. If $(u, v) \in U$, then

$$v = v1 = v(\alpha\bar{a} + \beta\bar{b}) = -(\alpha b + \beta a)u.$$

Since $n(v) \neq 0$, it follows from this that $n(\alpha b + \beta a) \neq 0$ and

$$(\alpha b + \beta a)u\bar{a} = bu, \ (\alpha b + \beta a)u\bar{b} = au. \tag{5.7}$$

Hence, $(\alpha b+\beta a)u(\alpha\bar{b}+\beta\bar{a}) = (\beta b+\alpha a)u = u$, and multiplying this by $\alpha b+\beta a$ from the right, we obtain

$$n(\alpha b+\beta a)(\alpha b+\beta a)u = u(\alpha b+\beta a) \tag{5.8}$$

and so $n(\alpha b+\beta a)(\alpha b+\beta a) = u(\alpha b+\beta a)u^{-1}$. Thus, $n(\alpha b+\beta a)(\alpha b+\beta a)$ and $\alpha b+\beta a$ have the same norm and trace. This implies

$$n(\alpha b+\beta a) = \pm 1,\ \ (n(\alpha b+\beta a)-1)t(\alpha b+\beta a) = 0. \tag{5.9}$$

If $n(\alpha b+\beta a) = 1$, from (5.7) we get $n(a) = n(b)$. Therefore, by (5.9) it must be that $n(\alpha b+\beta a) = -1$ and $t(\alpha b+\beta a) = 0$. Hence, 1 and $\alpha b+\beta a$ are linearly independent, so $Fa+Fb = F1+F(\alpha b+\beta a)$. Linearizing $x^2 = t(x)x - n(x)1$ and using (5.8), we have

$$\begin{aligned} 0 &= (\alpha b+\beta a)u + u(\alpha b+\beta a) \\ &= t(\alpha b+\beta a)u + t(u)(\alpha b+\beta a) - 2(\alpha b+\beta a, u)1 \\ &= t(u)(\alpha b+\beta a) - 2(\alpha b+\beta a, u)1, \end{aligned}$$

where $(\ ,\ )$ is the bilinear form associated with n. Since 1 and $\alpha b+\beta a$ are linearly independent, from this we get $(1,u) = 2t(u) = 0 = (\alpha b+\beta a, u)$, and hence u is orthogonal to $F1+F(\alpha b+\beta a) = Fa+Fb$. Now, $t(u\bar{a}) = 2(u,a) = 0$, so $u\bar{a} = -a\bar{u} = au$, and similarly, $u\bar{b} = bu$. Since u is invertible, this and (5.7) give

$$(\alpha b+\beta a)a = b,\ \ (\alpha b+\beta a)b = a,$$

which imply $n(a) = -n(b)$ $(\neq 0)$ and $\alpha b+\beta a = ba^{-1} = a^{-1}b$. But,

$$0 = t(\alpha b+\beta a) = t(ba^{-1}) = \frac{1}{n(a)}t(b\bar{a}) = \frac{1}{n(a)}t(\bar{a}b),$$

and therefore, condition (5.6) follows.

Conversely, if (5.5) holds, then $n(\alpha b+\beta a) = (\alpha^2+\beta^2)n(a) + 2\alpha\beta(a,b) = n(\alpha a+\beta b) = n(1) = 1$. For $u = 1$ and $v = -(\alpha b+\beta a)$, one gets

$$v\bar{a} = -(\alpha b+\beta a)\bar{a} = -\alpha b\bar{a} - \beta a\bar{a} = -\alpha b\bar{a} - \beta b\bar{b} = -b(\alpha\bar{a}+\beta\bar{b}) = -bu,$$

and also $v\bar{b} = -au$. Hence, $(u,v)\in U$ and $U\neq\emptyset$. On the other hand, if (5.6) holds, then

$$\begin{aligned} ba^{-1} &= ba^{-1}(\alpha a+\beta b) = \alpha b + \beta ba^{-1}b \\ &= \alpha b + \beta t(ba^{-1})b - \beta\bar{a}^{-1}\bar{b}b = \alpha b - \beta n(b)\bar{a}^{-1} \\ &= \alpha b + \beta a\bar{a}\bar{a}^{-1} = \alpha b+\beta a, \end{aligned}$$

since $t(ba^{-1}) = n(a)^{-1}t(b\bar{a}) = 0$. Similarly, we have $ab^{-1} = \alpha b + \beta a$. Since $n(a+b) = 2(a,b) - (n(a)+n(b)) = 0$, Q is split and so $Q \cong M_2(F)$. In this case, $V = (Fa+Fb)^\perp$ is a subspace of Q of dimension 2 such that $t(V) = (1, V) = 0$, since $1 \in Fa + Fb$. Hence, we can choose an $u \in V$ with $n(u) = -1$ (note, $n(x) = \det x$). If we let $v = -(\alpha b + \beta a)$, then

$$v\bar{a} = -(\alpha b + \beta a)u\bar{a} = -(\alpha b + \beta a)au = -ba^{-1}au = -bu,$$

using $ba^{-1} = \alpha b + \beta a$. Similarly, one gets $v\bar{b} = -au$, so $(u, v) \in U$ and $U \neq \emptyset$. □

If we let $\tau(x) = \bar{x}$, the standard involution of Q, then U equals the set of elements $(u, v) \in Q^\times \times Q^\times$ satisfying condition (1.5). Hence, if $U \neq \emptyset$ and $(u, v) \in U$, then as in (1.6) the map

$$\tau(u,v) : Q(a,b) \to Q(a,b) \ : \ x \mapsto u\bar{x}v^{-1} \tag{5.10}$$

is an automorphism of $Q(a, b)$, and so the set $T(Q)$ defined in Proposition 1.3 is nonempty with $T(Q) \cap \Phi(H) = \emptyset$, where H is given by

$$H = \{(u,v) \in Q^\times \times Q^\times : au = va, bu = vb\} \tag{5.11}$$

and the map $\Phi : H \to \operatorname{Aut} Q(a, b)$ by (1.3) and (1.4).

Lemma 5.6. *Let Q be a generalized quaternion algebra over F and let a, b be elements of Q with $\alpha a + \beta b = 1$ for some $\alpha, \beta \in F$. Then,*

$$Z_Q(Fa + Fb) = Z_Q(\bar{a}b).$$

Proof. Since $1 \in Fa + Fb$, $\bar{a} = t(a)1 - a \in Fa + Fb$. Hence, if $u \in Z_Q(Fa + Fb)$, then $u\bar{a}b = \bar{a}ub = \bar{a}bu$ and so $u \in Z_Q(\bar{a}b)$. Conversely, if $u\bar{a}b = \bar{a}bu$, then

$$\begin{aligned} ua &= u(\alpha\bar{a} + \beta\bar{b})a = \alpha n(a)u + \beta u\bar{b}a \\ &= \alpha n(a)u + \beta t(\bar{a}b)u - \beta u\bar{a}b \\ &= (\alpha n(a) + \beta t(\bar{a}b) - \beta\bar{a}b)u = (\alpha\bar{a} + \beta\bar{b})au = au. \end{aligned}$$

Similarly, $ub = bu$, so $u \in Z_Q(Fa + Fb)$. □

Proposition 5.7. *Let Q be a generalized quaternion algebra over F, and let a, b be elements of Q such that $a \neq b$ and $\alpha a + \beta b = 1$ for some $\alpha, \beta \in F$. Then,* $\operatorname{Aut} Q(a, b) = \operatorname{Aut}_{\bar{a}b} Q$, *unless either (5.5) or (5.6) holds. In either case of the latter,* $\operatorname{Aut}_{\bar{a}b} Q$ *is a subgroup of* $\operatorname{Aut} Q(a, b)$ *of index* 2.

Proof. Note first that since $\alpha a + \beta b = 1$, $R^0(a, b) = 0$ and so $Q(a, b)$ is simple. Thus, Q and $Q(a, b)$ satisfy the conditions of Theorem 3.15. Let U be the set defined by (5.4). If $U = \emptyset$, then as in the proof of Theorem 3.15(ii) $Q(a,b)$ cannot have an automorphism described by (3.13) and (3.15) and hence by Theorem 3.15 $\operatorname{Aut} Q(a,b) = \Phi(H)$ where H is given by (5.11). On the other hand, if $(u,v) \in H$, then $u = (\alpha a + \beta b)u = v(\alpha a + \beta b) = v$ and hence by Lemma 5.6

$$\begin{aligned} H &= \{(u,u) \in Q^\times \times Q^\times : u \in Z_Q(Fa + Fb)^\times\} \\ &= \{(u,u) \in Q^\times \times Q^\times : u \in Z_Q(\bar{a}b)^\times\}. \end{aligned}$$

Therefore, $\Phi(H) = \{\Phi_{u,u} \in \operatorname{Aut} Q(a,b) : u \in Z_Q(\bar{a}b)^\times\} = \{i_u : u \in Z_Q(\bar{a}b)^\times\} = \operatorname{Aut}_{\bar{a}b} Q$ (see (1.3)). This shows that if $U = \emptyset$, then $\operatorname{Aut} Q(a,b) = \operatorname{Aut}_{\bar{a}b} Q$. If $U \neq \emptyset$, then the automorphism $\tau(u,v)$ of $Q(a,b)$ given by (5.10) is in $T(Q)$, and as in Theorem 3.15(ii) we have

$$\operatorname{Aut} Q(a,b) = (\operatorname{Aut}_{\bar{a}b} Q) \cup (\tau(u,v) \circ \operatorname{Aut}_{\bar{a}b} Q) \text{ (disjoint)} \tag{5.12}$$

for any $(u,v) \in U$. Thus, $[\operatorname{Aut} Q(a,b) : \operatorname{Aut}_{\bar{a}b} Q] = 2$, and the result follows from Lemma 5.5. □

From the proof of Lemma 5.5, the map $\tau(u,v)$ above can be explicitly expressed according as case (5.5) or (5.6). If $n(a) = n(b)$, then from that proof we find a pair $(u,v) = (-1, -(\alpha b + \beta a)) \in U$ and so by (5.10) the map

$$\tau(u,v)(x) = -\bar{x}(\alpha b + \beta a)^{-1} = -\bar{x}\overline{(\alpha b + \beta a)} = -\overline{(\alpha b + \beta a)x}$$

gives an automorphism of $Q(a,b)$, since $n(\alpha b + \beta a) = 1$. But, as above, $\tau(u,v) \notin \operatorname{Aut}_{\bar{a}b} Q$. If $n(a) = -n(b) \neq 0 = t(\bar{a}b)$, then from the same proof we find a pair $(u,v) = (u, -(\alpha b + \beta a)u) \in U$ such that $u \in (Fa + Fb)^\perp$ and $n(u) = -1$. Since $n(\alpha b + \beta a) = -1$ and $t(u) = 0$ in this case, the map

$$\tau(u,v)(x) = u\bar{x}v^{-1} = -u\bar{x}(\alpha b + \beta a)$$

is an automorphism of $Q(a,b)$, where we have used $t(\alpha b + \beta a) = 0$ (see (5.9)). In both cases, every element of $\operatorname{Aut} Q(a,b)$ is given by decomposition (5.12).

Let Q be a generalized quaternion algebra over F, and let $p, q \in Q$ be such that $p \neq q$ and $Q(p,q)$ is simple. Then, as in (5.1), there is an invertible element s of the form $s = \alpha p + \beta q$ for some $\alpha, \beta \in F$. Let

$$a = s^{-1}p, \quad b = s^{-1}q \tag{5.13}$$

so that $\alpha a + \beta b = 1$. We have $\bar{a}b = \overline{s^{-1}p}s^{-1}q = \bar{p}\bar{s}^{-1}s^{-1}q = n(s)^{-1}\bar{p}q$ and hence

$$Z_Q(\bar{a}b) = Z_Q(\bar{p}q).$$

We now prove our principal result on $\operatorname{Aut} Q(p,q)$.

Theorem 5.8. *Let A be a generalized quaternion algebra over F, and let $p, q \in Q$ be such that $p \neq q$ and $Q(p,q)$ is simple. Let $s = \alpha p + \beta q$ be an invertible element in $Fp + Fq$ (such an element s exists).*

(i) *If w is any invertible element of Q contained in $Fp+Fq$, then $R_s^{-1} \circ \sigma \circ R_s = R_w^{-1} \circ \sigma \circ R_w$ for all $\sigma \in \operatorname{Aut}_{\bar{p}q} Q$.*

(ii) *Either $\operatorname{Aut} Q(p,q) = R_s^{-1} \circ (\operatorname{Aut}_{\bar{p}q} Q) \circ R_s \cong \operatorname{Aut}_{\bar{p}q} Q$, or $R_s^{-1} \circ (\operatorname{Aut}_{\bar{p}q} Q) \circ R_s$ is a subgroup of $\operatorname{Aut} Q(p,q)$ of index 2.*

Proof. (i) Let $w = \mu p + \nu q \in Fp + Fq$ be invertible in Q, and let $\sigma \in \operatorname{Aut}_{\bar{p}q} Q$. We note that

$$\begin{aligned} \bar{s}w &= \alpha\mu n(p) + \alpha\nu\bar{p}q + \beta\mu\bar{q}p + \beta\nu n(q) \\ &= (\alpha\mu n(p) + \beta\nu n(q) + \beta\mu t(\bar{p}q))1 + (\alpha\nu - \beta\mu)\bar{p}q \end{aligned}$$

and so $\bar{s}w$ is fixed by σ. It follows from this that

$$\begin{aligned} \sigma(w)w^{-1} &= n(s)^{-1}\sigma(s\bar{s}w)w^{-1} = n(s)^{-1}\sigma(s)\sigma(\bar{s}w)w^{-1} \\ &= n(s)^{-1}\sigma(s)\bar{s}ww^{-1} = \sigma(s)s^{-1}. \end{aligned}$$

Thus, $\sigma(xw)w^{-1} = \sigma(x)\sigma(w)w^{-1} = \sigma(x)\sigma(s)s^{-1} = \sigma(xs)s^{-1}$ for any $x \in Q$, showing part (i).

(ii) Let a and b be given by (5.13). By (5.2), the map

$$\operatorname{Aut} Q(a,b) \to \operatorname{Aut} Q(p,q) \;:\; s \mapsto R_s^{-1} \circ \sigma \circ R_s \tag{5.14}$$

is an isomorphism of $\operatorname{Aut} Q(a,b)$ onto $\operatorname{Aut} Q(p,q)$. Note also that $\operatorname{Aut}_{\bar{a}b} Q = \operatorname{Aut}_{\bar{p}q} Q$ since $\bar{a}b = n(s)^{-1}\bar{p}q$. Hence, by Proposition 5.7 we have that either $\operatorname{Aut} Q(a,b) = \operatorname{Aut}_{\bar{p}q} Q$, or $\operatorname{Aut}_{\bar{p}q} Q$ is a subgroup of $\operatorname{Aut} Q(a,b)$ of index 2. Therefore, the result follows from this and the isomorphism (5.14). □

Corollary 5.9. *Let Q be a generalized quaternion algebra over F, and let p, q be such that $Q(p,q)$ is a division algebra. Then, $\operatorname{Aut} Q(p,q) = R_s^{-1} \circ (\operatorname{Aut}_{\bar{p}q} Q) \circ R_s$, where R_s is the same as in Theorem 5.8.*

Proof. It follows from Theorem II.4.11 that Q is a division algebra and $n(p) \neq n(q)$, so $p \neq \pm q$. If $n(p) = -n(q) \neq 0 = t(\bar{p}q)$, then $n(p+q) = 0$ and so $p + q = 0$, a contradiction. Hence, the result follows from Theorem 5.8. □

Assume $s = \alpha p + \beta q$ is invertible. Then, $\bar{p}q \in F$ if and only if $s^{-1}p = n(s)^{-1}\bar{s}p = n(s)^{-1}(\alpha n(p) + \beta\bar{q}p) \in F$ and $s^{-1}q = n(s)^{-1}(\alpha\bar{p}q + \beta n(q)) \in F$. Thus, by Theorem II.2.7(vi), conditions (i)–(v) of Corollary II.2.3 are equivalent if and only if $\bar{p}q \in F$ if and only if $Q(p,q)$ is flexible. It follows from this and Theorem II.2.8 that if $p - q$ is invertible, then $\bar{p}q \in F$ if and only if $Q(p,q)$ has a unit element. We use this to prove:

Corollary 5.10. *Assume that $Q(p,q)$ is a division algebra over F. Then $\operatorname{Aut} Q(p,q) \cong \operatorname{Aut} Q$ if and only if $Q(p,q)$ has a unit element. Otherwise, $\operatorname{Aut} Q(p,q) \cong S^\times/F^\times$, where $S = F1 + F\bar{p}q$ is a quadratic extension field of F.*

Proof. Since $s = p - q$ is invertible, by Corollary 5.9

$$\operatorname{Aut} Q(p,q) = R_s^{-1} \circ (\operatorname{Aut}_{\bar{p}q} Q) \circ R_s \cong \operatorname{Aut}_{\bar{p}q} Q. \tag{5.15}$$

But, by Corollary 5.4 $\operatorname{Aut} Q = \operatorname{Aut}_{\bar{p}q} Q$ if and only if $\bar{p}q \in F$ if and only if $Q(p,q)$ has a unit element. By the remarks following Corollary 5.4, we find that $\bar{p}q \notin F$ if and only if $Z_Q(\bar{p}q) = F1 + F\bar{p}q$ is a subalgebra of dimension 2. The result now follows from Corollary 5.4 and (5.15). □

We turn our attention to $\operatorname{Der} Q(p,q)$. For fixed elements $a, b \in Q$, as in (1.8), let

$$L = \{(u,v) \in Q \times Q : au = va, bu = vb\}.$$

Lemma 5.11. *Let Q be a generalized quaternion algebra over F, and let $a, b \in Q$ be such that $a \neq b$ and $\alpha a + \beta b = 1$ for some $\alpha, \beta \in F$. Then, $\operatorname{Der} Q(a,b) = \operatorname{Der}_{\bar{a}b} Q$.*

Proof. Since $\alpha a + \beta b = 1$, $R^0(a,b) = 0$ and $Q(a,b)$ is simple. As in the proof of Proposition 5.7, we have $L = \{(u,u) \in Q \times Q : u \in Z_Q(\bar{a}b)\}$ using Lemma 5.6. Hence, by Theorem 4.6

$$\operatorname{Der} Q(a,b) = \rho(L) = \{d_{u,u} : u \in Z_Q(\bar{a}b)\} = \{ad_u : u \in Z_Q(\bar{a}b)\} = \operatorname{Der}_{\bar{a}b} Q,$$

where $d_{u,u}$ and ρ are the maps defined by (4.9) and (4.10). Here, we have used $\operatorname{Der} Q = \operatorname{IDer} Q$. □

We use this lemma to prove our main result on $\operatorname{Der} Q(p,q)$.

Theorem 5.12. *Let Q be a generalized quaternion algebra over F, and let p, q be elements of Q such that $p \neq q$ and $Q(p,q)$ is simple. Let $s = \alpha p + \beta q$ be an invertible element of Q for some $\alpha, \beta \in F$.*

(i) *If w is any invertible element of Q contained in $Fa + Fb$, then $R_s^{-1} \circ \partial \circ R_s = R_w^{-1} \circ \partial \circ R_w$ for all $\partial \in \operatorname{Der}_{\bar{p}q} Q$.*

(ii) $\operatorname{Der} Q(p,q) = R_s^{-1} \circ (\operatorname{Der}_{\bar{p}q} Q) \circ R_s \cong \operatorname{Der}_{\bar{p}q} Q$.

Proof. (i) The same calculation as in Theorem 5.8(i) shows that $\partial(\bar{s}w) = 0$ for any $\partial \in \operatorname{Der}_{\bar{p}q} Q$ and hence $\partial(xw)w^{-1} = \partial(xs)s^{-1}$ for $x \in Q$.

(ii) Let a and b be the same as in (5.13). By (5.2), the map

$$\operatorname{Der} Q(a,b) \to \operatorname{Der} Q(p,q) \ : \ \partial \mapsto R_s^{-1} \circ \partial \circ R_s$$

is a Lie algebra isomorphism of $\operatorname{Der} Q(a,b)$ onto $\operatorname{Der} Q(p,q)$. Since $\operatorname{Der}_{\bar{a}b} Q = \operatorname{Der}_{\bar{p}q} Q$, using this isomorphism the result follows from Lemma 5.11. □

Theorems 5.8(i) and 5.12(i) show that the isomorphisms (5.14) and (5.16) are independent of the choice of $s \in (Fp+Fq)\cap Q^{\times}$. When $Q(p,q)$ is a division algebra, we obtain a result similar to Corollary 5.10 for $\operatorname{Der} Q(p,q)$.

Corollary 5.13. *Assume that $Q(p,q)$ is a division algebra over F. Then, $\operatorname{Der} Q(p,q) \cong \operatorname{Der} Q$ if and only if $Q(p,q)$ has a unit element. Otherwise, $\operatorname{Der} Q(p,q)$ is a one–dimensional Lie algebra over F.*

Proof. Since $s = p - q$ is invertible, from Theorem 5.12(ii) we obtain $\operatorname{Der} Q(p,q) \cong \operatorname{Der}_{\bar{p}q} Q$. By Corollary 5.4, $\operatorname{Der} Q = \operatorname{Der}_{\bar{p}q} Q$ if and only if $\bar{p}q \in F$ if and only if $Q(p,q)$ has a unit element, as argued in Corollary 5.10. The remaining proof is immediate from the remarks following Corollary 5.4. □

A special case of interest is when $Q(p,q)$ is a division algebra over the field $\mathbb{R}$ of real numbers. Thus, Q is the division algebra $\mathbb{H}$ of real quaternions. If $\bar{p}q \notin \mathbb{R}$, then $\mathbb{R} + \mathbb{R}\bar{p}q$ is the field $\mathbb{C}$ of complex numbers, and hence by Corollary 5.4, Theorem 5.12 and (5.15), $\operatorname{Aut} \mathbb{H}(p,q) \cong \mathbb{C}^{\times}/\mathbb{R}^{\times}$ and $\operatorname{Der} \mathbb{H}(p,q) \cong \mathbb{C}^{-}/\mathbb{R} \cong \mathbb{R}$.

Assume that $\bar{p}q \in \mathbb{R}$. By the same results as above, we have $\operatorname{Aut} \mathbb{H}(p,q) \cong \mathbb{H}^{\times}/\mathbb{R}^{\times} \cong SU(2)$ and $\operatorname{Der} \mathbb{H}(p,q) \cong \mathbb{H}^{-}/\mathbb{R} \cong su(2)$. Moreover, we can let $q = \lambda p$ with $p \neq 0$ and $\lambda \neq \pm 1$, since $n(p) \neq n(q)$. Since $\mathbb{H}(p) \cong \mathbb{H}$ and $\mathbb{H}(p,q) = \mathbb{H}(p)(1,\lambda)$, for $\mathbb{H}(p,q)$ it suffices to determine $\mathbb{H}(1,\lambda)$ with multiplication $x * y = xy - \lambda yx$. Let $\{1, u_1, u_2, u_3\}$ be a standard basis of $\mathbb{H}$. Letting

$$e = (1-\lambda)^{-1}1, \ e_i = (1+\lambda)^{-1}u_i, \ i = 1,2,3,$$

we have that e is a unit element of $\mathbb{H}(1,\lambda)$ and the multiplication in $\mathbb{H}(1,\lambda)$ is given by Table IV.1.

TABLE IV.1

	e	e_1	e_2	e_3
e	e	e_1	e_2	e_3
e_1	e_1	$-(1-\lambda)^2 e$	e_3	$-e_2$
e_2	e_2	$-e_3$	$-(1-\lambda)^2 e$	e_1
e_3	e_3	e_2	$-e_1$	$-(1-\lambda)^2 e$

We note that $\mathbb{H}(1,\lambda)$ is a real flexible division algebra which also appears in the classification of real division algebras of dimension 4 having $su(2)$ as its derivation algebra [**BO1**,**BBO2**]. In fact, $\mathbb{H}(1,\lambda)$ is isomorphic to the algebra given by [**BO1**,(6.2), **BBO2**,(3.4)] with $\eta = \zeta = 1$ and $\beta > 0$.

6. Composition algebras: The octonion case

Let C be a Cayley–Dickson algebra over F. For fixed elements $p, q \in C$, let $C(p,q)$ denote the (left) (p,q)–mutation of C. The main concern in this section is to determine $\operatorname{Aut} C(p,q)$ and $\operatorname{Der} C(p,q)$, when $p+q$ is invertible in C. We recall some basic facts from Chapter III and Section I.4. Any Cayley–Dickson algebra C over F has the decomposition

$$C = F1 \oplus V = F1 \oplus [C,C] \tag{6.1}$$

where $V = [C,C] = \{x \in C : t(x) = 0\}$. Moreover, if the characteristic of F is not three, then V becomes a 7–dimensional central simple, non–Lie, Malcev algebra over F under the product $[\ ,\ ]$ (see Table I.2). For $u, v \in V$, we have $u \circ v = \frac{1}{2}(uv+vu) = -\frac{1}{2}(u\bar{v} + v\bar{u}) = -\frac{1}{2}t(u\bar{v})1 = -(u,v)1$, using (III.2.9), and hence

$$uv = \tfrac{1}{2}[u,v] - (u,v)1, \tag{6.2}$$

$$(\alpha 1 + u) \circ (\beta 1 + v) = (\alpha\beta - (u,v))1 + \alpha v + \beta u, \tag{6.3}$$

$$(\alpha 1 + u)(\beta 1 + v) = (\alpha\beta - (u,v))1 + \alpha v + \beta u + \tfrac{1}{2}[u,v] \tag{6.4}$$

for $u, v \in V$ and $\alpha, \beta \in F$, where $(\ ,\ )$ is the symmetric bilinear form associated with the norm n of C.

If $u, v \in V$, then $[u,[u,v]] = u^2v + vu^2 - 2uvu = -2n(u)v - 2t(uv)u + 2n(u)\bar{v} = -4n(u)v + 4(u,v)u$, i.e.,

$$[u,[u,v]] = -4n(u)v + 4(u,v)u. \tag{6.5}$$

Here, we have used (III.2.10) and the fact that $\bar{u} = -u$ and $\bar{v} = -v$. Linearizing (6.5), we obtain

$$[u,[w,v]] + [w,[u,v]] = -8(u,w)v + 4(w,v)u + 4(u,v)w \tag{6.6}$$

for $u, v, w \in V$.

Lemma 6.1. *Let C be a Cayley–Dickson algebra over F with decomposition (6.1).*

(i) *The bilinear form* $(\,,\,)$ *is* $\operatorname{Aut} V$*–invariant as well as* $\operatorname{Der} V$*–invariant; that is,*

$$(6.7) \qquad (\varphi(u), \varphi(v)) = (u, v), \quad (d(u), v) + (u, d(v)) = 0$$

for all $u, v \in V$, $\varphi \in \operatorname{Aut} V$ *and* $d \in \operatorname{Der} V$.

(ii) $\operatorname{Aut} C = \operatorname{Aut}_1 C^-$ *and* $\operatorname{Der} C = \operatorname{Der}_1 C^-$.

(iii) $\operatorname{Aut} V = \{\varphi|_V : \varphi \in \operatorname{Aut} C^-\} = \{\varphi|_V : \varphi \in \operatorname{Aut} C\}$.

(iv) $\operatorname{Der} V = \{d|_V : d \in \operatorname{Der} C^-\} = \{d|_V : d \in \operatorname{Der} C\}$.

(v) *If* φ *is an automorphism of* C^- *with* $\varphi(1) = -1$, *then* $-\varphi$ *is an anti–automorphism of* C.

Proof. (i) The first relation is immediate from (6.5). For the $\operatorname{Der} V$–invariance of $(\,,\,)$, we apply d to (6.5) and use (6.6) to obtain $-8(u, d(u))v + 4((d(u), v) + (u, d(v)))u = 0$ for any $u, v \in V$, hence the relation follows.

(ii) Clearly, $\operatorname{Aut} C \subseteq \operatorname{Aut}_1 C^-$ and $\operatorname{Der} C \subseteq \operatorname{Der}_1 C^-$. If $\varphi \in \operatorname{Aut} C^-$ with $\varphi(1) = 1$, then $\varphi \in \operatorname{Aut} C$, using (6.4) and the $\operatorname{Aut} V$–invariance of $(\,,\,)$. The second relation similarly follows from the $\operatorname{Der} V$–invariance of $(\,,\,)$.

(iii) If $\varphi \in \operatorname{Aut} V$, then we extend φ to a map $\varphi : C \to C$ by putting $\varphi(1) = 1$. It readily follows from (6.7) and (6.4) that $\varphi \in \operatorname{Aut} C$. If $\varphi \in \operatorname{Aut} C$, then $\varphi \in \operatorname{Aut} C^-$. Since $F1$ and V are the only proper ideals of C^- (Theorem I.4.17), we have $\varphi(V) = V$ and so $\varphi|_V \in \operatorname{Aut} V$.

(iv) Using the $\operatorname{Der} V$–invariance of $(\,,\,)$ and (6.4), any $d \in \operatorname{Der} V$ is extended to a derivation of C or C^- by setting $d(1) = 0$. Conversely, if $d \in \operatorname{Der} C$ (or $d \in \operatorname{Der} C^-$), then $d(V) = d([C, C]) \subseteq [d(C), C] + [C, d(C)] \subseteq V$ and so $d|_V$ is a derivation of V.

(v) Let $\varphi \in \operatorname{Aut} C^-$ with $\varphi(1) = -1$. Then, $\varphi|_V \in \operatorname{Aut} V$ and so $(\,,\,)$ is $\varphi|_V$–invariant. If we let $\sigma = -\varphi$, then $\sigma(1) = 1$ and from (6.4)

$$\begin{aligned}
\sigma((\alpha 1 + u)(\beta 1 + v)) &= (\alpha\beta - (u, v))1 + \alpha\sigma(v) + \beta\sigma(u) - \frac{1}{2}[\varphi(u), \varphi(v)] \\
&= (\alpha\beta - (u, v))1 + \alpha\sigma(v) + \beta\sigma(u) + \frac{1}{2}[\sigma(v), \sigma(u)] \\
&= \sigma(\beta 1 + v)\sigma(\alpha 1 + u),
\end{aligned}$$

since $(\,,\,)$ is σ–invariant. Hence, $\sigma = -\varphi$ is an anti–automorphism of C. □

Let $O(V)$ and $o(V)$ respectively denote the orthogonal group and algebra of V relative to $(\,,\,)$, i.e., $O(V)$ and $o(V)$ are the group and the Lie algebra of maps $\varphi \in GL(V)$ and $d \in Hom_F V$ satisfying the invariant relations of (6.7). It follows immediately from (6.3) that $\varphi \in \operatorname{Aut} C(1, -1) = \operatorname{Aut} C^+$ if and only if $\varphi(1) = 1$ and $\varphi|_V \in O(V)$, and $d \in \operatorname{Der} C(1, -1) = \operatorname{Der} C^+$ if and only if $d(1) = 0$ and

$d|_V \in o(V)$. Thus, $\operatorname{Aut} C(1,-1) \cong O(V)$ and $\operatorname{Der} C(1,-1) \cong o(V)$. This illustrates that $\operatorname{Aut} C(p,q)$ and $\operatorname{Der} C(p,q)$ are in general considerably larger than $\operatorname{Aut} C$ and $\operatorname{Der} C$. Note that $\operatorname{Der} C$ is a central simple Lie algebra over F of type G_2 of dimension 14 and $\operatorname{Der} C(1,-1) \cong o(V)$ is a central simple Lie algebra over F of type B_3 (of dimension 21).

The main task is to obtain the results for $\operatorname{Aut} C(p,q)$ and $\operatorname{Der} C(p,q)$ which are analogous to Theorems 5.8 and 5.12. For this, we need the restriction that $p+q$ is invertible in C. Thus, by Proposition III.2.2 there is an algebra isomorphism

$$\sigma : C(p+q) \to C, \tag{6.8}$$

which is instrumental for our discussion. Moreover, if n is nondegenerate on the subalgebra S of C generated by p and q (so $\dim S \leq 4$), then σ can be taken to be an extension of the isomorphism

$$\sigma : S(p+q) \to S : x \mapsto x(p+q). \tag{6.9}$$

This follows from the fact that S and $S(p+q)$ are composition algebras with norms n and $x \mapsto n(x(p+q))$, respectively, and hence C and $C(p+q)$ are regarded as Cayley–Dickson algebras obtained from S and $S(p+q)$ by the Cayley–Dickson process.

Lemma 6.2. *Let C be a Cayley–Dickson algebra over F, and let p,q be elements of C such that $s \equiv p+q$ is invertible. Then, the map σ of (6.8) becomes an algebra isomorphism*

$$\sigma : C(p,q) \to C(a,b), \tag{6.10}$$

where $a = \sigma(s^{-1}ps^{-1})$ and $b = \sigma(s^{-1}qs^{-1})$, with $a+b=1$. Moreover, if n is nondegenerate on the subalgebra S generated by p and q, then σ can be taken in such a way that $a = s^{-1}p$ and $b = s^{-1}q$.

Proof. Denote by "$\cdot$" the multiplication in $C(s)$. For $x,y \in C$, we find

$$(xp)y = (x(ss^{-1}ps^{-1}s))y = (((xs)(s^{-1}ps^{-1}))s)y = (x \cdot (s^{-1}ps^{-1})) \cdot y,$$

using the Moufang identity (III.2.2). Hence, by (6.8)

$$\sigma((xp)y) = (\sigma(x)\sigma(s^{-1}ps^{-1}))\sigma(y) = (\sigma(x)a)\sigma(y)$$

with $a = \sigma(s^{-1}ps^{-1})$ and similarly, $\sigma((yq)x) = (\sigma(y)b)\sigma(x)$ with $b = \sigma(s^{-1}qs^{-1})$. Therefore, the map σ of (6.8) is an isomorphism of $C(p,q)$ to $C(a,b)$, and $a + b = \sigma(s^{-1}) = 1$, since s^{-1} is the unit element of $C(s)$ (Proposition III.2.1). If σ is an extension of the map of (6.9), then since $s^{-1}ps^{-1}, s^{-1}qs^{-1} \in S$, we have $a = \sigma(s^{-1}ps^{-1}) = s^{-1}ps^{-1}s = s^{-1}p$ and also, $b = s^{-1}q$. □

Lemma 6.3. *Let C be a Cayley–Dickson algebra over F, and let $a, b \in C$ be such that $a \neq b$ and $a + b = 1$. Then, $\operatorname{Aut} C(a,b) = \operatorname{Aut}_{a-b} C$, unless $t(a) = 1$. In the latter case, $\operatorname{Aut}_{a-b} C$ is a subgroup of $\operatorname{Aut} C(a,b)$ of index 2.*

Proof. Note first that $\operatorname{Aut}_{a-b} C \subseteq \operatorname{Aut} C(a,b)$, since $\operatorname{Aut}_{a-b} C \subseteq \operatorname{Aut} C(a-b)^+ = \operatorname{Aut} C(a,b)^+$ and $\operatorname{Aut}_{a-b} C \subseteq \operatorname{Aut} C^- = \operatorname{Aut} C(a,b)^-$. Conversely, let φ be any automorphism of $C(a,b)$. Then, $\varphi \in \operatorname{Aut} C(a,b)^- = \operatorname{Aut} C(a+b)^- = \operatorname{Aut} C^-$. Hence, by Lemma 6.1(iii), $\varphi|_V \in \operatorname{Aut} V$ and $\varphi(1) = \alpha 1$ for some $\alpha \neq 0$ in F, so that

$$\varphi(a-b) = \varphi(1 * 1) = \varphi(1) * \varphi(1) = \alpha^2(a-b). \tag{6.11}$$

We proceed the proof by cases.

(i) Assume $a \in F$. Since $a + b = 1$, $a - b \in F$ and so $\varphi(a-b) = \alpha(a-b)$, which by (6.11) implies $\alpha = 1$. Hence, by Lemma 6.1(ii) $\varphi \in \operatorname{Aut} C$ and so $\varphi \in \operatorname{Aut}_{a-b} C$. In this case, we have $\operatorname{Aut} C(a,b) = \operatorname{Aut}_{a-b} C$. Note also that $t(a) \neq 1$, since if $t(a) = 1$ then $a = b = \frac{1}{2}$.

(ii) Assume $a \notin F$ and $t(a) \neq 1$. Since $a + b = 1$, it must be that $a - b \notin F1$ and $a - b \notin V$. Since $\varphi(a-b) = \alpha^2(a-b)$, it follows from this that $\varphi(1) = \alpha^2 = \alpha$, hence $\alpha = 1$. Therefore, as above, $\varphi \in \operatorname{Aut}_{a-b} C$ and $\operatorname{Aut} C(a,b) = \operatorname{Aut}_{a-b} C$.

(iii) Assume $t(a) = 1$, so that $a \notin F$. Then, $b = 1 - a = \bar{a}$ and $a - b = a - \bar{a} \in V$. We can choose an $x \in V$ such that $\{\varphi(x), a - b\}$ is linearly independent and $n(x) = n(\varphi(x)) \neq 0$ (see (6.7)). From (III.2.10) and (6.11) we obtain

$$\begin{aligned}
\varphi(x * x) &= \varphi(x(a-b)x) = \varphi(t(x(a-b))x + n(x)(a-b)), \\
&= t(x(a-b))\varphi(x) + \alpha^2 n(x)(a-b), \\
\varphi(x * x) &= \varphi(x) * \varphi(x) = \varphi(x)(a-b)\varphi(x) \\
&= t(\varphi(x)(a-b))\varphi(x) + n(\varphi(x))(a-b).
\end{aligned}$$

Hence, $\alpha^2 n(x) = n(\varphi(x)) = n(x)$ and $\alpha^2 = 1$, so $\alpha = 1$ or -1. If $\alpha = 1$, then by Lemma 6.1 $\varphi \in \operatorname{Aut}_{a-b} C = \operatorname{Aut}_a C$. If $\alpha = -1$, then $\varphi(1) = -1$ and by Lemma 6.1(v), $-\varphi$ is an anti–automorphism of C with $\varphi(a-b) = a - b$. Conversely, any $\varphi \in \operatorname{Aut}_a C$ belongs to $\operatorname{Aut} C(a,b)$ since $\varphi(b) = b$ also. If $\varphi : C \to C$ satisfies the conditions that $\varphi(a-b) = a - b$ and $-\varphi$ is an anti–automorphism of C, then $\varphi \in \operatorname{Aut} C(a,b)$ since $\varphi(a+b) = \varphi(1) = -1 = -a - b$ and so $\varphi(a) = -b$, $\varphi(b) = -a$. Moreover, if φ_1 and φ_2 are any two such maps, then $\varphi_1 \circ \varphi_2 = (-\varphi_1) \circ (-\varphi_2) \in \operatorname{Aut} C$ and $\varphi_1 \circ \varphi_2(a-b) = a - b$, so $\varphi_1 \circ \varphi_2 \in \operatorname{Aut}_{a-b} C$. On the other hand, by the assumption, we have $a = \frac{1}{2} + u$ for some $0 \neq u \in V$, so $b = \frac{1}{2} - u$. Define the map τ by

$$\tau : C \to C \ : \ x \mapsto -\bar{x}. \tag{6.12}$$

Then, $-\tau$ is an automorphism of C such that $\tau(a-b) = 2\tau(u) = -2\bar{u} = 2u = a - b$, hence, $\tau \in \operatorname{Aut} C(a,b)$ but $\tau \notin \operatorname{Aut}_{a-b} C$. Therefore, in this case, $[\operatorname{Aut} C(a,b) : \operatorname{Aut}_{a-b} C] = 2$. □

Let $a \neq b$ be elements of C such that $a+b=1$ and $t(a)=1$. Thus, $a=\frac{1}{2}+u$ for some $0 \neq u \in V$, and we have shown that

$$\operatorname{Aut} C(a,b) = \operatorname{Aut}_u C \cup \tau \circ (\operatorname{Aut}_u C) \quad \text{(disjoint)},$$

where τ is the map given by (6.12). When $p+q$ is invertible, using Lemmas 6.2 and 6.3 we determine $\operatorname{Aut} C(p,q)$ as follows:

Theorem 6.4. *Let C be a Cayley–Dickson algebra over F, and let $p \neq q$ be elements of C such that p+q is invertible in C. Let $G = \sigma^{-1} \circ (\operatorname{Aut}_{\sigma(\overline{p-q})} C) \circ \sigma$ where σ is the isomorphism of (6.10). Then, either $\operatorname{Aut} C(p,q) = G \cong \operatorname{Aut}_{\sigma(\overline{p-q})} C$ or $\operatorname{Aut} C(p,q)$ is an extension group of G by index 2. Moreover, if n is nondegenerate on the subalgebra generated by p and q, then $\operatorname{Aut}_{\sigma(\overline{p-q})} C = \operatorname{Aut}_{\bar{p}q} C$.*

Proof. Let $s=p+q$. As in Lemma 6.2, $\sigma : C(p,q) \to C(a,b)$ is an isomorphism with $a=\sigma(s^{-1}ps^{-1})$ and $b=\sigma(s^{-1}qs^{-1})$, so $\sigma(s^{-1})=1$. From this, we have a group isomorphism

$$\operatorname{Aut} C(a,b) \to \operatorname{Aut} C(p,q) \ : \ \varphi \mapsto \sigma^{-1} \circ \varphi \circ \sigma.$$

But, since

$$\begin{aligned} a-b=\sigma(s^{-1}(p-q)s^{-1}) &= \sigma\left[t(s^{-1}(p-q))s^{-1} - n(s^{-1})(\overline{p-q})\right] \\ &= t(s^{-1}(p-q)) - n(s^{-1})\sigma(\overline{p-q}) \end{aligned}$$

using (III.2.10), we have $\operatorname{Aut}_{a-b} C = \operatorname{Aut}_{\sigma(\overline{p-q})} C$. Thus, using the isomorphism above, the assertion follows from Lemma 6.3. If n is nondegenerate on the subalgebra generated by p and q, then as in Lemma 6.2 one has $a-b=s^{-1}(p-q)=n(s)^{-1}\bar{s}(p-q)=n(s)^{-1}(n(p)-n(q)+t(\bar{q}p)-2\bar{p}q)$, hence $\operatorname{Aut}_{a-b} C = \operatorname{Aut}_{\bar{p}q} C$. □

Corollary 6.5. *Let C be a Cayley–Dickson algebra over F, and let p,q be elements of C such that $C(p,q)$ is a division algebra. Then, $\operatorname{Aut} C(p,q) = \sigma^{-1} \circ (\operatorname{Aut}_{\bar{p}q} C) \circ \sigma \cong \operatorname{Aut}_{\bar{p}q} C$.*

Proof. By Theorem III.5.8, C is a division algebra and $n(p) \neq n(q)$. Hence, n is nondegenerate on the subalgebra generated by p and q. In view of Lemma 6.3 and Theorem 6.4, it suffices to verify $t(a) \neq 1$ for $a=(p+q)^{-1}p$. But, if $t(a)=1$ then since $a+b=1$, $b=\bar{a}$ and so $n(a)=n(b)$. This implies $n(p)=n(q)$, a contradiction. □

Therefore, if $C(p,q)$ is a division algebra, then for $\operatorname{Aut} C(p,q)$ it suffices to determine the isotropy subgroup $\operatorname{Aut}_u C$ of $\operatorname{Aut} C$. Let C be a Cayley–Dickson algebra over F. For $u \in C$, let $K = F1 + Fu$ be the unital subalgebra of C generated by u. Assume that n is nondegenerate on K. If $u \in F$, then clearly $\operatorname{Aut}_u C = \operatorname{Aut} C$.

Thus, we may assume that $u \neq 0$ and $u \in V$. The nondegeneracy of n on K is equivalent to the condition $n(u) \neq 0$, which is the case for any $u \neq 0$ if C is a division algebra. Thus, $K = F1 + Fu$ is a composition algebra with norm n and $\{1, u\}$ is completed to a standard basis $1, u = e_1, \ldots, e_7$ of C given by Table I.1 (Chapter I). If we let $\lambda = -n(u) \neq 0$, then $u^2 = \lambda$. Since $u \in V$, it follows easily that if $\sqrt{\lambda} \notin F$, then K is a quadratic extension field of F, and if $\sqrt{\lambda} \in F$ then $K \cong F \oplus F$.

Let $W = K^\perp = \{x \in C : (x, K) = 0\}$. Then, $V = Fu \oplus W$, and W is regarded as a (left) K–module under the composition $1x = x$ and $ux = \frac{1}{2}[u, x]$ for $x \in W$ (see (6.2)). Define an F–bilinear map

$$\langle\, , \rangle : W \times W \to K \ : \ (x, y) \mapsto \lambda(x, y) - u(ux, y). \tag{6.13}$$

Applying (6.2) to the flexible identity $(xy)x = x(yx)$, one easily sees that $(\, ,)$ is an invariant form on V, i.e., $([x, y], z) + (y, [x, z]) = 0$ for all $x, y, z \in V$. Using this, it is readily seen that $\langle\, , \rangle$ is a nondegenerate K–hermitian form on W. In fact, for $x, y \in W$ one has

$$\begin{aligned} \langle y, x\rangle = \lambda(y, x) - u(uy, x) &= \lambda(y, x) - \frac{1}{2}u([u, y], x) = \lambda(x, y) + \frac{1}{2}u([u, x], y) \\ &= \lambda(x, y) + u(ux, y) = \overline{\langle x, y\rangle}, \end{aligned}$$

$$\begin{aligned} \langle ux, y\rangle &= \lambda(ux, y) - u(u^2x, y) = u^2(ux, y) - \lambda u(x, y) = \bar{u}\langle x, y\rangle, \\ \langle x, uy\rangle &= \lambda(x, uy) - u(ux, uy) = \frac{1}{2}\lambda(x, [u, y]) - \frac{1}{2}u(ux, [u, y]) \\ &= -\frac{1}{2}\lambda([u, x], y) + \frac{1}{2}u([u, ux], y) = -u^2(ux, y) + u(u^2x, y) = u\langle x, y\rangle. \end{aligned}$$

Hence, $\langle\, , \rangle$ is K–hermitian and is nondegenerate on W, since $u \notin F$. It follows from Table I.1 and (6.13) that W is a free K–module with a K–basis $\{e_2, e_4, e_6\}$ relative to which $\langle\, , \rangle$ has a nonsingular diagonal matrix J in $M_3(F)$.

It is straightforward to see that the map

$$M_3(K) \to M_3(K) \ : \ B \mapsto B^* = J^{-1}\bar{B}^t J \tag{6.14}$$

becomes an involution of second kind. It is shown in [**J1**] that if $\sqrt{\lambda} \notin F$, so K is a field, then

$$\operatorname{Aut}_u C \cong SU(3, K) = \{B \in M_3(K) : B^* = B^{-1} \text{ and } \det B = 1\}, \tag{6.15}$$

and if $\sqrt{\lambda} \in F$, so $K \cong F \oplus F$, then

$$\operatorname{Aut}_u C \cong SL(3, F) = \{B \in GL(3, F) : \det B = 1\}.$$

Note that if C is a division algebra and $u \notin F$, then (6.15) is the only possibility.

We utilize these remarks to obtain an analogous result of Corollary 5.10 for a division mutation algebra $C(p, q)$.

Corollary 6.6. *Let C be a Cayley–Dickson algebra over F, and let p, q be elements of C such that $C(p,q)$ is a division algebra. Then,* $\operatorname{Aut} C(p,q) \cong \operatorname{Aut} C$ *if and only if $C(p,q)$ has a unit element.*

Proof. Since $C(p,q)$ is a division algebra, so is C with $n(p) \neq n(q)$. Assume $\operatorname{Aut} C(p,q) \cong \operatorname{Aut} C$. Since $\operatorname{Aut} C(p,q) \cong \operatorname{Aut}_{\bar{p}q} C$ by Corollary 6.5, we must have $\bar{p}q \in F$ from the remarks above. This implies $(p-q)^{-1}p = n(p-q)^{-1}(\overline{p-q})p \in F$, so $p \in F(p-q)$, and similarly, $q \in F(p-q)$. Thus, for any $x \in C$, $\{x, p, q\}$ is contained in the associative algebra generated by x and $p-q$. Since $(p-q)^{-1}p = p(p-q)^{-1} \in F$, $(p-q)^{-1}px = x(p-q)^{-1}p = xp(p-q)^{-1}$ and so $px(p-q) = (p-q)xp$, implying $p(xq) = q(xp)$. We conclude from Theorem III.3.5 that $C(p,q)$ has a unit element.

Conversely, suppose that $C(p,q)$ has a unit element. From Theorem III.3.5, we have $p(xq) = q(xp)$ for any $x \in C$, in particular, $pq = qp$. Hence, p and q generate a commutative associative subalgebra S of C. Since C is a division algebra, n is nondegenerate on S and so $\dim_F S \leq 2$. It follows from this that S is generated by a single element u. Thus, for any $x \in C$, $\{x, p, q\}$ is contained in the associative subalgebra generated by x and u. This together with $pxq = qxp$ implies $(p-q)xp = px(p-q)$ and so $xp(p-q)^{-1} = (p-q)^{-1}px$ for any $x \in C$. Therefore, $(p-q)^{-1}p \in F$ and so $\bar{p}q \in F$. □

Finally, we determine $\operatorname{Der} C(p,q)$ for $p+q$ invertible. We need first:

Lemma 6.7. *Let C be a Cayley–Dickson algebra over F, and let a, b be elements of C such that $a \neq b$ and $a+b=1$. Then,* $\operatorname{Der} C(a,b) = \operatorname{Der}_{a-b} C$.

Proof. Since $\operatorname{Der}_{a-b} C \subseteq \operatorname{Der} C^{-} = \operatorname{Der} C(a,b)^{-}$ and $\operatorname{Der}_{a-b} C \subseteq \operatorname{Der} C(a-b)^{+} = \operatorname{Der} C(a,b)^{+}$, $\operatorname{Der}_{a-b} C \subseteq \operatorname{Der} C(a,b)$. Let d be any derivation of $C(p,q)$. Since $d \in \operatorname{Der} C^{-}$, in view of Lemma 6.1(ii) it suffices to verify $d(a-b) = d(1) = 0$. Since $d(F1) \subseteq F1$, we can let $d(1) = \alpha \in F$. Then, $d(a-b) = d(1*1) = d(1)*1 + 1*d(1) = 2\alpha(a-b)$. There exists an $x \in V$ such that $n(x) \neq 0 = t(x(a-b))$ and $\{x, \overline{a-b}\}$ is linearly independent. Using this and (III.2.10), we find

$$
\begin{aligned}
d(x*x) &= d(x(a-b)x) = d(t(x(a-b))x - n(x)(\overline{a-b})) = -n(x)d(\overline{a-b}), \\
d(x*x) &= d(x)*x + x*d(x) = (d(x)(a-b))x + (x(a-b))d(x) \\
&= t(x(a-b))d(x) + t(d(x)(a-b))x - 2(x, d(x))(\overline{a-b}) \\
&= t(d(x)(a-b))x
\end{aligned}
$$

by (III.2.11). Hence, $-n(x)d(\overline{a-b}) = t(d(x)(a-b))x$. But, since $d(\overline{a-b}) = d(t(a-b)) - d(a-b) = \alpha t(a-b) - 2\alpha(a-b)$, we have either $\alpha = 0$ or $a-b \in F$. In the latter case, $d(a-b) = (a-b)d(1) = \alpha(a-b) = 2\alpha(a-b)$, so $\alpha = 0$ also. Thus, $d(1) = d(a-b) = 0$ and $d \in \operatorname{Der}_{a-b} C$, showing $\operatorname{Der} C(a,b) \subseteq \operatorname{Der}_{a-b} C$. □

Our principal result on $\operatorname{Der} C(p,q)$ follows easily from this lemma, the isomorphism σ of (6.10) and the same arguments as in Theorem 6.4.

Theorem 6.8. *Let C be a Cayley–Dickson algebra over F, and let p, q be elements of C such that $p \neq q$ and $p+q$ is invertible in C. Then,*

$$\operatorname{Der} C(p,q) = \sigma^{-1} \circ \operatorname{Der}_{\sigma(\overline{p-q})} C \circ \sigma \cong \operatorname{Der}_{\sigma(\overline{p-q})} C$$

where σ is the isomorphism of (6.10). Moreover, if n is nondegenerate on the subalgebra generated by p and q, then $\operatorname{Der}_{\sigma(\overline{p-q})} C = \operatorname{Der}_{\bar{p}q} C$.

Proof. The isomorphism σ of (6.10) gives a Lie algebra isomorphism

$$\operatorname{Der} C(a,b) \longrightarrow \operatorname{Der} C(p,q) \;:\; d \mapsto \sigma^{-1} \circ d \circ \sigma$$

where $a = \sigma(s^{-1}ps^{-1})$ and $b = \sigma(s^{-1}qs^{-1})$ with $s = p+q$ and $a+b = 1$. As in the proof of Theorem 6.4, we have $\operatorname{Der}_{a-b} C = \operatorname{Der}_{\sigma(\overline{p-q})} C$. Therefore, the result follows from Lemma 6.7 and the same arguments as in Theorem 6.4. □

For $u \in C$, let $K = F1 + Fu$ be the unital subalgebra of C generated by u. If $u \in F$, then $\operatorname{Der}_u C = \operatorname{Der} C$. Thus, as in the remarks following Corollary 6.5 we assume that $0 \neq u \in V$ and n is nondegenerate on K, or equivalently, $n(u) \neq 0$. As before, let $\lambda = -n(u)$, and consider the K–hermitian form $\langle\,,\,\rangle$ on W given by (6.13) and the map defined by (6.14). Let

$$su(3,K) = \{B \in M_3(K) : B^* = -B \text{ and } \operatorname{trace} B = 0\}. \tag{6.16}$$

It is proven in **[EMy1]** that if $\sqrt{\lambda} \notin F$, then $\operatorname{Der}_u C \cong su(3,K)$ and if $\sqrt{\lambda} \in F$ then $\operatorname{Der}_u C \cong sl(3,F)$. Using this, we have an analogous result of Corollary 6.6 for $\operatorname{Der} C(p,q)$.

Corollary 6.9. *Let C be a Cayley–Dickson algebra over F, and let p, q be elements of C such that $C(p,q)$ is a division algebra. Then, $\operatorname{Der} C(p,q) \cong \operatorname{Der} C$ if and only if $C(p,q)$ has a unit element.*

Proof. Since C is a division algebra, n is nondegenerate on the subalgebra generated by p and q. In view of Theorem 6.8 and the remarks above, we find that $\operatorname{Der} C(p,q) \cong \operatorname{Der}_{\bar{p}q} C \cong \operatorname{Der} C$ if and only if $\bar{p}q \in F$. But, the same arguments as for Corollary 6.6 show that $\bar{p}q \in F$ if and only if $C(p,q)$ has a unit element. □

Proposition 6.10. *Let C be a Cayley–Dickson algebra over F such that $C(p,q)$ is a division algebra, and let $K = F1 + F\bar{p}q$.*

(i) *If $\bar{p}q \in F$, then $\operatorname{Aut} C(p,q) \cong \operatorname{Aut} C$ and $\operatorname{Der} C(p,q) \cong \operatorname{Der} C$ (non–split G_2).*

(ii) *If* $\bar{p}q \notin F$, *then* $\operatorname{Aut} C(p,q) \cong SU(3,K)$ *and* $\operatorname{Der} C(p,q) \cong su(3,K)$.

Proof. Since n is nondegenerate on K, part (i) follows from Corollary 6.5 and Theorem 6.8. If $\bar{p}q \notin F$, then K is a quadratic extension field of F since C is a division algebra. Hence, part (ii) is a consequence of (6.15) and (6.16). □

Applying Proposition 6.10 to the special case where $F = \mathbb{R}$, we have:

Corollary 6.11. *If $C(p,q)$ is a real division algebra, then*

(i) $\operatorname{Aut} C(p,q) \cong SU(3)$ *or compact* G_2,

(ii) $\operatorname{Der} C(p,q) \cong su(3)$ *or compact* G_2. □

Corollary 6.11 can be used to determine real division mutations $C(p,q)$ of C. In this case, C is the Cayley algebra and $n(p) \neq n(q)$. Assume $\operatorname{Der} C(p,q) =$ compact G_2, or equivalently, $\bar{p}q \in \mathbb{R}$. Hence $q = \lambda p$ with $p \neq 0$ and $\lambda \neq \pm 1$ in $\mathbb{R}$. Since $C(p) \cong C$ and $C(p,q) = C(p)(1,\lambda)$, we may assume that $C(p,q) = C(1,\lambda)$ with multiplication $x * y = xy - \lambda yx$. As for the quaternion case, let $\{1, u_1, \ldots, u_7\}$ be a standard basis of C. Thus, C has the multiplication given by Table I.1 with $\alpha = \beta = \gamma = -1$. Let

$$e = (1-\lambda)^{-1}1, \quad e_i = (1+\lambda)^{-1}u_i, \; i = 1, \ldots, 7.$$

It is easy to see that e is a unit element of $C(1,\lambda)$ and the multiplication of $C(1,\lambda)$ relative to $\{e, e_1, \ldots, e_7\}$ is the same as the standard multiplication of C, except $e_i * e_i = -(1-\lambda)^2 e$, $i = 1, \ldots, 7$. We note from this that $C(1,\lambda)$ is isomorphic to an algebra given by [**BO1**, (2.1)] with $\eta = \zeta = 1$ and $\beta > 0$, which arises from the classification of real division algebras of dimension 8 with compact G_2 as its derivation algebra.

Real division algebras of dimension 8 with derivation algebra $su(3)$ have been determined in [**BO1**]. Such division algebras are divided into two types. The first type consists of Lie–admissible algebras whose minus algebras are isomorphic to $su(3)$. Since $C(1,\lambda)^- = C(1+\lambda)^-$ is a non–Lie, Malcev algebra, $C(1,\lambda)$ cannot be an algebra of this type. The second type comprises real division algebras A such that A is the direct sum $A = U \oplus V$ of a 2–dimensional trivial $su(3)$–module U and a 6–dimensional irreducible $su(3)$–module V. The multiplication between U and V has been determined in [**BO1**, (4.2)] in a rather crude form. In case $\operatorname{Der} C(p,q) \cong su(3)$, it is not known which of these algebras A are isomorphic to division mutations $C(p,q)$ of C.

The discussion in this section is based on the results from [**EMMy1**,**Mo1**].

CHAPTER V

THE MUTATION VARIETY AND STRICT REPRESENTATIONS

1. Basic concepts

In this chapter we study identities satisfied by all mutations of associative algebras and its special representations which are consequences of one–sided modules of associative algebras. The approach we employ here reveals that the identities satisfied by all mutation algebras have much greater diversity than those identities resulting from Jordan– or Lie–admissibility and that the complete analysis of these identities appears to be a formidable task. Accordingly, our discussion focuses on establishing a general technique to determine the identities which is based on a certain gradation of the T–ideal of identities satisfied by all mutation algebras. As a consequence, we determine all identities of degree ≤ 4 in this T–ideal, and in particular, the degree 4 case results in two new identities which are not consequences of Jordan– and Lie–admissibilities. Some detail of the degree 4 case is omitted here which involves lengthy calculation. In Section V.3, we extend some of the results in Section II.2 for mutation algebras to the variety defined by Lie–, Jordan–admissibilities and these new identities.

Since there lacks a complete description of the variety generated by mutations of all associative algebras, it seems difficult to develop a representation theory for these algebras in the general framework based on the theory of varieties. However, (one–sided) modules for an associative algebra give rise to a certain special type of representations for mutation algebras, called here "strict" representations. The primary concern in Section V.4 is to determine all irreducible strict representations for mutations of any simple artinian associative algebra.

Let F be a field of characteristic $\neq 2$ and let $X = \{x_1, \ldots, x_n, \ldots\}$ be a countable set of indeterminates. Let $F[X]$ denote the free (nonassociative) algebra on X over F. Thus, any map $\varphi : X \to A$ for an F–algebra A extends to a unique algebra homomorphism of $F[X]$ into A. Each $f \in F[X]$ may be denoted by $f(x_1, \ldots, x_m)$ since it is contained in the subalgebra of $F[X]$ generated by $x_1, \ldots, x_m$ for some $m \geq 1$. If A is an algebra over F and $a_1, \ldots, a_m \in A$, then $f(a_1, \ldots, a_m)$ stands for the image of $f(x_1, \ldots, x_m)$ by any homomorphism of $F[X]$ into A that sends x_i to a_i, $i = 1, \ldots, m$. An element $f = f(x_1, \ldots, x_m)$ (or $= 0$) in $F[X]$ is called an *identity* for an algebra B over F if $f \in \cap_\varphi \ker \varphi$ where φ runs over the set of all homomorphisms of $F[X]$ into B, or equivalently, $f(a_1, \ldots, a_m) = 0$ for all $a_1, \ldots, a_m \in B$. Let $I(B)$ denote the set of all identities for B. If U is a class of F–algebras, then we let

$$I(U) = \cap_{B \in U} I(B),$$

that is, $I(U)$ is the set of identities satisfied by all algebras in U. An element of $I(U)$ is also called a U*–identity*.

Definition 1.1. A class U of algebras over F is called a *variety* if there is a subset S of $F[X]$ such that U equals the set of algebras satisfying all $f \in S$. In this case, U is called the *variety defined by* S or the S–variety, and is denoted by $V(S)$. □

Thus, any subset of $F[X]$ defines a variety. A class U of F–algebras is termed a *closed class* if U contains all subalgebras, all homomorphic images of any algebra in U and complete direct products of any algebras in U. It is clear that any variety is a closed class. The converse of this, which is also true, is known as Birkhoff's theorem [**Os1**]:

Theorem 1.1. *A class of F–algebras is a variety if and only if it is a closed class.* □

The intersection of any collection of closed classes is a closed class. Hence, for any class U of F–algebras, there exists the smallest closed class containing U which, by Birkhoff's theorem, is a variety. This variety is called the *variety generated by* U and is denoted by $\langle U\rangle$. As shown by Example II.4.6, not all subalgebras or quotient algebras of a mutation of an associative algebra are mutation algebras. Thus, the class Mut of all mutations of associative algebras over F does not form a variety.

Definition 1.2. The variety $\langle Mut\rangle$ generated by Mut is called the *mutation variety* (of associative algebras over F). □

An ideal of $F[X]$ is called a T–ideal if it is stable under any endomorphism of $F[X]$. It is clear that for any class U of F–algebras, $I(U)$ is a T–ideal of $F[X]$, and that any subset S of $F[X]$ generates a T–ideal of $F[X]$, denoted by $[S]$, i.e., the smallest T–ideal of $F[X]$ containing S. The relationship between varieties and T–ideals is given by the duality theorem [**Os1**,p.173]:

Theorem 1.2. *The map $U \longrightarrow I(U)$ is a bijective correspondence between the set of all varieties of F–algebras and the set of all T–ideals of $F[X]$ such that $V(I(U)) = U$ and $I(V(S)) = S$ for any variety U and T–ideal S of $F[X]$.* □

Using Theorems 1.1 and 1.2, the following can be easily proven.

Proposition 1.3. *Let U, W be classes of F–algebras and let S be a subset of $F[X]$.*

(i) $I(\langle U\rangle) = I(U)$ *and* $V(I(U)) = \langle U\rangle$.

(ii) *If* $I(U) = I(W)$, *then* $\langle U\rangle = \langle W\rangle$.

(iii) *There exists an algebra* $A \in \langle U\rangle$ *such that* $\langle A\rangle = \langle U\rangle$ *and* $I(A) = I(U)$.

(iv) $V(S) = V([S])$ *and* $I(V(S)) = [S]$.

Proof. Clearly, $I(\langle U\rangle) \subseteq I(U)$ and the reverse inclusion follows from Theorem 1.1. Hence, by Theorem 1.2 $V(I(U)) = V(I(\langle U\rangle)) = \langle U\rangle$, verifying (i). Part (ii) is an immediate consequence of part (i) and Theorem 1.2. Part (iii) follows from [**Os1**, Theorem 2.2 and Lemma 2.3], using Theorem 1.1 and part (i). For (iv), clearly $V([S]) \subseteq V(S)$. For any $B \in V(S)$, let $\varphi : F[X] \to B$ be a homomorphism and let η be any endomorphism of $F[X]$. If $f \in S$, then $\varphi(\eta(f)) = (\varphi \circ \eta)(f) = 0$ since $\varphi \circ \eta : F[X] \to B$ is a homomorphism and $B \in V(S)$. Hence, $\eta(f) \in \ker \varphi$ and so $B \in V([S])$, showing $V([S]) = V(S)$. The second part follows from this and Theorem 1.2. □

We say that an $f \in F[X]$ is a consequence of S if f is a $V(S)$–identity, or equivalently, $f \in [S]$ by Proposition 1.3(iv). We now turn our attention to a gradation of $F[X]$ that plays a central role for the analysis of the mutation variety and $I(Mut)$. For $n_1, \dots, n_m \in \mathbb{Z}^+$, a monomial in $F[X]$ is said to be of degree $[n_1, \dots, n_m]$ if it is of degree n_i in x_i for $i = 1, \dots, m$ and of degree 0 in x_i for $i > m$. An $f \in F[X]$ of degree $[n_1, \dots, n_m]$ is a linear combination of monomials of degree $[n_1, \dots, n_m]$ and hence is homogeneous of degree n_i in x_i for $i = 1, \dots, m$ and of degree 0 in x_i for $i > m$. If we let $F^{[n_1,\dots,n_m]}[X]$ denote the linear span of all elements in $F[X]$ of degree $[n_1, \dots, n_m]$, then we have the standard gradation for $F[X]$:

$$F[X] = \sum_{m\geq 1} \sum_{n_1,\dots,n_m \geq 0} \bigoplus F^{[n_1,\dots,n_m]}[X]. \tag{1.1}$$

Denote the multiplication in $F[X]^+$ and $F[X]^-$ by $u \circ v = \frac{1}{2}(uv + vu)$ and $[u, v] = uv - vu$. To refine the gradation (1.1), for $(r, s) \in \mathbb{Z} \times \mathbb{Z}$ we introduce the subspace $F_{(r,s)}[X]$ of $F[X]$ recursively as follows. If $r < 0$ or $s < 0$, then we put $F_{(r,s)}[X] = 0$. Let $F_{(0,0)}[X] = FX$, the linear span of X. For $r \geq 0$ and $s \geq 0$ we define $F_{(r,s)}[X]$ by

$$\begin{aligned} F_{(r,s)}[X] &= \sum_{\substack{i+j+1=r\\ k+l=s}} F_{(i,k)}[X] \circ F_{(j,l)}[X] \\ &\quad + \sum_{\substack{i+j=r\\ k+l+1=s}} \left[F_{(i,k)}[X], F_{(j,l)}[X]\right]. \end{aligned} \tag{1.2}$$

For example,

$$\begin{aligned} F_{(r,0)}[X] &= \sum_{i+j+1=r} F_{(i,0)}[X] \circ F_{(j,0)}[X], \\ F_{(0,s)}[X] &= \sum_{k+l+1=s} \left[F_{(0,k)}[X], F_{(0,l)}[X]\right], \\ F_{(1,1)}[X] &= FX \circ [FX, FX] + [FX, FX \circ FX]. \end{aligned}$$

Thus, $F_{(r,s)}[X]$ is the span of elements in $F[X]$ which are formed by applying r times the product $\circ$ and s times the product $[\ ,\ ]$ to X in any order and association.

The subspaces $F_{(r,s)}[X]$ yield a gradation for $F[X]$:

Proposition 1.4. $F[X] = \sum_{r,s\geq 0} \oplus F_{(r,s)}[X]$.

Proof. Since $uv = u\circ v + \frac{1}{2}[u,v]$ for $u, v \in F[X]$, $F[X]$ is the sum of all $F_{(r,s)}[X]$. To show that the sum is direct, for indeterminates a and b not in X, we consider the free nonassociative algebra $F[Y]$ on $Y = X \cup \{a, b\}$. We define a new multiplication $*$ on $F[Y]$ by

$$u * v = \frac{1}{2}\left((u(a+b))v + (v(a-b))u\right), \tag{1.3}$$

that is, $(F[Y], *)$ is the left $(\frac{1}{2}(a+b), -\frac{1}{2}(a-b))$–mutation of $F[Y]$ (see Definition I.4.4), Let $F^{\sharp}[X]$ denote the algebra obtained from $F[X]$ by adjoining a unit element, i.e., $F^{\sharp}[X] = F1 \oplus F[X]$. Then, there exist the unique homomorphisms σ and τ:

$$\sigma : F[X] \to (F[Y], *),$$
$$\tau : F[Y] \to F^{\sharp}[X]$$

such that $\sigma(x_i) = x_i$ for all i and $\tau(a) = \tau(b) = 1$, $\tau(x_i) = x_i$ for all i.

It is clear that $\tau : (F[Y], *) \to F^{\sharp}[X]$ is also a homomorphism such that $\tau(\sigma(w)) = w$ for all $w \in F[X]$, and hence σ is injective. Since $\sigma(u \circ v) = \frac{1}{2}(\sigma(u) * \sigma(v) + \sigma(v) * \sigma(u)) = \frac{1}{2}((\sigma(u)a)\sigma(v) + (\sigma(v)a)\sigma(u))$ and $\sigma([u,v]) = \sigma(u) * \sigma(v) - \sigma(v) * \sigma(u) = (\sigma(u)b)\sigma(v) - (\sigma(v)b)\sigma(u)$ using (1.3), we find that $\sigma(F_{(r,s)}[X])$ comprises the elements of $\sigma(F[X])$ which are homogeneous of degree r in a and of degree s in b. Therefore, if $\sum_{r,s\geq 0} f_{(r,s)} = 0$ for $f_{(r,s)} \in F_{(r,s)}[X]$, then $\sum \sigma(f_{(r,s)}) = 0$ and hence $\sigma(f_{(r,s)}) = 0$, so $f_{(r,s)} = 0$ for all $r, s \geq 0$ since σ is injective. This shows that the sum $\sum_{(r,s)} F_{(r,s)}[X]$ is direct. □

For $n_1, \ldots, n_m \geq 0$ and for $r, s \geq 0$ such that $r + s + 1 = n_1 + \cdots + n_m$, let

$$F^{[n_1,\ldots,n_m]}_{(r,s)}[X] = F^{[n_1,\ldots,n_m]}[X] \cap F_{(r,s)}[X]. \tag{1.4}$$

Each element of $F^{[n_1,\ldots,n_m]}_{(r,s)}[X]$ is said to be of *type* $\binom{[n_1,\ldots,n_m]}{(r,s)}$. It follows easily from Proposition 1.4 and (1.1) that

$$F[X] = \sum_{\substack{m\geq 1 \\ n_1,\ldots,n_m\geq 0}} \ \sum_{\substack{r,s\geq 0 \\ r+s+1=n_1+\cdots+n_m}} \bigoplus F^{[n_1,\ldots,n_m]}_{(r,s)}[X]. \tag{1.5}$$

2. Identities for the mutation variety

Let F be a field of characteristic $\neq 2$. Let Ass and LIE respectively denote the varieties of associative algebras and of Lie algebras over F. Thus, we have $Mut = \{A(p,q) : A \in Ass,\ p,q \in A\}$, and consider the classes:

$$\begin{aligned} Mut^{\pm} &= \{A(p,q)^{\pm} : A(p,q) \in Mut\}, \\ Ass^{\pm} &= \{A^{\pm} : A \in Ass\}. \end{aligned}$$

The class, denoted by SJ, of special Jordan algebras over F consists of subalgebras of A^+ for any $A \in Ass$. Hence,

$$Mut^+ \subseteq Ass^+ \subseteq SJ, \quad Mut^- \subseteq Ass^- \subseteq LIE, \tag{2.1}$$

since $A(p,q)^+ = A(p-q)^+$ and $A(p,q)^- = A(p+q)^-$ for $A(p-q), A(p+q) \in Ass$.

Recall that each $A(p,q) \in Mut$ is Lie– and Jordan–admissible, that is, $A(p,q)^-$ and $A(p,q)^+$ satisfy the Jacobi and Jordan identities

$$[[x,y]^*,z]^* + [[y,z]^*,x]^* + [[z,x]^*,y]^* = 0, \tag{2.2}$$

$$\{\{\{x,x\}^*,y\}^*,x\}^* = \{\{x,x\}^*,\{y,x\}^*\}^* \tag{2.3}$$

where $x*y = xpy - yqx$, $[x,y]^* = x*y - y*x$ and $\{x,y\}^* = \frac{1}{2}(x*y + y*x)$. As noted in Chapter II, certain well known identities considered in the theory of nonassociative algebras, such as flexibility and third power–associativity, are not identities for Mut, although they are satisfied by some $A(p,q) \in Mut$ under a strong restriction on p and q.

For the set $X = \{x_1, \ldots, x_n, \ldots\}$ of indeterminates, let $Ass[X]$ denote the free associative algebra on X over F. For two indeterminates p and q not in X, we consider the (p,q)–mutation $Ass[X \cup \{p,q\}](p,q)$ of the free associative algebra $Ass[X \cup \{p,q\}]$ on $X \cup \{p,q\}$. Let $Mut[X]$ be the subalgebra of $Ass[X \cup \{p,q\}](p,q)$ generated by X. Then, there exists a unique surjective homomorphism

$$\psi : F[X] \to Mut[X]$$

such that $\psi(x_i) = x_i$ for all $i = 1, 2, \ldots$. We first show:

Proposition 2.1. $\ker \psi = I(Mut) = I(Mut[X])$.

Proof. We first verify $I(Mut) = I(Mut[X])$. Since Mut contains $Ass[X \cup \{p,q\}](p,q)$ and $Mut[X]$ is a subalgebra of this, we have $I(Mut) \subseteq I(Mut[X])$. Conversely, let $f = f(x_1, \ldots, x_n) \in I(Mut[X])$. For any $A(\tilde{p},\tilde{q}) \in Mut$ and for any $a_1, \ldots, a_n \in A$, there is a unique homomorphism

$$\sigma : Ass[X \cup \{p,q\}] \to A$$

such that $\sigma(x_i) = a_i$ for $i = 1, \dots, n$, $\sigma(x_i) = 0$ for $i > n$ and $\sigma(p) = \tilde{p}$, $\sigma(q) = \tilde{q}$. Hence, the restriction of σ to $Mut[X]$ gives a homomorphism $\tilde{\sigma} : Mut[X] \to A(\tilde{p}, \tilde{q})$, and

$$f(a_1, \dots, a_n) = f(\tilde{\sigma}(x_1), \dots, \tilde{\sigma}(x_n)) = \tilde{\sigma}(f(x_1, \dots, x_n)) = 0$$

since $f(x_1, \dots, x_n) = 0$ in $Mut[X]$. This proves $I(Mut) = I(Mut[X])$.

If $A(c,d) \in Mut$, then let $\varphi : F[X] \to A(c,d)$ be any homomorphism. Hence, there is a unique homomorphism $\varphi_1 : Ass[X \cup \{p,q\}] \to A$ such that $\varphi_1(x_i) = \varphi(x_i)$ for all i, $\varphi_1(p) = c$ and $\varphi_1(q) = d$. As above, φ_1 induces a homomorphism $\tilde{\varphi}_1$ of $Mut[X]$ into $A(c,d)$ with $\tilde{\varphi}_1(x_i) = \varphi(x_i)$ for all i. For $f = f(x_1, \dots, x_n) \in F[X]$, we have

$$\begin{aligned}(\tilde{\varphi}_1 \circ \psi)(f) &= \tilde{\varphi}_1(\psi(f(x_1, \dots, x_n))) \\ &= \tilde{\varphi}_1(f(\psi(x_1), \dots, \psi(x_n))) \\ &= f(\tilde{\varphi}_1(x_1), \dots, \tilde{\varphi}_1(x_n)) \\ &= f(\varphi(x_1), \dots, \varphi(x_n)) = \varphi(f).\end{aligned}$$

Thus, $\tilde{\varphi}_1 \circ \psi = \varphi$ and so $\ker \psi \subseteq \ker \varphi$. This implies $\ker \psi \subseteq I(Mut) = I(Mut[X])$. Since $I(Mut[X]) \subseteq \ker \psi$, this completes the proof. □

Therefore, by Proposition 1.3(i),(ii), $Mut[X]$ generates the mutation variety $\langle Mut \rangle$, and $I(\langle Mut \rangle) = \ker \psi$. A class U of F–algebras is called *homogeneous* if $I(U)$ is a homogeneous ideal of $F[X]$ relative to the gradation (1.5). In view of Proposition 2.1, Mut is homogeneous if and only if $\ker \psi$ is homogeneous, or equivalently, $\ker \psi$ has the decomposition

$$\ker \psi = \sum_{\substack{m \geq 1 \\ n_1, \dots, n_m \geq 0}} \sum_{\substack{r,s \geq 0 \\ r+s+1 = n_1 + \cdots + n_m}} \bigoplus \Psi_{(r,s)}^{[n_1, \dots, n_m]} \tag{2.4}$$

where we set

$$\Psi_{(r,s)}^{[n_1, \dots, n_m]} = F_{(r,s)}^{[n_1, \dots, n_m]}[X] \cap \ker \psi = \ker \left(\psi|_{F_{(r,s)}^{[n_1, \dots, n_m]}[X]} \right). \tag{2.5}$$

The second equality in (2.5) follows easily.

Let $a = p - q$ and $b = p + q$. Then, $Ass[X \cup \{p,q\}] = Ass[X \cup \{a,b\}]$ and

$$x * y = xpy - yqx = \tfrac{1}{2}[x(a+b)y + y(a-b)x], \tag{2.6}$$

$$\{x,y\}^* = \tfrac{1}{2}(xay + yax), \quad [x,y]^* = xby - ybx \tag{2.7}$$

for $x, y \in Mut[X]$. As in (1.2), we can define the subspace $Mut_{(r,s)}[X]$ of $Mut[X]$ by

$$Mut_{(r,s)}[X] = \sum_{\substack{i+j+1=r \\ k+l=s}} \left\{ Mut_{(i,k)}[X], Mut_{(j,l)}[X] \right\}^* \tag{2.8}$$

$$+\sum_{\substack{i+j=r\\k+l+1=s}}\left[Mut_{(i,k)}[X], Mut_{(j,l)}[X]\right]^*$$

with $Mut_{(r,s)}[X]=0$ for $r<0$ or $s<0$.

We note from (2.7) and (2.8) that each element of $Mut_{(r,s)}[X]$ is homogeneous of degree r in a and of degree s in b, viewed as an element of $Ass[X\cup\{a,b\}]$. Hence, as in the proof of Proposition 1.4, it follows that

$$Mut[X]=\sum_{r,s\geq 0}\bigoplus Mut_{(r,s)}[X]. \tag{2.9}$$

The decomposition (2.9) also shows that $Mut_{(r,s)}[X]$ is the set of all elements in $Mut[X]$ which are homogeneous of degree r in a and of degree s in b. It is clear from (1.2) and (2.8) that

$$\psi(F_{(r,s)}[X])=Mut_{(r,s)}[X] \tag{2.10}$$

for all (r,s). For $n_1,\ldots,n_m\geq 0$, let $Mut^{[n_1,\ldots,n_m]}[X]$ be the subspace of all elements in $Mut[X]$ which are homogeneous of degree n_i in x_i for $1\leq i\leq m$ and degree 0 in x_j for $j>m$, and let

$$Mut^{[n_1,\ldots,n_m]}_{(r,s)}[X]=Mut^{[n_1,\ldots,n_m]}[X]\cap Mut_{(r,s)}[X]$$

for all types $\binom{[n_1,\ldots,n_m]}{(r,s)}$.

Since $Mut[X]$ is the direct sum of all $Mut^{[n_1,\ldots,n_m]}[X]$, it follows easily from (2.9) and (2.10) that

$$Mut[X]=\sum_{\substack{m\geq 1\\n_1,\ldots,n_m\geq 0}}\ \sum_{\substack{r,s\geq 0\\r+s+1=n_1+\cdots+n_m}}\bigoplus Mut^{[n_1,\ldots,n_m]}_{(r,s)}[X], \tag{2.11}$$

$$\psi\left(F^{[n_1,\ldots,n_m]}_{(r,s)}[X]\right)=Mut^{[n_1,\ldots,n_m]}_{(r,s)}[X]. \tag{2.12}$$

Thus, ψ is a grade preserving homomorphism. The homogeneity of Mut now follows easily from this.

Proposition 2.2. *Mut is a homogeneous class of F-algebras.*

Proof. By Proposition 2.1 it suffices to verify the decomposition (2.4). The sum on the right side of (2.4) is clearly contained in $\ker\psi$ by (2.5). But, if $f\in\ker\psi$, then $0=\psi(f)$ and by (2.12) the components of $\psi(f)$ in (2.11) must be zero since the sum in (2.11) is direct. This shows that f is contained in the sum on the right side of (2.4). □

Therefore, for $I(Mut)$ it suffices to determine all $\Psi^{[n_1,\ldots,n_m]}_{(r,s)}$. The primary objective in this section is to determine these subspaces of $\ker\psi$ for types $\binom{[n_1,\ldots,n_m]}{(r,s)}$ with

$r+s+1 = n_1 + \cdots + n_m \le 4$. A complete analysis of the degree 4 case is somewhat complicated, but two fundamental identities appear for Mut of types $\binom{[3,1]}{(2,1)}$ and $\binom{[2,2]}{(2,1)}$ which are not consequences of Lie– and Jordan–admissibilities (2.2) and (2.3). We begin the analysis by constructing an identity of type $\binom{[4]}{(2,1)}$ for $Mut[X]$ which is implied by one of the two identities above (for type $\binom{[3,1]}{(2,1)}$), but not by (2.2) and (2.3).

Let x be any element of $Mut[X]$. All 4th powers of x in $Mut[X]$ form a 5×1 matrix S given by

$$S = \begin{pmatrix} ((x*x)*x)*x \\ x*(x*(x*x)) \\ (x*(x*x))*x \\ x*((x*x)*x) \\ (x*x)*(x*x) \end{pmatrix}.$$

Expansion of S using (2.6) is expressed by the matrix equation $S = DT$, where D is a 5×8 matrix over F and T is an 8×1 matrix over $Ass[X \cup \{p,q\}]$, given by

$$D = \begin{pmatrix} 1 & 0 & 0 & 0 & -3 & 0 & 3 & -1 \\ 1 & -3 & 0 & 3 & 0 & 0 & 0 & -1 \\ 1 & 0 & -2 & 0 & -1 & 2 & 1 & -1 \\ 1 & -1 & -2 & 1 & 0 & 2 & 0 & -1 \\ 1 & -1 & -1 & 1 & -1 & 1 & 1 & -1 \end{pmatrix},$$

$$T = \begin{pmatrix} xpxpxpx \\ xpxpxqx \\ xpxqxpx \\ xpxqxqx \\ xqxpxpx \\ xqxpxqx \\ xqxqxpx \\ xqxqxqx \end{pmatrix}.$$

Hence, the identity $S = DT$ is satisfied by $Mut[X]$.

Noting that $\operatorname{rank} D = 3$, we obtain two solutions $\Lambda_1 = (1,-1,-3,3,0)$ and $\Lambda_2 = (1,1,1,1,-4)$ to the equation

$$(\lambda_1, \lambda_2, \lambda_3, \lambda_4, \lambda_5)D = (0,0,0,0,0),$$

so that all other solutions are linear combinations of Λ_1 and Λ_2. In particular, $\Lambda_i S = \Lambda_i(DT) = (\Lambda_i D)T = 0$ for $i = 1,2$. Expanding $\Lambda_1 S = \Lambda_2 S = 0$ yields the following two identities

(2.13) $\qquad ((x*x)*x)*x - x*(x*(x*x)) - 3(x*(x*x))*x$

$$+3x*((x*x)*x)=0,$$

(2.14) $$4(x*x)*(x*x)=((x*x)*x)*x+x*(x*(x*x)) +(x*(x*x))*x+x*((x*x)*x)$$

which are identities for $Mut[X]$ and hence for Mut. Identity (2.14) is just the expansion of

$$\{\{x,x\}^*,\{x,x\}^*\}^*=\{\{\{x,x\}^*,x\}^*,x\}^*$$

which is a consequence of Jordan–admissibility (2.3).

On the other hand, identity (2.13) is not a consequence of (2.2) and (2.3). To illustrate this, let B be an algebra over F with a basis $\{a,b,c,d\}$ and with multiplication given by

$$a^2=b,\ ab=2c,\ ac=2d,\ b^2=d$$

and all other products are zero. We note that the multiplication in B^+ is given by

$$a\circ a=b,\ a\circ b=c,\ a\circ c=d,\ b\circ b=d$$

with other products equal to zero. Thus, B^+ is the commutative associative algebra generated by the element a with $a^{\circ 5}=0$. Since the nonzero products in B^- are given by $[a,b]=2c$ and $[a,c]=2d$, it follows that B^- is a Lie algebra. Therefore, B is Lie– and Jordan–admissible. However, (2.13) is not an identity for B since

$$(a^2a)a=(aa^2)a=a(a^2a)=0,\ a(aa^2)=4d\neq 0.$$

We now express (2.13) in terms of the multiplications in $Mut[X]^+$ and $Mut[X]^-$. Thus,

$$\begin{aligned}
&((x*x)*x)*x-x*(x*(x*x))-3(x*(x*x))*x-3x*((x*x)*x)\\
&=2[((x*x)*x)*x+x*((x*x)*x)]\\
&\quad-2[x*(x*(x*x))+(x*(x*x))*x]\\
&\quad-[((x*x)*x)*x+(x*(x*x))*x]+[x*(x*(x*x))+x*((x*x)*x)]\\
&=4\{(x*x)*x,x\}^*-4\{x*(x*x),x\}^*-2\{x*x,x\}^**x+2x*\{x*x,x\}^*\\
&=4\{[\{x,x\}^*,x]^*,x\}^*-2[\{\{x,x\}^*,x\}^*,x]^*=0.
\end{aligned}$$

This shows that

(2.15) $$[x\circ(x\circ x),x]+2x\circ[x,x\circ x]$$

is an element of $\Psi^{[4]}_{(2,1)}$ (strictly speaking, $x=x_1\in X$).

Let $F_+[X]$ and $F_-[X]$ denote the subspaces of $F[X]$ given by

$$F_+[X] = \sum_{\substack{m\geq 1 \\ n_1,\ldots,n_m\geq 0}} \sum_{\substack{r\geq 0 \\ r+1=n_1+\cdots+n_m}} \bigoplus F_{(r,0)}^{[n_1,\ldots,n_m]}[X],$$

$$F_-[X] = \sum_{\substack{m\geq 1 \\ n_1,\ldots,n_m\geq 0}} \sum_{\substack{s\geq 0 \\ s+1=n_1+\cdots+n_m}} \bigoplus F_{(0,s)}^{[n_1,\ldots,n_m]}[X].$$

It follows easily that $F_\pm[X]$ is the subalgebra of $F[X]^\pm$ generated by X and hence is the free commutative or anticommutative algebra over X. Moreover, if $\psi_\pm$ denotes the restriction of ψ to $F_\pm[X]$, then

$$\ker\psi_+ = \sum_{\substack{m\geq 1 \\ n_1,\ldots,n_m\geq 0}} \sum_{\substack{r\geq 0 \\ r+1=n_1+\cdots+n_m}} \bigoplus \Psi_{(r,0)}^{[n_1,\ldots,n_m]},$$

$$\ker\psi_- = \sum_{\substack{m\geq 1 \\ n_1,\ldots,n_m\geq 0}} \sum_{\substack{s\geq 0 \\ s+1=n_1+\cdots+n_m}} \bigoplus \Psi_{(0,s)}^{[n_1,\ldots,n_m]},$$

and $\psi_\pm : F_\pm[X] \to Mut[X]^\pm$ is a unique surjective homomorphism such that $\psi_\pm(x_i) = x_i$ for all i. From the same arguments as in the proof of Proposition 2.1, we deduce

Lemma 2.3. $\ker\psi_\pm = I(Mut[X]^\pm) = I(Mut^\pm)$. □

Theorem 2.4. (i) $I(SJ) = I(Ass^+) = I(Mut^+) = \ker\psi_+$.

(ii) $I(LIE) = I(Ass^-) = I(Mut^-) = \ker\psi_-$.

Proof. (i) By (2.1) we have $I(SJ) \subseteq I(Ass^+) \subseteq I(Mut^+)$, and since any $f \in I(Ass^+)$ is satisfied by all subalgebras of any $A^+ \in Ass^+$, $I(SJ) = I(Ass^+)$. Let $A^+ \in Ass^+$. If A has a unit element 1, then $A^+ = A(1,0)^+ \in Mut^+$. Otherwise, we let $A^\sharp = F1 \oplus A$, the algebra obtained from A by adjoining a unit element. Hence, A^+ is a subalgebra of $(A^\sharp)^+ = A^\sharp(1,0)^+ \in Mut^+$. This shows $I(Ass^+) = I(Mut^+)$, and the last equality follows from Lemma 2.3.

(ii) From (2.1) we note $I(LIE) \subseteq I(Ass^-) \subseteq I(Mut^-)$. But, by the Poincaré–Birkhoff–Witt Theorem, any Lie algebra is a subalgebra of some $A^- \in Ass^-$ and hence $I(LIE) = I(Ass^-)$. Since $A(c,d)^- = A(c+d)^-$ for $A(c,d) \in Mut$, as above, any $A^- \in Ass^-$ is a subalgebra of some $B(1,0)^- \in Mut^-$, implying $I(Ass^-) = I(Mut^-)$. □

Since LIE is the variety defined by the Jacobian $J(x,y,z)$:

$$J(x,y,z) = [[x,y],z] + [[y,z],x] + [[z,x],y],$$

any element f of $\ker \psi_- = I(Mut^-)$ is a consequence of $J(x,y,z)$, i.e., f is contained in the T–ideal of $F_-[X]$ generated by $J(x,y,z)$ (Proposition 1.3(iv)). In contrast, this is not true for $\ker \psi_+ = I(SJ) = I(Mut^+)$, that is, not every element of $\ker \psi_+$ is a consequence of the Jordan identity

$$j(x,y) = (x \circ x) \circ (y \circ x) - ((x \circ x) \circ y) \circ x,$$

or equivalently, SJ does not generate the variety of Jordan algebras defined by $j(x,y)$ (Proposition 1.3(ii)). Clearly, SJ is closed under subalgebras and complete direct sums (such a class of algebras is called a quasi–variety), but not under homomorphic images. In fact, due to Cohn's theorem, there exists a free special Jordan algebra of three generators whose homomorphic image is exceptional [**J3**,**ZSSS1**]. As is well known in the Jordan theory, there is an identity of degree 8, called the Glennie identity, which is satisfied by all homomorphic images of any algebra in SJ but not by Albert algebras of dimension 27 (see Section I.5). The Glennie identity belongs to $\Psi_{(8,0)}^{[3,2,3]}$ [**Ja3**].

To facilitate calculation for the remaining types (i.e., $rs > 0$), we consider the unique homomorphism $u \mapsto u^{\cdot}$ of $Ass[X \cup \{a,b\}]$ to the opposite algebra $Ass[X \cup \{a,b\}]^{op}$ such that $x_i^{\cdot} = x_i$ for all $x_i \in X$ and $p^{\cdot} = -q$, $q^{\cdot} = -p$. Thus, $u \mapsto u^{\cdot}$ is an involution on $Ass[X \cup \{a,b\}]$. Since $a = p - q$ and $b = p + q$, $a^{\cdot} = p^{\cdot} - q^{\cdot} = a$ and $b^{\cdot} = -q - p = -b$. Let $H[X]$ be the subspace of $Ass[X \cup \{a,b\}]$ spanned by X and by all elements of the form

$$y_1 z_1 y_2 z_2 \cdots y_{r-1} z_{r-1} y_r + (y_1 z_1 y_2 z_2 \cdots y_{r-1} z_{r-1} y_r)^{\cdot}$$

for $r = 1, 2, \ldots$, where $y_i \in X$ and $z_i \in \{a,b\}$ for all $i = 1, \ldots, r$. Hence, every element of $H[X]$ is $\cdot$ – symmetric, and for $u, v \in H[X]$ we have $(u * v)^{\cdot} = (upv)^{\cdot} - (vqu)^{\cdot} = vp^{\cdot}u - uq^{\cdot}v = u * v$. Moreover,

$$\begin{aligned} u * v &= \frac{1}{2}(u * v + (u * v)^{\cdot}) \\ &= \frac{1}{4}\left[u(a+b)v + (u(a+b)v)^{\cdot} + v(a-b)u + (v(a-b)u)^{\cdot}\right] \in H[X]. \end{aligned}$$

This proves that $H[X]$ is a subalgebra of $Ass[X \cup \{p,q\}](p,q)$ containing X and hence $Mut[X] \subseteq H[X]$.

Therefore, the canonical homomorphism $\psi : F[X] \to Mut[X]$ is regarded as a unique homomorphism

$$\psi : F[X] \to H[X]$$

such that $\psi(x_i) = x_i$ for all $x_i \in X$ (ψ is not necessarily surjective). As for $Mut[X]$, we can define the subspaces $H_{(r,s)}[X]$ (see (2.8)) and $H_{(r,s)}^{[n_1,\ldots,n_m]}[X]$ which yield the gradation for $H[X]$:

$$H[X] = \sum_{\substack{m\geq 1\\ n_1,\ldots,n_m\geq 0}} \sum_{\substack{r,s\geq 0\\ r+s+1=n_1+\cdots+n_m}} \bigoplus H^{[n_1,\ldots,n_m]}_{(r,s)}[X],$$

where

(2.16) $$\psi : F^{[n_1,\ldots,n_m]}_{(r,s)}[X] \to H^{[n_1,\ldots,n_m]}_{(r,s)}[X].$$

We first treat the case of degree ≤ 3.

Theorem 2.5. *All identities of degree ≤ 3 in $I(Mut)$ are consequences of Lie-admissibility, i.e., the Jacobian $J(x,y,z)$.*

Proof. It suffices to determine $S = \Psi^{[n_1,\ldots,n_m]}_{(r,s)}$ for $1 \leq r+s+1 = n_1+\cdots+n_m \leq 3$. It is clear that there are no identities for Mut of degree 2. Since there are no identities of degree 3 for SJ, in view of Theorem 2.4 we have $S = 0$ if $(r,s) = (2,0)$ while $J(x,y,z)$ spans S for $(r,s) = (0,2)$. Thus, we are left with the types $\binom{[1,1,1]}{(1,1)}$, $\binom{[2,1]}{(1,1)}$ and $\binom{[3]}{(1,1)}$. By direct calculation one has the relations

(2.17) $$\begin{aligned}&\psi([x,y]\circ z - [y,z]\circ x - [x,z]\circ y + [x\circ y, z] - [y\circ z, x] - [x\circ z, y])\\ &\quad = xbyaz - zaybx = xbyaz + (xbyaz)^{\cdot} = -(zaybx)^{\cdot} - zaybx\end{aligned}$$

for all $x,y,z \in X$, which clearly span $H^{[n_1,\ldots,n_m]}_{(1,1)}[X]$. Hence, the map ψ of (2.16) is surjective. It also follows from (2.17) that both spaces in (2.16) have dimension 6 for type $\binom{[1,1,1]}{(1,1)}$, 3 for type $\binom{[2,1]}{(1,1)}$ and 1 for type $\binom{[3]}{(1,1)}$. Therefore, ψ is injective and so $S = 0$ in these cases. □

For the degree 4 case, F is assumed to be of characteristic $\neq 3$. We omit a full detail of analysis which involves tedious calculations and which may be found in [**Mo2**]. Our discussion focuses on the following two identities for Mut and their linearizations:

$$\begin{aligned} f(x,y) &= -[x^2\circ y, x] + x^2\circ [y,x] - x\circ [y,x^2] + [x\circ y, x^2] - y\circ [x,x^2],\\ g(x,y) &= -2y\circ(x\circ[x,y]) - 6x\circ(y\circ[y,x]) + 2[y\circ x^2, y] + [x\circ y^2, x]\\ &\quad -2[y\circ(y\circ x), x] - 2x\circ[x,y^2] - 4x\circ[y\circ x, y] - 2(x\circ y)\circ[x,y].\end{aligned}$$

It is shown in [**Mo2**] that $g(x,y) \in \Psi^{[2,2]}_{(2,1)}$ and that $f(x,y)$ forms a basis of $\Psi^{[3,1]}_{(2,1)}$ together with the following two identities:

$$\begin{aligned} f_1(x,y) &= [x^2\circ y, x] + 2[x\circ(x\circ y), x] - 4x\circ[x\circ y, x]\\ &\quad +2x\circ[y,x^2] + [x^2\circ x, y] + 2y\circ[x,x^2],\\ f_2(x,y) &= -x^2\circ[y,x] + 2x\circ(x\circ[y,x]) - 2x\circ[y,x^2] - [x^2\circ x, y].\end{aligned}$$

We note that $-f(x,x)$ is just identity (2.15) and hence $f(x,y)$ is not a consequence of Lie– or Jordan–admissibility. An algebra is constructed in [**Mo2**] to show that the same is true for $g(x,y)$. Let $f^* = \psi(f)$ and $g^* = \psi(g)$, so that $f^*(x,y) = g^*(x,y) = 0$ for any $x, y \in Mut[X]$. Therefore,

$$\begin{aligned} 2f^*(x,y) &= -2(y*(x*x))*x + 2x*((x*x)*y) \\ &\quad -2(x*x)*(x*y) + 2(y*x)*(x*x) \\ &\quad -y*(x*(x*x)) + y*((x*x)*x) \\ &\quad -(x*(x*x))*y + ((x*x)*x)*y = 0, \end{aligned} \tag{2.18}$$

$$\begin{aligned} g^*(x,y) &= -2\{y,\{x,[x,y]^*\}^*\}^* - 6\{x,\{y,[y,x]^*\}^*\}^* \\ &\quad +2\,[\{y,\{x,x\}^*\}^*,y]^* + [\{x,\{y,y\}^*\}^*,x]^* \\ &\quad -2\,[\{y,\{y,x\}^*\}^*,x]^* - 2\{x,[x,\{y,y\}^*]^*\}^* \\ &\quad -4\{x,[\{y,x\}^*,y]^*\}^* - 2\{\{x,y\}^*,[x,y]^*\}^* = 0 \end{aligned} \tag{2.19}$$

for all $x, y \in Mut[X]$. In fact, identity (2.18) is just the expansion of $2f^*(x,y) = 0$.

It is well known in the Jordan theory that $\Psi^{[n_1,\dots,n_m]}_{(3,0)}$ is spanned by Jordan–admissibility (i.e., $j(x,y)$), $j(x,x)$ and its linearizations (also see Theorem 2.4(i)). On the other hand, $\Psi^{[n_1,\dots,n_m]}_{(0,3)}$ is spanned by homogeneous elements of degree 4 in the T–ideal of $F_-[X]$ generated by $J(x,y,z)$ (see Theorem 2.4(ii)). For the remaining types, i.e., the case $(r,s) = (2,1)$ or $(1,2)$, $\Psi^{[n_1,\dots,n_m]}_{(2,1)}$ is shown to be spanned by $f(x,y)$, $g(x,y)$ and their linearizations with substitution of variables, while $\Psi^{[n_1,\dots,n_m]}_{(1,2)}$ has a basis consisting of elements of the form $J(x\circ y,z,t)$ and $J(x,y,z)\circ t$ (for detail, see [**Mo2**]). Hence, for the case $(r,s) = (1,2)$, identities for Mut are consequences (in $F[X]$) of Lie–admissibility.

For explicit forms of these identities, we first consider a full linearization of $f(x,y)$:

$$\begin{aligned} C(x,y,z,t) &= -[(x\circ z)\circ y,t] - [(z\circ t)\circ y,x] - [(x\circ t)\circ y,z] \\ &\quad +(z\circ t)\circ[y,x] + (x\circ t)\circ[y,z] + (x\circ z)\circ[y,t] \\ &\quad -x\circ[y,z\circ t] - z\circ[y,x\circ t] - t\circ[y,x\circ z] \\ &\quad +[x\circ y,z\circ t] + [z\circ y,x\circ t] + [t\circ y,x\circ z] \\ &\quad -y\circ[x,z\circ t] - y\circ[z,x\circ t] - y\circ[t,x\circ z]. \end{aligned}$$

We note that $C(x,y,x,x) = 3f(x,y)$ and $C(x,y,z,t) \in \Psi^{[1,1,1,1]}_{(2,1)}$.

For a full linearization of $g(x,y)$, let

$$\begin{aligned} D_1(x,y,z,t) &= \frac{1}{2}\left([x\circ(y\circ z),t] + [z\circ(y\circ x),t] - [(z\circ x)\circ y,t]\right), \\ D_2(x,y,z,t) &= \frac{1}{2}\left(x\circ(y\circ[z,t]) - y\circ(x\circ[z,t]) + (y\circ x)\circ[z,t]\right) \end{aligned}$$

$$\begin{aligned} E(x,y,z,t) &= 2D_1(x,y,z,t) - D_1(t,y,x,z) + D_2(x,y,z,t) \\ &\qquad -D_2(t,y,x,z) - D_2(z,y,x,t), \\ G(x,y,z,t) &= x\circ(t\circ[y,z]) - x\circ(y\circ[z,t]) - x\circ(z\circ[y,t]) \\ &\qquad +x\circ[y\circ z,t] - x\circ[z\circ t,y] - x\circ[y\circ t,z]. \end{aligned}$$

It is proven in [**Mo2**] that $g(x,y)$ is linearized to

$$H(x,y,z,t) = G(x,y,z,t) + G(t,z,y,x) + E(x,t,z,y) + E(t,x,y,z)$$

with $4H(x,y,y,x) = g(x,y)$, and that $f_1(x,y) = 0$ and $f_2(x,y) = 0$ are consequences of $H(x,y,z,t) = 0$ and $f(x,y) = 0$.

The identities f, g, C, H, j and J essentially determine the structure of $\Psi^{[n_1,\ldots,n_m]}_{(r,s)}$ for degree 4 which we summarize in Table V.1 for the cases $(r,s) = (2,1)$ and $(1,2)$ (for detail, see [**Mo2**]).

The following result now follows from Theorem 2.5 and Table V.1 (also, see [**Mo2**]).

Theorem 2.6. *The identities for Mut of degree ≤ 4 are consequences of $J(x,y,z), j(x,y), f(x,y)$ and $g(x,y)$. Moreover, no one of these four identities is a consequence of the remaining three.* □

TABLE V.1

Type	$\dim \Psi_{(r,s)}^{[n_1,\ldots,n_m]}$	Basis of $\Psi_{(r,s)}^{[n_1,\ldots,n_m]}$
$\begin{pmatrix}[4]\\(2,1)\end{pmatrix}$	1	$[x\circ x^2, x] + 2x\circ [x, x^2]$
$\begin{pmatrix}[3,1]\\(2,1)\end{pmatrix}$	3	$f(x,y), f_1(x,y), f_2(x,y)$
$\begin{pmatrix}[2,1,1]\\(2,1)\end{pmatrix}$	6	$C(x,y,x,z), C(x,z,x,y), H(x,y,z,x),$ $H(y,x,x,z), H(x,y,x,z), H(x,z,x,y)$
$\begin{pmatrix}[2,2]\\(2,1)\end{pmatrix}$	4	$C(x,y,x,y), H(x,y,y,x) = \frac{1}{4}g(x,y),$ $H(y,x,x,y), H(x,y,x,y)$
$\begin{pmatrix}[1,1,1,1]\\(2,1)\end{pmatrix}$	9	$C(x,y,z,t), C(y,x,z,t), C(x,z,y,t),$ $H(x,y,z,t), H(y,x,z,t), H(x,y,t,z),$ $H(t,y,z,x), H(z,y,x,t), H(x,t,z,y)$
$\begin{pmatrix}[4]\\(1,2)\end{pmatrix}$	0	
$\begin{pmatrix}[3,1]\\(1,2)\end{pmatrix}$	1	$J(x^2, x, y)$
$\begin{pmatrix}[2,1,1]\\(1,2)\end{pmatrix}$	4	$J(x^2,y,z), J(x\circ y,x,z), J(x\circ z,y,x),$ $J(x,y,z)\circ x$
$\begin{pmatrix}[2,2]\\(1,2)\end{pmatrix}$	1	$J(x\circ y, x, y)$
$\begin{pmatrix}[1,1,1,1]\\(1,2)\end{pmatrix}$	10	$J(x\circ y,z,t), J(z\circ y,t,x), J(x\circ t,y,z),$ $J(x\circ z,y,t), J(t\circ y,z,x), J(t\circ z,y,x),$ $J(x,y,z)\circ t, J(x,y,t)\circ z, J(x,z,t)\circ y,$ $J(y,z,t)\circ x$

3. Additional properties of the mutation variety

As noted earlier, the mutation variety $\langle Mut\rangle$ properly contains Mut. Let $\mathcal{M}$ denote the variety defined by the four identities $J(x,y,z), j(x,y), f(x,y)$ and $g(x,y)$, and let LJ be the variety defined by $J(x,y,z)$ and $j(x,y)$. Since each term in $f(x,y)$ and $g(x,y)$ involves both the products $[\,,\,]$ and $\circ$, any Lie and Jordan algebras are contained in $\mathcal{M}$. But, since Albert algebras do not belong to $\langle Mut\rangle$ (see the remarks following Theorem 2.4), combining this with the discussion in Section V.2, we obtain

$$Mut \subset \langle Mut\rangle \subset \mathcal{M} \subset LJ.$$

In this section, we extend some of the results in Section II.2 obtained for mutation algebras to algebras in the variety $\mathcal{M}$. In particular, we show that third power–associativity is equivalent to power–associativity for algebras in $\mathcal{M}$. Let $*$ denote the multiplication in any algebra in $\mathcal{M}$.

If B is an algebra over F with multiplication xy, then for $x \in B$ we define $x^1 = x$ and $x^{n+1} = x^n x$ for $n = 1, 2, \ldots$. The following lemma is due to Albert [**A2**].

Lemma 3.1. *Let B be a third power–associative algebra over F. Let x be an element of B and n be a positive integer with $n \geq 4$ such that $x^r x^s = x^{r+s}$ for $r+s < n$. Then,*

$$n[x^{n-1}, x] = 0, \quad [x^{n-r}, x^r] = r[x^{n-1}, x]$$

for $r = 1, \ldots, n-1$.

Proof. Linearizing $u^2u = uu^2$ gives

$$[x \circ y, z] + [y \circ z, x] + [z \circ x, y] = 0 \tag{3.1}$$

for all $x, y, z \in B$. Putting $y = x^r$ and $z = x^{n-(r+1)}$ in (3.1), we obtain

$$[x^{n-(r+1)}, x^{r+1}] = [x^{n-r}, x^r] + [x^{n-1}, x],$$

and by induction on r, this implies $[x^{n-r}, x^r] = r[x^{n-1}, x]$ for $r = 1, \ldots, n-1$. When $r = n-1$, this becomes $n[x^{n-1}, x] = 0$. □

For any algebra $(A, *)$ in the variety $\mathcal{M}$, let $[\,,\,]^*$ and $\{\,,\,\}^*$ denote the multiplications in $(A,*)^-$ and $(A,*)^+$. Thus, the identities $f^*(x,y) = g^*(x,y) = 0$ given by (2.18) and (2.19) hold for any $(A,*) \in \mathcal{M}$. The following result extends Theorem II.2.1 to $\mathcal{M}$.

Theorem 3.2. *An algebra $(A, *) \in \mathcal{M}$ is power–associative if and only if it is third power–associative.*

Proof. Let x be any element of A. Assume that $(A, *)$ is third power–associative and that $x^{*r} * x^{*s} = x^{*(r+s)}$ for $r+s < n$ with $n \geq 4$. It suffices to verify $x^{*r} * x^{*s} = x^{*n}$ for $r + s = n$. If $y = x^{*(n-3)}$ in the identity $f^*(x, y) = 0$, then we obtain

$$\begin{aligned} 0 &= -2x^{*(n-1)} * x + 2x * x^{*(n-1)} - 2x^{*2} * x^{*(n-2)} + 2x^{*(n-2)} * x^{*2} \\ &= -2\left([x^{*(n-1)}, x]^* - [x^{*(n-2)}, x^{*2}]^*\right). \end{aligned}$$

It follows from this and Lemma 3.1 that $[x^{*(n-1)}, x]^* = 0$, which in turn implies $[x^{*r}, x^{*s}]^* = 0$ for $r + s = n$, using Lemma 3.1. Since $(A, *)^+$ is a Jordan algebra and hence is power–associative, from this we obtain

$$x^{*r} * x^{*s} = \{x^{*r}, x^{*s}\}^* = \{x^{*(n-1)}, x\}^* = x^{*n}$$

for any r, s with $r + s = n$. Therefore, $(A, *)$ is power–associative. □

In view of Theorem 3.2, the following corollary follows from the same proof as for Corollary II.2.3.

Corollary 3.3. *Let $(A, *)$ be an algebra in $\mathcal{M}$. Consider the conditions:*

(i) *$(A, *)$ is flexible;*

(ii) *$(A, *)$ is a noncommutative Jordan algebra;*

(iii) *$(A, *)$ is third power–associative;*

(iv) *$(A, *)$ is power–associative.*

Then, we have the implications: (i) $\Leftrightarrow$ (ii) $\Rightarrow$ (iii) $\Leftrightarrow$ (iv). □

As shown in Theorem II.2.6, the conditions (i)–(iv) are equivalent for algebras $A(p, q) \in Mut$ under a certain natural restriction such as: A contains an element which is not a zero divisor in A, or $A(p, q)$ is simple for artinian A (so A is simple). But, this is no longer true for algebras in $\mathcal{M}$. For example, let $(A, *)$ be the twisted mutation algebra in Example II.4.6. Then, $(A, *)$ is a quotient algebra and a subalgebra of a mutation algebra, hence $(A, *) \in \mathcal{M}$ but $(A, *) \notin Mut$. We can let $A = F \oplus F$ with multiplication $*$ given by

$$\begin{aligned} (a, b) * (c, d) &= (a, b)\begin{pmatrix}1\\0\end{pmatrix}(c, d) - (c, d)\begin{pmatrix}0\\1\end{pmatrix}(a, b) \\ &= (a(c-d), (a-b)d). \end{aligned}$$

Then, $((a, b)*(a, b))*(a, b) = (a-b)^2(a, b) = (a, b)*((a, b)*(a, b))$, hence $(A, *)$ is third power–associative. But, $((1, 0) * (0, 1)) * (1, 0) = (-1, 0)$ and $(1, 0) * ((0, 1) * (1, 0)) = (0, 0)$, so $(A, *)$ is not flexible. We note that $(A, *)$ is simple.

However, as for mutation algebras (see Theorem II.2.8), the existence of a unit element in an algebra of $\mathcal{M}$ implies the equivalence of the conditions (i)–(iv) in Corollary 3.3.

Lemma 3.4. *If B is a third power–associative algebra over F such that B^+ has a unit element e, then e is also a unit element of B.*

Proof. If $y = z = e$ in (3.1), then $[x \circ e, e] + [e \circ e, x] + [e \circ x, e] = 0$, so $[x, e] = 0$ for all $x \in B$. Hence, $ex = e \circ x + \frac{1}{2}[e, x] = e \circ x = x$ and similarly, $xe = x$ for any $x \in B$. □

Theorem 3.5. *If $(A, *) \in \mathcal{M}$, then the following are equivalent:*

(i) *$(A, *)$ has a unit element;*

(ii) *$(A, *)^+$ has a unit element and $(A, *)$ is flexible;*

(iii) *$(A, *)^+$ has a unit element and $(A, *)$ is third power-associative.*

Proof. Let e be the unit element of $(A, *)$. From the identity $g^*(x, y) = 0$ of (2.19), we obtain

$$\begin{aligned} 0 = g^*(x+e, y) &= g^*(x, y) + 2\{y, [x, y]^*\}^* + [\{y, y\}^*, x]^* \\ &= 2\{y, [x, y]^*\}^* + [\{y, y\}^*, x]^* \end{aligned}$$

for all $x, y \in A$. This is expanded to the identity

$$\begin{aligned} 0 &= y * (x * y - y * x) + (x * y - y * x) * y + (y * y) * x - x * (y * y) \\ &= -(y, x, y)^* + (x, y, y)^* + (y, y, x)^*. \end{aligned}$$

Linearizing this gives

$$-(y, x, z)^* - (z, x, y)^* + (x, y, z)^* + (x, z, y)^* + (y, z, x)^* + (z, y, x)^* = 0$$

for all $x, y, z \in A$. Putting $y = x$ in this, one has $2(x, z, x)^* = 0$ and hence $(A, *)$ is flexible, implying (ii). Clearly, (ii) implies (iii) and the implication (iii)⇒(i) follows from Lemma 3.4. □

The identities $f(x, y)$ and $g(x, y)$ may be used for the study of power–associative Jordan–admissible algebras. Let L be any anticommutative algebra with multiplication $[x, y]$ over F and $\tau : L \to F$ be any linear functional. Let $(L, *)$ denote the algebra defined on the vector space L with multiplication

$$x * y = \frac{1}{2}[x, y] + \frac{1}{2}(\tau(x)y + \tau(y)x), \tag{3.2}$$

so that $[x,y]^* = [x,y]$ and $\{x,y\}^* = \frac{1}{2}(\tau(x)y + \tau(y)x)$. Hence, $(L,*)^-$ has the same structure as L for any τ while $(L,*)^+$ depends only on τ. Since $x^{*m} = \tau(x)^{m-1}x$, $x^{*m} * x^{*n} = \tau(x)^{m+n-1}x = x^{*(m+n)}$ and

$$\begin{aligned}\{\{\{x,x\}^*,y\}^*,x\}^* &= \frac{1}{2}\tau(x)^2\{x,y\}^* + \frac{1}{2}\tau(x)^2\tau(y)x \\ &= \{\{x,x\}^*,\{y,x\}^*\}^*\end{aligned}$$

for any $x, y \in L$, we find that $(L,*)$ is a power–associative Jordan–admissible algebra.

The algebra $(L,*)$ arises in the study of certain power–associative products defined on Lie and Malcev algebras [**My4**]. When L is a Lie algebra, $(L,*)$ has been termed a *pseudo–Lie algebra* [**Mo5,6**]. It is proven in [**Mo5**] that for any anticommutative algebra, $(L,*)$ given by (3.2) satisfies the identities $f(x,y)$ and $g(x,y)$, and hence belongs to $\mathcal{M}$ if L is a Lie algebra.

Let A be a finite–dimensional power–associative Jordan–admissible algebra over F of characteristic 0. It has been shown in [**Mo5**] that if A is a non–nil, simple algebra satisfying the identities $f(x,y)$ and $g(x,y)$, then A is either an algebra $(L,*)$ given by (3.2) or a noncommutative Jordan algebra. The former case occurs when A has a unique idempotent, and otherwise, A is flexible (so noncommutative Jordan) [**Mo5**]. Thus, if $A \in \mathcal{M}$ is non–nil and simple, then A is pseudo–Lie or noncommutative Jordan. We notice that the algebra $(A,*)$ of Example II.4.6 is a pseudo–Lie algebra $(L,*)$ with $L = (A,*)^-$ and $\tau((a,b)) = a - b$ for $(a,b) \in A$.

4. Strict representations of mutation algebras

For a variety of F–algebras there is a standard approach to the theory of representations and modules. Let A be an algebra over F in a variety and let V be a vector space over F. A *representation* of A in V is a pair (λ, ρ) of linear maps

$$\lambda, \rho : A \to Hom_F V$$

such that the vector space direct sum $M = V \oplus A$ becomes an algebra in the given variety under the multiplication

$$(v_1 + a_1)(v_2 + a_2) = \rho(a_2)v_1 + \lambda(a_1)v_2 + a_1a_2$$

for $v_i \in V$ and $a_i \in A$, $i = 1, 2$. In this case, (V, λ, ρ) is called a *module* for A.

In general, this approach does not lead to a meaningful representation theory if a variety is too large, or if a variety lacks the intrinsic characterization by means of a convenient set of defining identities. There are two obvious reasons that this approach is not applicable for mutation algebras. The first is that the mutation algebras do not form a variety or even a quasi–variety. Secondly, as noted in Section V.2, it seems impossible to identify a minimal set of generators for $I(Mut) = I(\langle Mut\rangle) = I(Mut[X])$, although we know by Theorem 2.6 that the identities $J(x,y,z), j(x,y), f(x,y)$ and $g(x,y)$ are contained in such a minimal set. Accordingly, there lacks structural information for the mutation variety $\langle Mut\rangle$ and even for simple algebras in $\langle Mut\rangle$ or in Mut.

The primary concern in this section is to study a "special" type of representations and modules for $A(p,q) \in Mut$ which are induced from left or right modules for the associative algebra A. Let V be a left module for A with composition $(x,v) \mapsto xv$ $(x \in A, v \in V)$. We define the linear maps λ, ρ by

$$\lambda, \rho : A(p,q) \to Hom_F V, \tag{4.1}$$

$$\lambda(x)v = x * v = xpv, \quad \rho(x)v = v * x = -xqv \tag{4.2}$$

for $x \in A$ and $v \in V$. If V is a right A–module with composition $(x,v) \mapsto vx$, then the maps λ, ρ of (4.1) are defined by

$$\lambda(x)v = x * v = -vqx, \quad \rho(x)v = v * x = vpx. \tag{4.3}$$

It is straightforward to check that (4.2) or (4.3) satisfies the relations

$$\begin{aligned} \lambda(x * y) &= \lambda(x)\lambda(y) + \rho(y)\lambda(x), \\ \rho(x * y) &= \lambda(x)\rho(y) + \rho(y)\rho(x) \end{aligned} \tag{4.4}$$

for all $x, y \in A$. Moreover, it is readily seen that any linear maps λ, ρ of (4.1) satisfying (4.4) yield the relations

$$(\lambda - \rho)([x,y]^*) = [(\lambda - \rho)(x), (\lambda - \rho)(y)]. \tag{4.5}$$

$$(\lambda + \rho)(\{x,y\}^*) = (\lambda + \rho)(x) \circ (\lambda + \rho)(y) \tag{4.6}$$

for all $x, y \in A$. Thus, (4.5) holds if and only if $\lambda - \rho$ is a representation of the Lie algebra $A(p,q)^- = A(p+q)^-$ in V, while the validity of (4.6) means that $\lambda + \rho$ is a special representation of the special Jordan algebra $A(p,q)^+ = A(p-q)^+$ (or an associative specialization of $A(p,q)^+$ in $Hom_F V$ [**J3**]). A pair (λ, ρ) of linear maps λ, ρ of (4.1) or (V, λ, ρ) satisfying (4.5) and (4.6) is called here a *special representation* of $A(p,q)$ or a *special module* for $A(p,q)$. Thus, if (λ, ρ) satisfies (4.4), then it is a special representation of $A(p,q)$.

Definition 4.1. As in Section V.3, let $\mathcal{M}$ denote the variety defined by $J(x,y,z), j(x,y), f(x,y)$ and $g(x,y)$, and let $(B,*) \in \mathcal{M}$. A pair (λ, ρ) of linear maps $\lambda, \rho : (B, *) \longrightarrow Hom_F V$ for a vector space V over F is called a *strict representation* of $(B, *)$ in V if (λ, ρ) and $(B, *)$ satisfy (4.4). In this case, (V, λ, ρ) is said to be a *strict module* for $(B, *)$. □

Therefore, for any left or right module V for an associative algebra A over F, the pair (λ, ρ) defined by (4.2) or (4.3) is a strict representation of $A(p, q)$ for any fixed $p, q \in A$. More generally, let V be a bimodule for A with compositions $(x, v) \mapsto xv$ and $(x, v) \mapsto vx$, so that $M = V \oplus A$ becomes an associative algebra over F under the product $(v_1 + a_1)(v_2 + a_2) = v_1a_2 + a_1v_2 + a_1a_2$ for $a_i \in A$ and $v_i \in V$, $i = 1, 2$. For fixed $p, q \in A$, we consider the (p, q)–mutation $M(p, q)$ of M, and let $\lambda(x)$ and $\rho(x)$ $(x \in A)$ denote the restrictions to V of the left and right multiplications in $M(p, q)$ by x, respectively. Hence,

$$\lambda(x)v = x * v = (xp)v - v(qx),$$
$$\rho(x)v = v * x = v(px) - (xq)v \tag{4.7}$$

for $v \in V$ and $x \in A$, and moreover,

$$(v_1 + a_1) * (v_2 + a_2) = \lambda(a_1)v_2 + \rho(a_2)v_1 + a_1 * a_2. \tag{4.8}$$

We notice that if V is a left or right module for A, then V is made into a bimodule for A via $vx = 0$ or $xv = 0$ for all $x \in A$ and $v \in V$. Thus, the strict representation (λ, ρ) of $A(p, q)$ obtained by (4.2) or (4.3) is a special case of (4.7). For any maps λ, ρ of (4.1), denote

$$\mu(x) = \lambda(x) + \rho(x), \quad x \in A. \tag{4.9}$$

Let V be a bimodule for A and let λ, ρ be given by (4.7). Then, by (4.8) $[x, v]^* = (\lambda - \rho)(x)v$, and since $M(p, q)^-$ is a Lie algebra, it follows easily that $\lambda - \rho$ satisfies (4.5). Using $\{x, v\}^* = \frac{1}{2}\mu(x)v = \frac{1}{2}[x(p - q)v + v(p - q)x]$, we compute from (4.8)

$$\mu(\{x, y\}^*)v - \mu(x) \circ \mu(y)v = x(p - q)v(p - q)y + y(p - q)v(p - q)x$$

for all $x, y \in A$ and $v \in V$. Hence, (λ, ρ) is a special representation of $A(p, q)$ if and only if

$$x(p - q)v(p - q)y + y(p - q)v(p - q)x = 0 \tag{4.10}$$

for all $x, y \in A$ and $v \in V$. From this it is easy to construct bimodules for A which do not give rise to special representations of $A(p, q)$ given by (4.7). For instance, for a semiprime associative algebra A with unit element 1, let $V = A$ be the (regular) bimodule given by the product in A. If $p - q \neq 0$ and $x = y = 1$, then the left side of (4.10) is not zero for some $v \in V$ (it is erroneous in [**MS1**] that any bimodule for A induces a special module for $A(p, q)$ via (4.7)).

In the remainder of this section, we focus on strict representations for $A(p,q)$. The main objective is to determine all irreducible strict modules for $A(p,q)$ when A is simple artinian with $p-q$ invertible. They are shown to be strict modules for $A(p,q)$ given by (4.2) or (4.3). Strict modules also arise from Peirce decompositions relative to idempotents in $A(p,q)$ [**Os2**].

Let e be an idempotent in $A(p,q)$, so that $e*e = e(p-q)e = e$. In particular, e is an idempotent of the $(p-q)$–homotope $A(p-q)$ of A and $e(p-q)$ and $(p-q)e$ are idempotents of A. If c is an idempotent of A, then A has the Peirce decomposition

$$A = A_{11}(c) \oplus A_{10}(c) \oplus A_{01}(c) \oplus A_{00}(c), \tag{4.11}$$

$$A_{ij}(c)A_{kl}(c) \subseteq \delta_{jk}A_{il}(c) \tag{4.12}$$

for $i,j,k,l = 0,1$ [**Sc3**], where the $A_{ij}(c) = \{x \in A : cx = ix, xc = jx\}$ are the Peirce spaces relative to c. For an idempotent e in $A(p,q)$, we consider the subspace

$$\tilde{A}_{ij}(e) = \tilde{A}_{ij} = \{x \in A(p,q) : e(p-q)x = ix,\ x(p-q)e = jx\}$$

for $i,j = 0,1$. Denote:

$$\begin{aligned} B_i(e) &= A_{i1}(e(p-q)) + A_{i0}(e(p-q)), \\ C_j(e) &= A_{1j}((p-q)e) + A_{0j}((p-q)e), \quad i,j = 0,1. \end{aligned}$$

Lemma 4.1. *If e is an idempotent of $A(p,q)$, then for $i,j,k,l = 0,1$*

$$\tilde{A}_{ij} = B_i(e) \cap C_j(e), \tag{4.13}$$

$$A(p,q) = \tilde{A}_{11} \oplus \tilde{A}_{10} \oplus \tilde{A}_{01} \oplus \tilde{A}_{00}, \tag{4.14}$$

$$B_i(e) = \tilde{A}_{i1} + \tilde{A}_{i0}, \quad C_j(e) = \tilde{A}_{1j} + \tilde{A}_{0j}, \tag{4.15}$$

$$\tilde{A}_{ij}A\tilde{A}_{kl} \subseteq A_{il}. \tag{4.16}$$

Proof. For (4.13), let $x_i = y_j$ for $x_i \in B_i(e)$, $y_j \in C_j(e)$. Then $e(p-q)x_i = ix_i$ but $x_i(p-q)e = y_j(p-q)e = jy_j = jx_i$, hence $B_i(e) \cap C_j(e) \subseteq \tilde{A}_{ij}$. The reverse inclusion is immediate from (4.11) with $c = e(p-q)$ or $c = (p-q)e$. If $x \in A(p,q)$, then

$$\begin{aligned} x = {} & e(p-q)x(p-q)e \quad \in \tilde{A}_{11} \\ & + e(p-q)x - e(p-q)x(p-q)e \quad \in \tilde{A}_{10} \\ & + x(p-q)e - e(p-q)x(p-q)e \quad \in \tilde{A}_{01} \\ & + x - e(p-q)x - x(p-q)e + e(p-q)x(p-q)e \quad \in \tilde{A}_{00}. \end{aligned}$$

If $\sum x_{ij} = 0$ for $x_{ij} \in \tilde{A}_{ij}$, $i,j = 0,1$, then multiplying $e(p-q)$ and $(p-q)e$ to this from the left and right, we have all $x_{ij} = 0$. This verifies (4.14).

For (4.15), we note that $\tilde{A}_{i1} + \tilde{A}_{i0} \subseteq B_i(e)$ by (4.13) and the reverse inclusion is obtained from (4.14). The second relation follows similarly. From (4.12), we have $B_i(e)A \subseteq B_i(e)$ and $AC_l(e) \subseteq C_l(e)$, hence

$$\tilde{A}_{ij}A\tilde{A}_{kl} \subseteq B_i(e)AC_l(e) \subseteq B_i(e) \cap C_l(e) = \tilde{A}_{il},$$

using (4.13). This implies (4.16). □

Corollary 4.2. *Each $\tilde{A}_{ij}$ is a subalgebra of A as well as of $A(p,q)$, and hence is contained in the variety $\mathcal{M}$.*

Proof. From (4.12) and (4.15) we find that $\tilde{A}_{i1} + \tilde{A}_{i0}$ and $\tilde{A}_{1j} + \tilde{A}_{0j}$ are subalgebras of A for all $i, j = 0, 1$. This with (4.14) implies

$$\begin{aligned}\tilde{A}_{i1}\tilde{A}_{i1} &\subseteq (\tilde{A}_{i1} + \tilde{A}_{i0}) \cap (\tilde{A}_{11} + \tilde{A}_{01}) = \tilde{A}_{i1},\\ \tilde{A}_{i0}\tilde{A}_{i0} &\subseteq (\tilde{A}_{i1} + \tilde{A}_{i0}) \cap (\tilde{A}_{10} + \tilde{A}_{00}) = \tilde{A}_{i0},\end{aligned}$$

hence $\tilde{A}_{ij}$ is a subalgebra of A. Moreover, by (4.16)

$$\tilde{A}_{ij} * \tilde{A}_{kl} \subseteq \tilde{A}_{ij}A\tilde{A}_{kl} + \tilde{A}_{kl}A\tilde{A}_{ij} \subseteq \tilde{A}_{il} + \tilde{A}_{kj}, \tag{4.17}$$

and, in particular, $\tilde{A}_{ij}$ is a subalgebra of $A(p,q)$. □

For $x \in A$, let $[x]_{ij}$ denote the component of x in $\tilde{A}_{ij}$. If $x_{ij}, y_{ij} \in \tilde{A}_{ij}$, then by (4.16) and (4.17)

$$[x_{ij} * y_{kl}]_{il} = x_{ij}py_{kl}, \quad [x_{ij} * y_{kl}]_{kj} = -y_{kl}qx_{ij}. \tag{4.18}$$

The following result due to Osborn [**Os2**] shows that the subspaces $\tilde{A}_{11}$ and $\tilde{A}_{10}+\tilde{A}_{01}$ become strict modules for certain algebras in $\mathcal{M}$.

Theorem 4.3. *Let e be an idempotent in $A(p,q)$. Then,*

(i) *$\tilde{A}_{10} + \tilde{A}_{01}$ is a strict module for $\tilde{A}_{11}$ with the maps λ, ρ given by*

$$\begin{aligned}\lambda(x)(v_1 + v_2) &= [x * v_1]_{10} + [x * v_2]_{01},\\ \rho(x)(v_1 + v_2) &= [v_1 * x]_{10} + [v_2 * x]_{01}\end{aligned}$$

where $x \in \tilde{A}_{11}$, $v_1 \in \tilde{A}_{10}$ and $v_2 \in \tilde{A}_{01}$.

(ii) *$\tilde{A}_{11}$ is a strict module for $\tilde{A}_{10}$ under the maps λ, ρ given by*

$$\lambda(v)x = [v * x]_{11}, \quad \rho(v)x = [x * v]_{11}$$

for $x \in \tilde{A}_{11}$ and $v \in \tilde{A}_{10}$.

Proof. (i) By (4.18) we obtain

$$\lambda(x)(v_1+v_2) = xpv_1 - v_2qx,\ \rho(x)(v_1+v_2) = -xqv_1 + v_2px,$$

from which it follows that

$$\begin{aligned}\lambda(x*y)(v_1+v_2) &= (x*y)pv_1 - v_2q(x*y)\\ &= xpypv_1 - yqxpv_1 - v_2qxpy + v_2qyqx\\ &= \lambda(x)(ypv_1 - v_2qy) + \rho(y)(xpv_1 - v_2qx)\\ &= (\lambda(x)\lambda(y) + \rho(y)\lambda(x))(v_1+v_2).\end{aligned}$$

Similarly, (4.4) is obtained for $\rho(x*y)$.

(ii) If $v, w \in \tilde{A}_{10}$ and $x \in \tilde{A}_{11}$, then by (4.8)

$$\lambda(v)x = vpx,\ \rho(v)x = -vqx$$

and hence $\lambda(v*w)x = vpwpx - wqvpx = (\lambda(v)\lambda(w) + \rho(w)\lambda(v))x$. □

Since $\tilde{A}_{10}(p-q)\tilde{A}_{10} = \tilde{A}_{10}(p-q)(e(p-q))\tilde{A}_{10} = (\tilde{A}_{10}(p-q)e)(p-q)\tilde{A}_{10} = 0$, $\{\tilde{A}_{10}, \tilde{A}_{10}\}^* = 0$ and so $(\tilde{A}_{10}, *)$ is a Lie algebra. Similarly, $\{\tilde{A}_{11}, \tilde{A}_{10}\}^* \subseteq \tilde{A}_{10}$ and hence $\{\tilde{A}_{11}, \tilde{A}_{10}\}^*_{11} = 0$. Therefore, the strict representation of $\tilde{A}_{10}$ in $\tilde{A}_{11}$ given by Theorem 4.3(ii) is a Lie representation of $(\tilde{A}_{10}, *)$.

The following two lemmas are instrumental for our main result.

Lemma 4.4. *Assume that A is an associative algebra over F with unit element 1 and $p \neq q$ are fixed elements of A such that $p-q$ is invertible with inverse $u = (p-q)^{-1}$. Let (V, λ, ρ) be a strict module for $A(p,q)$, and denote $\mu = \lambda + \rho$ as in (4.9) and $E = \mu(u)$. Then,*

(i) *E is an idempotent in $Hom_F V$ and for $x \in A$*

$$\mu(x) = E\mu(x) = \mu(x)E,\ \lambda(x) = E\lambda(x),\ \rho(x) = E\rho(x). \tag{4.19}$$

In particular, $E(V) = (\lambda(A) + \rho(A))V$.

(ii) *$\lambda(x) = \mu(x)\lambda(u)$ and $\rho(x) = \mu(x)\rho(u)$ for any $x \in A$.*

Proof. (i) Since $\{u, x\}^* = \frac{1}{2}(u(p-q)x + x(p-q)u) = x$, u is the unit element of $A(p,q)^+$. Since (λ, ρ) is a special representation of $A(p,q)$, $\mu : A(p,q)^+ \to (Hom_F V)^+$ is an algebra homomorphism, hence

$$\mu(x) = \mu(\{u, x\}^*) = E\mu(x)$$

for $x \in A$. In particular, $E = \mu(u)$ is an idempotent in $Hom_F V$. This relation is expressed as

$$(I - E)\mu(x) = \mu(x)(E - I) \tag{4.20}$$

where I is the identity map on V. If we let $G = I - E$, then G is also an idempotent and from (4.20)

$$G\mu(x) = -\mu(x)G = -\mu(x)G^2 = G\mu(x)G = -G^2\mu(x) = -G\mu(x).$$

Hence, $G\mu(x) = \mu(x)G = 0$, and this gives the first relation of (4.19). Using (4.4), we obtain $2\lambda(x) = \lambda(x * u) + \lambda(u * x) = \lambda(x)\lambda(u) + \rho(u)\lambda(x) + \lambda(u)\lambda(x) + \rho(x)\lambda(u) = \mu(x)\lambda(u) + E\lambda(x)$, hence $2G\lambda(x) = (G\mu(x))\lambda(u) + (GE)\lambda(x) = 0$ since $G\mu(x) = GE = 0$. This gives the second relation of (4.19). Similarly, by (4.4) one has $2\rho(x) = \mu(x)\rho(u) + E\rho(x)$ which implies $\rho(x) = E\rho(x)$.

(ii) From the above, it follows that $2\lambda(x) = \mu(x)\lambda(u) + E\lambda(x) = \mu(x)\lambda(u) + \lambda(x)$, hence $\lambda(x) = \mu(x)\lambda(u)$ and similarly, $\rho(x) = \mu(x)\rho(u)$. □

Lemma 4.5. *If there hold the same hypotheses as in Lemma 4.4, then for any* $x, y \in A$

$$[\mu(x), \mu(y)](\lambda(u) - \rho(u))E = \mu([x, y]^*) + \mu(x)\mu([y, u]^*) - \mu(y)\mu([x, u]^*).$$

Proof. From Lemma 4.4(ii) and (4.4), we have $\mu(x * y)\lambda(u) = \lambda(x * y) = \lambda(x)\lambda(y) + \rho(y)\lambda(x) = \mu(x)\lambda(u)\mu(y)\lambda(u) + \mu(y)\rho(u)\mu(x)\lambda(u)$, which implies

$$(\mu(x * y) - \mu(x)\lambda(u)\mu(y) - \mu(y)\rho(u)\mu(x))\lambda(u) = 0.$$

Similarly, using $\mu(x)\rho(u) = \rho(x)$ and $\rho(x * y) = \lambda(x)\rho(y) + \rho(y)\rho(x)$, we get

$$(\mu(x * y) - \mu(x)\lambda(u)\mu(y) - \mu(y)\rho(u)\mu(x))\rho(u) = 0.$$

Since $E = \mu(u) = \lambda(u) + \rho(u)$ and $\mu(x)E = \mu(x)$ by (4.19), adding the two identities above gives

$$\mu(x * y) = \mu(x)\lambda(u)\mu(y) + \mu(y)\rho(u)\mu(x). \tag{4.21}$$

In particular, (4.21) shows that μ is an algebra homomorphism of $A(p, q)$ into the $(\lambda(u), -\rho(u))$–mutation $(Hom_F V)\,(\lambda(u), -\rho(u))$ of $Hom_F V$ and hence

$$\mu([x, y]^*) = \mu(x)(\lambda(u) - \rho(u))\mu(y) - \mu(y)(\lambda(u) - \rho(u))\mu(x) \tag{4.22}$$

for any $x, y \in A$. When $y = u$, (4.22) gives

$$\begin{aligned} \mu([x, u]^*) &= \mu(x)(\lambda(u) - \rho(u))E - E(\lambda(u) - \rho(u))\mu(x) \\ &= \mu(x)(\lambda(u) - \rho(u))E - (\lambda(u) - \rho(u))\mu(x)E \\ &= [\mu(x), \lambda(u) - \rho(u)]E \end{aligned} \tag{4.23}$$

by (4.19). We now rewrite (4.22) as

$$\begin{aligned}\mu([x,y]^*) &= \mu(x)[\lambda(u)-\rho(u),\mu(y)]+\mu(x)\mu(y)(\lambda(u)-\rho(u))\\ &\quad -\mu(y)[\lambda(u)-\rho(u),\mu(x)]-\mu(y)\mu(x)(\lambda(u)-\rho(u))\\ &= \mu(y)[\mu(x),\lambda(u)-\rho(u)]-\mu(x)[\mu(y),\lambda(u)-\rho(u)]\\ &\quad +[\mu(x),\mu(y)](\lambda(u)-\rho(u)).\end{aligned}$$

Multiplying E to this from the right and using (4.19), the Lemma follows from (4.23).

□

We now prove our main result in this section.

Theorem 4.6. *Let A be a simple artinian associative algebra over F, i.e., $A = M_n(D)$ with a division algebra D over F, and assume that $n \geq 1$ and if $n = 1$ then $Z(A)/F$ is a separable extension of F with $1 < \dim_{Z(A)} A < \infty$. Let p and q be fixed elements of A with $p-q$ invertible. Then, any irreducible strict module (V,λ,ρ) for $A(p,q)$ is given either by (4.2) with an irreducible left A–module V, or by (4.3) with an irreducible right A–module V.*

Proof. We first note that $\mu = \lambda+\rho : A(p,q)^+ = A(p-q)^+ \to (Hom_F V)^+$ is an algebra homomorphism. Since $p-q$ is invertible, $A(p-q) \cong A$ is simple. Therefore, by a known result in the theory of semisimple Jordan algebras [**JJ1**,**JR1**], we find that $\mu : A(p-q) \to Hom_F V$ is either a homomorphism or an anti–homomorphism. Hence, in the former V is made into a left $A(p-q)$–module under the composition $xv = \mu(x)v$, while in the latter V becomes a right $A(p-q)$–module via $vx = \mu(x)v$.

Assume that V is a left $A(p-q)$–module. Letting $\varphi(x) = \mu(x(p-q)^{-1})$, we have that $\varphi : A \to Hom_F V$ is a homomorphism of the associative algebra A and V is a left A–module via $xv = \varphi(x)v$, since $(xy)v = \varphi(xy)v = \mu(xy(p-q)^{-1})v = \mu(x(p-q)^{-1}(p-q)y(p-q)^{-1})v = \mu(x(p-q)^{-1})\mu(y(p-q)^{-1})v = (\varphi(x)\varphi(y))v = x(yv)$. Since (V,λ,ρ) is irreducible, in view of Lemma 4.4(i), $E(V) \neq 0$ and $E(V) = V$, hence

$$E = \mu(u) = \varphi(1) = I \tag{4.24}$$

(so φ is injective). Since $\mu(x) = \varphi(x(p-q))$, form Lemma 4.5 and (4.24) one has

$$\begin{aligned}&\varphi([x(p-q),y(p-q)])(\lambda(u)-\rho(u))\\ &= \varphi(x(p+q)y(p-q)-y(p+q)x(p-q))\\ &\quad +\varphi(x(p-q))\varphi(y(p+q)(p-q)^{-1}(p-q)-(p-q)^{-1}(p+q)y(p-q))\\ &\quad -\varphi(y(p-q))\varphi(x(p+q)(p-q)^{-1}(p-q)-(p-q)^{-1}(p+q)x(p-q))\\ &= \varphi(x(p-q)y(p+q)-y(p-q)x(p+q))\\ &= \varphi([x(p-q),y(p-q)])\varphi(u(p+q))\end{aligned}$$

for any $x, y \in A$. Hence,

$$\varphi([x(p-q), y(p-q)])(\lambda(u) - \rho(u) - \varphi(u(p+q))) = 0. \tag{4.25}$$

Assume that $\varphi([A,A])v = 0$ for $v \in V$. Since $[A,A] \neq 0$ by hypotheses, $[A,A] + A[A,A]$ is a nonzero ideal of A. Since A is simple, we conclude that $\varphi(A)v = 0$ and $v = 0$ by (4.24). Thus, it follows from (4.25) that $\lambda(u) - \rho(u) = \varphi(u(p+q))$. On the other hand, $\lambda(u) + \rho(u) = \mu(u) = \varphi(1) = \varphi(u(p-q))$. This implies $\lambda(u) = \varphi(up)$ and $\rho(u) = -\varphi(uq)$. Moreover, by Lemma 4.4(ii) $\lambda(x) = \mu(x)\lambda(u) = \varphi(x(p-q))\varphi(up) = \varphi(xp)$ for any $x \in A$, and similarly, $\rho(x) = -\varphi(xq)$. Therefore, (V, λ, ρ) is given by (4.2) for the left A–module V with $xv = \varphi(x)v$.

Finally, assume that V is a right $A(p-q)$–module. We note that $A(p,q) = A^{op}(-q,-p)$ for the opposite algebra A^{op} of A with product $x \cdot y = yx$, and $A^{op}(-q,-p)^+ = A^{op}(p-q)^+$. Since $\mu : A(p-q) \to Hom_F V$ is an anti–homomorphism, it is a homomorphism of $A^{op}(p-q)$ into $Hom_F V$. Letting $\mu(x) = \varphi(x \cdot (p-q)) = \varphi((p-q)x)$, $\varphi : A^{op} \to Hom_F V$ becomes a homomorphism which converts V into a right A–module via $vx = \varphi(x)v$. By the same calculation as above with $A^{op}(-q,-p)$, we have $\lambda(x) = \varphi(x \cdot (-q)) = -\varphi(qx)$ and $\rho(x) = -\varphi(x \cdot (-p)) = \varphi(px)$. Therefore, $\lambda(x)v = -\varphi(qx)v = -vqx$ and $\rho(x)v = \varphi(px)v = vpx$ for $x \in A$ and $v \in V$. Thus, (V, λ, ρ) is given by (4.3). □

The next example [**Os2**] shows that, even for well behaved mutation algebras such as simple ones, a strict module can be indecomposable while it is not irreducible.

Example 4.7. Let A be an associative algebra over F with unit element, let $p, q \in A$ be fixed elements with $p - q$ invertible. Let V be an irreducible left A–module with composition xv ($x \in A$, $v \in V$). For a vector space W over F, let $\phi : W \to V$ be an injective linear map. Let $U = V \oplus W$ and consider the linear maps $\lambda, \rho : A(p,q) \to Hom_F U$ given by

$$\begin{aligned} \lambda(x)(v+w) &= xpv + x(p-q)\phi(w), \\ \rho(x)(v+w) &= -xqv - x(p-q)\phi(w) \end{aligned} \tag{4.26}$$

for $x \in A$, $v \in V$ and $w \in W$. From this we have

$$\begin{aligned} \lambda(x*y)(v+w) &= (x*y)pv + (x*y)(p-q)\phi(w) \\ &= xpypv - yqxpv + xpy(p-q)\phi(w) - yqx(p-q)\phi(w) \\ &= \lambda(x)(ypv + y(p-q)\phi(w)) + \rho(y)(xpv + x(p-q)\phi(w)) \\ &= (\lambda(x)\lambda(y) + \rho(y)\lambda(x))(v+w), \end{aligned}$$

and similarly, the second relation of (4.4) follows. Hence, (V, λ, ρ) is a strict module for $A(p,q)$.

Clearly, V is a submodule of U for $A(p,q)$. Let $v+w$ be any nonzero element of U. By (4.26) we have

$$\mu(x)(v+w) = (\lambda(x)+\rho(x))(v+w) = x(p-q)v \tag{4.27}$$

for $x \in A$. In particular, $\mu(A)v = A(p-q)v = Av = V$ for any $v \neq 0$ in V since V is an irreducible left A–module, and hence V is an irreducible submodule for $A(p,q)$. If $v \neq 0$ in (4.27), then $\mu(A)(v+w) = A(p-q)v = V$ and V is contained in the submodule for $A(p,q)$ generated by $v+w$. If $v=0$, then $\mu(A)w = A(p-q)\phi(w) = A\phi(w) = V$ since $0 \neq \phi(w) \in V$, and again, V is contained in the submodule for $A(p,q)$ generated by $w = v+w$. Therefore, V is the unique minimal submodule for $A(p,q)$ and U is indecomposable. □

In Example 4.7, notice that any submodule of U for $A(p,q)$, which contains properly V, is indecomposable. This suggests that the class of indecomposable strict modules for any mutation algebra can be too large. This fact is a serious drawback for the strict representation theory of mutation algebras. Therefore, it seems desirable to find a different approach to representations of mutation algebras that is parallel to the standard representation theory of associative, Lie or Jordan algebras.

REFERENCES

[A] Albert, A. A.

1. On a certain algebra of quantum mechanics, *Ann. of Math.* **35** (1934), 65–73.
2. On the power–associativity of rings, *Brasil. Math.* **2** (1948), 1–13.
3. Power–associative rings, *Trans. Amer. Math. Soc.* **64** (1948), 552–593.

[AM] Atiyah, M. F. and Macdonald, I. G.

1. *Introduction to Commutative Algebra,* Addison–Wesley, Reading, MA, 1969.

[BBO] Benkart, G. M., Britten, D. J. and Osborn, J. M.

1. Real flexible division algebras, *Canad. J. Math.* **34** (1982), 550–588.

[BO] Benkart, G. M. and Osborn, J. M.

1. An investigation of real division algebras using derivations, *Pacific J. Math.* **96** (1981), 265–300.
2. Flexible Lie–admissible algebras, *J. Algebra* **71** (1981), 11–31.

[BK] Braun, H. and Koecher, M.

1. *Jordan–Algebren,* Springer–Verlag, New York, 1966.

[E] Elduque, A.

1. On supersolvable Malcev algebras, *Commun. Algebra* **14** (1986), 311–321.
2. On maximal subalgebras of central simple Malcev algebras, *J. Algebra* **103** (1986), 216–227.

3. Sobre algunas identidades en álgebras mutación (Spanish), Actas XV Jornadas Luso–Espanholas de Matemática, vol. I, 113–118. Evora, 1990.

[EGM] Elduque, A., González, S. and Martínez, C.

1. Unit element in mutation algebras, *Algebras Groups Geoom.* **1** (1984), 386–398.

[EM] Elduque, A. and Montaner, F.

1. On mutations of associative algebras, *J. Korean Math. Soc.* **28** (1991), 143–156.
2. A note on derivations of simple algebras, *J. Algebra,* to appear.

[EMMy] Elduque, A., Montaner, F. and Myung, H. C.

1. Automorphisms and derivations in a class of Jordan and Malcev admissible algebras, *Nova J. Algebra Geom.,* to appear.

[EMy] Elduque, A. and Myung, H. C.

1. Color algebras and affine connections on S^6, *J. Algebra* **149** (1992), 234–261.
2. The reductive pair (B_3, G_2) and affine connections on S^7, *J. Pure Appl. Algebra,* **86** (1992), 155–171.

[F] Filippov, V. T.

1. Semiprimary Mal'tsev algebras of characteristic 3, *Algebra i Logika* **14** (1975), 100–111 = *Algebra and Logic* **14** (1976), 64–71.
2. Central simple Mal'tsev algebras, *Algebra i Logika* **15** (1976), 235–242 = *Algebra and Logic* **15** (1977), 147–151.

[GM] González, S. and Martínez, C.

1. Noncommutative Jordan mutation algebras. A partial order relation, *Algebras Groups Geom.* **4** (1987), 119–127
2. Mutation and periodicity, *Hadronic Mechanics and Nonpotential Interactions,* M. Mijatovic, ed., Nova Science Publ., New York, (1990), 85–93.

[H] Herstein, I. N.

1. *Topics in Ring Theory,* University of Chicago Press, Chicago, 1969.

[Hu] Humphreys, J. E.

1. *Introduction to Lie Algebras and Representation Theory,* Springer–Verlag, New York, 1972.

[Hun] Hungerford, T. W.

1. *Algebra,* Holt, Rinehart and Winston, New York, 1974.

[JJ] Jacobson, F. D. and Jacobson, N.

1. Classification and representation of semisimple Jordan algebras, *Trans. Amer. Math. Soc.* **65** (1949), 141–169.

[J] Jacobson, N.

1. Composition algebras and their automorphisms, *Rend. Circ. Math. Palermo* **7**, Series II (1958), 55–80.
2. *Lie Algebras,* Interscience Tracts in Pure and Appl. Math., no. 10, Interscience, New York, 1962.
3. *Structure and Representations of Jordan Algebras,* Amer. Math. Soc. Colloq. Publ. Vol. 39, Amer. math. Soc., Providence, R.I., 1968.
4. *Basic Algebra I,* Second edition, Freeman and Company, New York, 1985.
5. *Basic Algebra II,* Freeman and Company, San Francisco, 1980.

[JR] Jacobson, N. and Rickart, C. E.

1. Jordan homomorphisms of rings, *Trans. Amer. Math. Soc.* **69** (1950), 479–502.

[JvonNW] Jordan, P., von Neumann, J. and Wigner, E. P.

1. On an algebraic generalization of the quantum mechanical formalism, *Ann. of Math.* **36** (1934), 29–64.

[K] Kass, S. N.

1. Explicit decompositions of some tensor products of modules for simple complex Lie algebras, *Commun. Algebra* **15** (1987), 2251–2261.

[Ku] Kuzmin, E. N.

1. Mal'cev algebras and their representations, *Algebra i Logika* **7** (1968), 48–69 = *Algebra and Logic* **7** (1968), 233–244.

2. Levi's theorem for Mal'cev algebras, *Algebra i Logika* **16** (1977), 424–431 = *Algebra and Logic* **16** (1978), 286–291.

[LT] Laufer, P. J. and Tomber, M. L.

1. Some Lie admissible algebras, *Canad. J. Math.* **14** (1962), 287–292.

[LS] Levin, J. J. and Shatz, S. S.

1. Riccati algebras, *Duke Math. J.* **30** (1963), 579–594.

[M] Malcev, A. A.

1. Analytic loops, *Mat. Sb.* **78** (1955), 569–578.

[Mc] McCrimmon, K.

1. Noncommutative Jordan rings, *Trans. Amer. Math. Soc.* **158** (1971), 1–33.
2. Homotopes of alternative algebras, *Math. Ann.* **191** (1971), 253–262.
3. Jordan algebras and their applications, *Bull. Amer. Math. Soc.* **84** (1978), 612–627.
4. The Russian revolution in Jordan algebras, *Algebras Groups Geom* **1** (1984), 1–64.

[MMP] McKay, W. G., Moody, R. V. and Patera, J.

1. Decomposition of tensor products of E_8 representations, *Algebras Groups Geom.* **3** (1986), 286–328.

[Mo] Montaner, F.

1. *Mutations of Associative and Alternative Algebras,* Doctoral thesis, University of Zaragoza, 1990.
2. Identities in mutations of associative algebras, *Commun. Algebra* **20** (1992), 55–67.
3. Prime and maximal ideals in mutations of associative algebras, *Nonassociative Algebraic Models,* S. González and H. C. Myung, eds., Nova Science Publ., New York, (1992), 213–221.
4. Mutations of alternative algebras, *Hadronic Mechanics and Nonpotential Interactions,* H. C. Myung, ed., Nova Science Publ., New York, (1992), to appear.

5. Power–associativity in mutations of associative algebras, *Commun. Algebra,* to appear.

6. On identities in mutation algebras, to appear.

[My] Myung, H. C.

1. Some classes of flexible Lie–admissible algebras, *Trans. Amer. Math. Soc.* **167** (1972), 79–88.

2. The exponentiation and deformations of Lie–admissible algebras, *Hadronic J.* **5** (1982), 771–903.

3. A Malcev–admissible mutation of an alternative algebra, *Bull. Korean Math. Soc.* **20** (1983), 37–43.

4. *Malcev–Admissible Algebras,* Birkhäuser, Boston–Basel–Stuttgart, 1986.

5. Lie–admissible algebra, *Soviet Encyclopaedia Math.* **5**, Kluwer Acad. Publ., (1990), 408–410.

[MS] Myung, H. C. and Sagle, A. A.

1. On Lie–admissible mutations of associative algebras, *Hadronic J.* **10** (1987), 35–51.

[MSh] Myung, H. C. and Shin, D. S.

1. Note on Malcev–admissible mutations of an alternative algebra, *Algebras Groups Geom.* **4** (1987), 139–143.

2. Semisimple Malcev–admissible mutation algebras, *J. Korean Math. Soc.* **24** (1987), 143–150.

[O] Okubo, S.

1. Non–associative quantum mechanics via flexible Lie–admissible algebras, *Current Problems in High Energy Particle Theory,* R. Casalbuoni, G. Domokos and S. Kövesi–Domokos, eds., Johns Hopkins Univ. Press, Baltimore, MD, (1979), 103–120.

[OM] Okubo, S. and Myung, H. C.

1. Adjoint operators in Lie algebras and the classification of simple flexible Lie–admissible algebras, *Trans. Amer. Math. Soc.* **264** (1981), 459–472.

[Os] Osborn, J. M.

1. Varieties of algebras, *Adv. Math.* **8** (1972), 163–369.

2. The Lie–admissible mutation $A(r, s)$ of an associative algebra A, *Hadronic J.* **5** (1982), 904–930.

[**P**] Pierce, R. S.

1. *Associative Algebras,* Springer–Verlag, New York, 1982.

[**R**] Racine, M. L.

1. On maximal subalgebras, *J. Algebra* **30** (1974), 155–180.

[**S**] Sagle, A. A.

1. Malcev algebras, *Trans. Amer. Math. Soc.* **101** (1961), 426–458.
2. On derivations of semisimple Malcev algebras, *Portugal. Math.* **21** (1962), 107–109.

[**SW**] Sagle, A. A. and Walde, R. E.

1. *Introduction to Lie Groups and Lie Algebras.* Academic Press, New York, 1973.

[**Sa**] Santilli, R. M.

1. *Lie–Admissible Approach to the Hadronic Structure,* Vols. I and II, Hadronic Press, Nonantum, MA, 1978 and 1982.

[**Sc**] Schafer, R. D.

1. Noncommutative Jordan algebras of characteristic 0, *Proc. Amer. Math. Soc.* **6** (1955), 472–475.
2. On noncommutative Jordan algebras, *Proc. Amer. Math. Soc.* **9** (1958), 110–117.
3. *An Introduction to Nonassociative Algebras,* Academic Press, New York, 1966.
4. Generalized standard algebras, *J. Algebra* **12** (1969), 386–417.

[**T**] Thedy, A.

1. Zum Wedderburnschen Zerlegungssatz, *Math. Z.* **113** (1970), 173–195.
2. On rings with completely alternative commutators, *Amer. J. Math.* **93** (1971), 42–51.

[**W**] Walcher, S.

1. On derivations of simple algebras, *Algebras Groups Geom.* **4** (1987), 379–382.

[Y] Yamaguti, K.

1. On the theory of Malcev algebras, *Kumamoto J. Sci.* **6** (1963), 9–45.

[Z] Zel'manov, E. I.

1. On prime Jordan algebras, *Algebra i Logika* **18** (1979), 162–175 = *Algebra and Logic* **18** (1979), 103–111.

[ZSSS] Zhevlakov, K. A., Slin'ko, A. M., Shestakov, I. P., and Shirshov, A. I.

1. *Rings That Are Nearly Associative,* (translated by H.F. Smith), Academic Press, New York, 1982.

SYMBOL INDEX

SUBJECT INDEX

N

O

P

Q

R

S